The Physics of Time Reversal

Robert G. Sachs

The Physics of Time Reversal

The University of Chicago Press | Chicago and London

THE UNIVERSITY OF CHICAGO PRESS, CHICAGO 60637
THE UNIVERSITY OF CHICAGO PRESS, LTD., LONDON

Printed in the United States of America
96 95 94 93 92 91 90 89 5432

∞ The paper used in this publication meets the minimum requirements of the American National Standard for Information Sciences—Permanence of Paper for Printed Library Materials, ANSI Z39.48-1984.

Library of Congress Cataloging in Publication Data

Sachs, Robert Green, 1916–
The physics of time reversal.

Bibliography: p.
Includes index.
1. Time reversal. 2. Symmetry (Physics) I. Title.
QC173.59.T53S23 1987 530.1 87-5826
ISBN 0-226-73330-0
ISBN 0-226-73331-9 (pbk.)

To Eugene Wigner in appreciation of his many magnificent contributions to physics, one, but only one, of them being his seminal work on time reversal

Contents

Preface

The time reversal transformation T has played an important role in physics at least during the time that the behavior of physical systems has been described in terms of equations of motion. Until the middle of the twentieth century this role was on the whole subliminal, although the notion that equations of motion should be invariant under the transformation T seems to have been implicit in attempts to formulate equations describing the motions of objects and fields at a fundamental level.

Explicit recognition of the invariance of classical equations of motion under time reversal did occur occasionally, the most important early example being Loschmidt's demonstration that Boltzmann's proof of the H theorem was in error because it was based on equations of motion that are invariant and ended with a consequence that is not.

The next important step came with Wigner's classic paper on the consequences of time reversal invariance in quantum mechanics. He apparently was motivated by his realization that the "Kramers degeneracy" is a consequence of time reversal invariance ("T invariance") and can be generalized to include all systems having half-integral spin. In this paper, Wigner also brought out many important general consequences of T invariance. However, the importance and usefulness of his results were not widely recognized or exploited until after World War II, when their impact on nuclear theory began to be recognized.

After a slow beginning in relating these results to nuclear theory, important applications to quantum field theory and particle physics were discovered and rapid development ensued, culminating in the ubiquitous CPT theorem. Thus T invariance is a very rich and powerful concept in physics, touching all subfields of the quantum theory of matter.

An entirely different aspect of the subject unfolded following the 1956 discovery of parity (P) nonconservation in weak interaction phenomena. The resulting displacement of parity conservation, or P invariance, by CP

invariance as the general principle to be satisfied by the equations of motion naturally led, through belief in the *CPT* theorem, to the notion that *T* invariance is a firm root in the foundations of physics. Nevertheless, the surprising experience with parity was taken as a warning, and some attention was given to the ways the validity of *T* invariance, *CP* invariance, and *CPT* invariance might be tested by experiments.

The discovery of *CP* violation in 1964 shattered the illusion concerning the fundamental nature of these invariances and opened up questions concerning its origin and the origin of the associated violation of *T* invariance. These questions have not been answered satisfactorily despite an enormous effort in experimental and theoretical physics.

The subject matter of this book ranges over all these basic ideas and their implications for physics. It includes a presentation of a systematic method for defining the time reversal transformation and the other two related improper transformations, space inversion, *P*, and charge conjugation, *C*, as "kinematically admissible transformations," which is a departure from standard practice but is needed to deal in a general way with interactions that are not invariant under one or more of these transformations.

The book has two separate foci of interest, one concerning the consequences of *T* invariance for those properties of matter that depend on electromagnetic and strong interactions, and even on the grosser features of the weak interactions, the other concerning the violation of *CP* invariance and *T* invariance in some special aspects of the weak interactions. The ability to separate these two aspects rests on the fact that the observed violation is an extremely small effect, not influencing in a (so far) measurable way even high-precision weak interaction measurements other than those specific, particularly sensitive ones by means of which the *CP* violation was discovered. Nevertheless, physicists believe firmly in the notion of a theory that unifies the electromagnetic, weak, and strong interaction phenomena at some level. At that level the separation of the phenomena into two classes must become meaningless. And the very fact that the observed violation occurs in such a limited way suggests that the level of unification at which *T* violation originates in a fundamental way must be very deep indeed. Therefore its ultimate elucidation may have profound implications for our understanding of the nature of physical theories. It may also have important implications for cosmology.

This ultimate unification is a hope for the future, but for the purposes of this book the two distinct classes of physical phenomena will be clearly separated. They will appear as those phenomena for which *T* invariance is a means of interpretation, on the one hand, and on the other, as those for which it is the background against which *T* violation is to be measured.

There is some mixing between the two classes, because for the first class of phenomena it is important to establish the limits of T invariance for all the interactions and for the second class it is necessary to identify the consequences of T invariance to establish the background.

The subject of time reversal threads its way through a rich variety of topics in physics, and it is my purpose to present an overall perspective on these subjects from the viewpoint of questions raised about T invariance. That perspective not only includes the well-known and widely accepted aspects of time reversal but also introduces some serious questions about our understanding of the subject in the context of recent developments.

Because I consider many topics of physics in following this thread, the interests and experiences of potential readers are expected to be correspondingly various. My intention is to provide a treatment that will be useful to workers in any one of the many subfields that are connected by this thread and to provide those associated with one subfield a view of physical methods and ideas that their work shares with other subfields.

Clearly, not everyone with an interest in some of the topics will be interested in all of them or will have the required theoretical background for all. To make the material accessible to as many readers as possible, I have taken up the subject in the order in which it developed historically, beginning with its foundations in classical physics and then in quantum mechanics and going on to quantum field theory. The classical foundations, given in chapters 1 and 2, are important for all readers who wish to gain an understanding of the fundamentals. The only theoretical background required for that purpose is some knowledge of Newtonian mechanics. Again, chapter 3 on the quantum mechanical foundations is important for understanding the role of time reversal in all aspects of the structure of matter, whether it be gross matter or matter at the atomic level, nuclear level, or quark level. Familiarity with the basic ideas and methods of quantum mechanics is assumed.

From that point on there are options depending on the interests of the reader, and the background requirements will vary. Chapters 4 and 5 are primarily concerned with straightforward applications to atomic and nuclear systems, using standard methods of quantum mechanics. I have given brief summaries of some of the more formal methods that are needed, but I have also tried to provide an adequate bibliography for the reader who feels the need for more detail. It is these two chapters that bring out how the subject threads its way through the many subfields of the quantum theory of matter. It has not been possible to mention every important application of T invariance to atomic, molecular, condensed matter and nuclear physics, but the important general principles are treated in some detail.

Insofar as specific applications to these properties of matter are concerned, I have limited my attention to simple, sometimes idealized examples that are intended to help the reader appreciate the richness of the subject and develop an insight into the physical meaning of the more general results. No attempt is made to derive general formulas covering all such examples, or even to write them down, although literature providing such formulas is cited. In my opinion, it is easier and more rewarding to arrive at the answer in a specific case by using physical insight and applying general principles in the way outlined here. Therefore I have not attempted to provide a comprehensive review of these very rich subjects, but instead have used the concept of time reversal as the instrument for presenting a way to view the physics.

In order to discuss the fundamentals of the theory, it is necessary to make use of quantum field theory. My treatment of these fundamentals in chapters 6 and 7 has been kept to the simplest possible level. I do not attempt to provide either generality or mathematical rigor but, instead, develop the subject by treating explicit examples of scalar, vector, tensor, and spinor fields, using material at the level of quantum field theory to which graduate students in physics should have been exposed. I do suggest the way the arguments can be generalized and provide references to more detailed and more rigorous treatments of the general subject of quantum field theory.

Chapter 8 is concerned with applications of the field theoretical operators for time reversal and other improper transformations to both general and specific examples. In particular Pauli's constructive method of proof of the CPT theorem is described in this chapter.

The next two chapters, chapters 9 and 10, deal explicitly with the phenomena of *CP* violation and *T* violation, which were referred to collectively as the "second focus" of the physics of time reversal. Chapter 9 presents a rather complete phenomenological treatment of the K°, $\bar{K}^\circ$ problem and the associated interpretation of the evidence for *CP* violation and *T* violation. The whole of chapter 10 is devoted to quark models, with some emphasis on the Kobayashi-Maskawa six-quark model because that model shows so much promise as a phenomenological description of the weak interaction processes, including *CP* violation. I discuss new opportunities for experiments that might extend our at present very limited information about *CP* violation, although attention is again limited to some of the simpler examples because this subject has become a very active industry among particle theorists and new suggestions regularly appear in the literature. I have given references to current literature having some comprehensive value but have not attempted an all-inclusive review of the subject. For background information I refer to some recent elementary textbooks on quarks, weak interactions, and quantum chromodynamics (QCD). Very little more than basic quantum mechanics is needed to follow the discussion of chapters 9 and 10, so a reader who has an interest in the phenomenological theory underlying the work on *CP* violation and its connection with *T* violation may find it desirable to bypass the material in chapters 6, 7, and 8 on a first reading of the book.

Since the questions raised about our current understanding of physics are the fuel for new physics and since speculating about them is part of the joy of the subject, I devote the last chapter, chapter 11, to some of the larger questions relating to the time reversal transformation and to speculation about the directions in which their answers may lie. First, the very old and much-discussed question of the relation between *T* invariance and the "Arrow of Time" is briefly reviewed from the Gibbs viewpoint, suggesting that it is a nonquestion in the sense that two distinctly different and unrelated aspects of physics are involved—equations of motion for the one and the statistics of initial conditions for the other.

This question also has cosmological implications, so that it leads to additional questions about the meaning of the time reversal transformation in general relativity. A much more recent cosmological issue concerning the role of *CP* violation in determining the matter-antimatter imbalance is also discussed briefly, but it clearly cannot be resolved before we have a clear understanding of the meaning of the transformations *T* and *CP* in the context of a satisfactory quantum relativity theory. The remainder of chapter 11 is devoted to questions and speculations about the role of *T* and *CP* in relation to recent developments that may or may not lead to such a satisfactory theory.

As I indicated earlier, the time reversal transformation and other improper transformations are defined here in terms of kinematics in order to avoid any presumption of invariance of the equations of motion. The need to raise this issue casts a new light on some of our concepts concerning the basic theory and suggests a number of questions on the application of conventional ideas about the behavior under improper transformations of these recent generalizations of quantum field theory—for example, non-Abelian gauge theories.

These questions seem to me to be important to our understanding of ultimate forms of the theory, and I hope that by raising them I shall stimulate others to give them some attention, even if only to show I am mistaken. If that is the case, so be it. To paraphrase a remark of Pauli's in an essay dedicated to Niels Bohr on his seventieth birthday, it is sometimes useful to start an argument without knowing the end of it.

A more prosaic question I have had to face is choosing a notation appropriate to the large variety of topics encompassed here. Each of these topics has associated with it its own more or less standard notation, and the same symbols may occur with entirely different meanings in connection with different topics. I have tried to use notation that is standard for a given topic and to explain my choice, where necessary, when it is introduced. At the same time, insofar as it is possible in these circumstances, I have made consistency of notation throughout the book one of my objectives. I have included in the index the key words that will locate the pages where frequently used symbols are defined, because I know from my own experience how annoying it can be to try to find the meaning of a symbol that suddenly appears at some point far removed from its definition.

I have benefited from discussions of time reversal with many colleagues over many years. Eugene Wigner called the subject to my attention in the early 1950s and stirred my original interest in it. During my years at the University of Wisconsin in Madison, my discussions with W. F. Fry about his experiments on K° mesons were very stimulating. In those years I also had the opportunity to spend a year at Princeton and began a lasting collaboration on the subject with Sam Treiman. At the University of Chicago I have had the benefit of many discussions with James Cronin and Bruce Winstein about *CP*-violation experiments and their interpretation. The person who pressed me for answers to some of the most cogent questions on time reversal and influenced me by his insistence on a clear, physical interpretation was my colleague and friend V. L. Telegdi.

The first draft of the manuscript was read by Lee Brekke, whose comments were very helpful. The comments of the two (anonymous) reviewers to whom the manuscript was sent by the University of Chicago Press were most helpful and led me to make a number of changes that certainly have improved the product. Other changes have been made in response to comments of K. C. Wali and R. H. Dalitz, both of whom read parts of the manuscript in its final stages.

I must also express my appreciation for the cooperativeness and patience of Fred Flowers, who, over a period of years, prepared, changed, and changed again the manuscript for this book. In addition, Roy Briere and Lynne Orr provided very substantial help by serving as independent proofreaders. Finally, I wish to acknowledge support over the years of my research from the Division of High Energy Physics of first the Atomic Energy Commission and then the Department of Energy, support that made it possible for me to study and learn the subject with which this book is concerned.

1 | Conventions, Symmetry Transformations, and Conservation Laws

The formulation of physical theories is greatly influenced by our perceptions about the space we live in and within which we observe the phenomena that are to be explained. Certain of these perceptions are universally taken for granted—possibly explicitly, but more often implicitly—and they may lead to the notion that some features of the description of physical systems are matters of convention rather than substance. But what appears to be a matter of convention sometimes turns out to be a substantive physical assumption, for example, an assumption about the structure of the empty space in which the physical system is embedded. As a result, the "convention" may have important physical consequences.

Among the features of the space that appear to be simply a question of convenience would be included the location of the origin of a coordinate system, because we assume that the space is uniform. Similarly, the orientation of the axes is assumed to be arbitrary because we believe the space to be isotropic. It is well known that both of these assumptions have important physical consequences; they lead to the laws of conservation of total (linear) momentum and total angular momentum, respectively, for isolated systems.

The ideas involved in formulating a time variable in physics are even more entangled with preconceptions, including a preconception about conventions, and these too lead to physical consequences. In order to elucidate the physical aspects of the nature of the time variable, it is convenient to consider analogies with the space variables, as far as they apply. Therefore I shall first summarize briefly some relevant aspects of space transformations.

1.1 Spatial Transformations

The conservation laws arise from the assumptions about the three-dimensional space through the spatial properties of the equations used to describe the motions of physical systems. I use the term "motion" here in its generic sense to describe motions of massive objects or the changes with time of either classical or quantum fields. The "equations of motion" make use of the reference frame that has been constructed to describe the system and, for isolated systems, incorporate the assumptions of uniformity and isotropy of the space as well as any other natural conventions. The implication is that the equations of motion are invariant under appropriate transformations of the coordinate system: uniformity of space implies translational invariance, isotropy implies rotational invariance, and these invariances lead inevitably to the conservation laws.

The conservation laws provide a means for experimental testing of the original assumptions, and the weight of the accumulated evidence in their favor for all aspects of physics is, of course, enormous. Nevertheless, it is always to be kept in mind that, in spite of their plausibility, such assumptions must be tested for their relevance to every aspect of physics.

The transformations of translation and rotation are referred to as "proper" transformations because they can be carried out continuously; they may be described in terms of continuously varying sets of parameters. Our interest here will focus on "improper" transformations, which cannot be so described. One such transformation is inversion or space reflection, the interchange of right-handed and left-handed reference frames.

Inversion enters into considerations of physics because of another common perception—that right- and left-handedness are matters of convention rather than of substance. Thus until Lee and Yang (1956) proposed and Wu and others (1957) confirmed a contradiction to this assumption, inversion was included among the transformations expressing the isotropy of space, and right- and left-handed coordinate systems were taken to be indistinguishable. Either implicitly or explicitly, the dynamic laws were assumed to be invariant under inversion.

Again, this invariance leads to a conservation law, in this case the conservation of parity for fields, such as the wave function of quantum mechanics. Because of its association with the parity quantum number, inversion is usually denoted by the symbol P. Although it has been well established since 1956 that parity conservation is not universal, the importance of the transformation P to physics has not been dimmed; rather, it has been enhanced. For one thing, our understanding of the structure of physical theories has been profoundly enriched by the discovery. For another, parity violation of the weak interactions makes possible methods for measuring properties of physical systems that otherwise would have been virtually inaccessible.

Because so much emphasis is being placed here on the symmetry (invariance) of the equations of motion under transformations, it is important to recognize that such symmetry does *not* imply corresponding symmetry of the motion itself. The motion is determined not only by the equations of motion but also by the initial or boundary conditions, which, being imposed externally, will usually violate the symmetry of the empty space, thereby causing a similar violation by the observed motion. We shall return to this question later.

1.2 Temporal Transformations

The nature and role of the time variable in physics are distinctly different from those of the space variable. The distinction between equations of motion and actual motion is of special importance for the time variable because our perceptions about time are influenced by the irreversibility of our everyday lives, which is a direct expression of motion rather than equations of motion, as we shall see.

The time variable is distinguished in other ways too: space and time variables clearly play distinct roles in Newtonian dynamics and in special relativity. That is explicit in the former case, and in the latter case the Lorentz transformations interrelating time and space variables make an invariant distinction between timelike and spacelike quantities. These distinctions mean not only that the time variable must be treated differently from the space variables, but also that the consequences of symmetry considerations may be quite different. In particular, we shall find that the time reflection transformation (time reversal) has consequences of a quite different kind from space reflection (inversion, P).

The qualitative meaning of the time variable is that of an ordering parameter in one-to-one correspondence to a sequence of events. As for all of the geometric variables defining the motion of a system, the quantitative definition of the time variable, t, is kinematic in nature—that is, it involves the noncausal relationships between time and motion, those relationships that are independent of the dynamics of systems. However, because the notion of time is inextricably intertwined with the dynamics (i.e., the actual motion) of systems, the complete separation of its kinematic and dynamic aspects is not possible, and one must instead be satisfied with a definition of t that appears to serve equally well for all dynamic systems. Thus the definition itself has physical content that is subject to experimental test to confirm its self-consistency. For example, we may choose a continuing sequence of well-established repetitive events as a starting point for the time ordering and choose the kind of event to be one associated with a simple dynamic system, such as the rotating earth, that *manifestly* has a constant period, then use the angle of rotation relative to the fixed stars as a scale of measurement (Whittaker 1944, p. 27). Of course, one is starting here with an assumption about dynamics based on a commonly held perception: the intervals associated with equal angles of rotation appear to be equal and reproducible when compared with intervals of time measured by an hourglass. But it is also a dynamic assumption that the hourglass measures equal time intervals. These assumptions are subject to continuing scrutiny of their consistency with the motions of other systems as described by the equations of motion based on the use of the chosen time variable. Comparing dynamic systems with one another not only tests consistency but also refines the method of measuring time using instruments (dynamic systems) for which the time intervals are much more closely reproducible than is the rotational period of the earth (or, for that matter, the interval determined by an hourglass).

Since the basic measure of time is taken to be the (constant) interval between full turns of the earth, only the time *interval* is defined, and the choice of time origin, $t = 0$, is a matter of convenience. But this again implies an assumption of uniformity in the time dimension, or invariance of all dynamics under time translations, which are proper transformations. The dynamic consequence (under certain restrictive assumptions about the dynamics such as the requirement that the forces are conservative) is again a conservation law for isolated systems, the conservation of energy.

The notion that time has a "direction" or "sense" is already built into this method of quantifying the time variable because, in describing a sequence of events, one necessarily presupposes that one event is known to occur before or after another. But again that supposition is based on a commonly held perception, in this case a perception of the sense of time resulting from our chemical, biological, and psychological experience, which is clearly irreversible.

How then does time reversal enter into physics? The answer is that it enters naturally as the result of the implicit assumption that the variable t, introduced to quantify the measure of time, is an algebraic variable; time intervals are additive. Time intervals are physically measurable quantities, and they can be assigned algebraic signs because we can add them to make longer intervals and subtract one from another to construct shorter intervals. Once the choice of an origin, $t = 0$, for the time variable has been made, the assignment of a sign, let us say, positive for t later than $t = 0$ and negative for t earlier than $t = 0$ appears to be merely a matter of convention. The "time-reversed" variable $t' = -t$ appears to have equal standing because time intervals $\Delta t = t_2 - t_1$ can just as well be expressed in terms of t', $\Delta t' = t_2' - t_1'$ without altering their essential algebraic properties.

However, we see immediately that

$$\Delta t' = -\Delta t, \tag{1.1}$$

which may make some difference in the appearance of equations describing motion. For example, if a particle goes from position $\vec{x}_1$ to position $\vec{x}_2$ in the interval (t_1, t_2), then its average velocity is $\vec{v} = (\vec{x}_2 - \vec{x}_1)/\Delta t$ as determined on the standard time scale. But if we use instead the time-reversed system t', then $\vec{v} = (\vec{x}_2 - \vec{x}_1)/\Delta t' = -\vec{v}$.

Another way to express these relationships is to introduce the time reversal transformation T where

$$T: \quad t \rightarrow t' = -t, \tag{1.2}$$

which is to be read as, "Under T, t transforms to $t' = -t$." It will be recognized that this is an improper transformation, a reflection akin to P. For the velocity it follows that

$$T: \quad v \rightarrow v' = -v, \tag{1.3}$$

and the transformation may be said to reverse the velocity. For this reason the transformation T is often referred to as "motion reversal" rather than "time reversal," but we shall find later that there are important aspects of the transformation having little to do with actual motion, and therefore we shall use the term "motion reversal" as a supplementary concept referring only to the transformation of variables describing motion, such as velocity, momentum, angular momentum, and so forth.

The reference system using t as a variable will henceforth be called the "standard system" and the one using t' as a variable, the "transformed" or "time-reversed" system. Since the transformation T from the one system to the other reverses velocities, it might appear that the two reference systems are physically distinguishable, a situation that would contradict our assertion that the difference between the systems is merely a matter of convention. In fact the reversal of velocities does *not* distinguish the reference systems because the sign of the velocity vector itself is merely a matter of convention. In prescribing the kinematic variables for particle motion we could just as well have defined the velocity by $-d\vec{x}/dt$ as by $d\vec{x}/dt$, and T simply causes a shift between the two.

The question now arises, If the difference between the two systems is merely a matter of convention, why are there any physical implications of the transformation T? The answer is that the physical implications arise from the dynamics, not the kinematics. The very statement that the choice of systems is "merely" a matter of convention implies a great deal about the dynamics of all systems because it means we cannot construct any dynamic system whatever that, by measurements of its motions, can reveal a difference between the two conventions. More explicitly, the assertion is that the (dynamic) equations of motion for all physical systems are unchanged (invariant) under the transformation T. Thus the acceptance of a convention led naturally to an implicit assumption of time reversal invariance (T invariance) despite the apparent irreversibility of nature, and this assumption was built into the formulation of physical theories until very recently, when the violation of parity conservation suggested that symmetry under improper transformations might not have the same "natural" standing as symmetry under proper transformations.

1.3 Time Reversal Invariance

Since time reversal invariance is a statement about the behavior of physical systems, it can be meaningful (in the sense of physics) only if it is subject to experimental test. As an assertion about dynamic behavior, it is in fact a statement about the forces and interactions between systems. Further, since in its original form it refers to all dynamic systems, it is an assertion about every force or interaction of nature and is subject to test for each such interaction—those already known and those that remain to be discovered.

To explain how this reversibility of time can be acceptable in our irreversible world, I repeat that it is important to recognize that the statement of invariance is a statement about the *equations* of motion, not about the actual motions of systems. Of course the allowed motions of systems are restricted by this invariance, but they are not themselves invariant, as we shall see much more closely in the next chapter.

Until parity violation was discovered, the broad assertion of universal time reversal invariance was generally accepted either implicitly or explicitly as a principle of physics, much like the assumption of isotropy of space leading to conservation of total angular momentum, or the translational uniformity of space leading to conservation of linear momentum. However, the principle of time reversal invariance, or "T invariance," does not lead directly to conservation laws, but it does restrict the form of the equations of motion, and these restrictions in turn affect the actual motion, which is given by the solution of those equations. These consequences depend on the nature of the physical system under consideration and may be quite subtle. Recognition of the principle has led to a useful tool for helping to find solutions to certain dynamic problems, and it has been used in exploring the dynamics of systems to circumscribe the mathematical form of acceptable interactions when the nature of the forces is unknown. T invariance is thus a powerful tool in elucidating the nature of the interactions between newly discovered physical systems.

The principle of universal T invariance therefore has an impact on all facets of physics, and its use provides a special opportunity to make predictions about interesting aspects of a great variety of physical systems. Furthermore, the principle is subject to continuing scrutiny by confronting these predictions with experimental tests. The tests may be direct, making use of the concept of motion reversal, or indirect, making use of the subtle connections between the transformation T and another fundamental transformation, CP, where C refers to charge conjugation (the interchange of all particles with their associated antiparticles) and P to inversion. CP is connected to T by the requirement of relativistic invariance under the continuous Lorentz transformations and by the structure of the theories of physics. This connection, the CPT theorem, makes possible an approach to the T-invariance question through tests of CP invariance, as we shall see. However, it must be kept in mind that the CPT theorem is based on physical assumptions and must itself be subjected to continuing experimental scrutiny.

Through this CP avenue, the surprising result of the landmark experiment of Christenson and others (1964) has led to the conclusion that T invariance is in fact not valid for certain special dynamic systems; therefore it is *not* a universal principle, as had been assumed before then. This conclusion, though generally accepted, has yet to be confirmed by a test of motion reversal, primarily because the degree of violation is very small in the domain of experiments accessible at present.

It should be apparent at this point that the study of time reversal is of interest not only because it provides a special insight into a great variety of physical problems, but also because the existence of a violation of T invariance demands explanation at a deeper level. Thus far T violation has proved to be associated only with very elusive phenomena, and its connections with the rest of physics are a great mystery. Its ultimate elucidation may very well have profound implications for both physics and cosmology.

1.4 Kinematically Admissible Transformations

The formulation of a time reversal transformation, or for that matter any of the transformations considered here, must avoid using properties of the forces or interactions that determine the dynamics, because it is the transformation properties of the dynamic equations that we seek to determine. Since the kinematics are those properties of the motion that are independent of the dynamics, we may accomplish this objective by requiring that the concept of an admissible transformation be formulated in kinematic terms.

In order to express explicitly the independence between the kinematics and the nature of the forces, we require that the transformations leave the equations of motion invariant *when all forces or interactions vanish.* Thus for a mass point in classical mechanics, a kinematically admissible transformation of the kinematic variables of position and time from $\vec{x}$ to $\vec{\xi}$ and from

t to τ, respectively, must transform the free-particle equation

$$\frac{d^2\vec{x}}{dt^2} = 0 \tag{1.4}$$

to the form

$$\frac{d^2\vec{\xi}}{d\tau^2} = 0. \tag{1.5}$$

If we take $\tau \equiv t$, the form of an admissible space transformation is found to be

$$x_i \to \xi_i = \sum_j a_{ij} x_j + b_i + c_i t, \qquad i = 1, 2, 3 \tag{1.6}$$

where x_i is the ith component of $\vec{x}$, a_{ij}, b_i and c_i are real numbers, and the matrix of the a_{ij} is nonsingular ($det\ a \neq 0$).

For $a_{ij} = \delta_{ij}$ and $c_i = 0$, eq. (1.6) describes a simple translation of coordinates by the constant vector $\vec{b}$. For $c_i \neq 0$, it is a Galilean transformation to a reference frame moving relative to the original frame with constant velocity $\vec{c}$. If a_{ij} is an orthogonal transformation with $det\ a = 1$, eq. (1.6) includes a proper rotation of axes. The case $a_{ij} = -\delta_{ij}$ corresponds to inversion:

$$P: \quad \vec{x} \to -\vec{x}. \tag{1.7}$$

Other forms of a_{ij} introduce a change of scale that, while kinematically admissible, is not expected to be of dynamic interest here. Thus we find that all the point transformations corresponding to our intuition about the uniformity and isotropy of the space are built into the way the kinematics have been formulated.

The kinematically admissible transformations of the time variable may be identified by noting that (Ames and Murnaghan 1929, p. 114) if $\tau = \tau(t)$,

$$\frac{d^2\vec{x}}{dt^2} = \frac{d^2\vec{x}}{d\tau^2}\left(\frac{d\tau}{dt}\right)^2 + \frac{d\vec{x}}{d\tau}\frac{d^2\tau}{dt^2}. \tag{1.8}$$

Thus the conditions eq. (1.4) and eq. (1.5), with $\vec{\xi} \equiv \vec{x}$, lead to

$$\frac{d^2\tau}{dt^2} = 0, \tag{1.9}$$

and admissible transformations are of the form

$$\tau = At + B. \tag{1.10}$$

It will be recognized that eq. (1.10) includes both time translations ($A = 1$) and time reversal ($A = -1$, $B = 0$), and again we find that our preconceptions are built into the kinematics, in this case the preconception of the uniformity and reversibility of time.

Since classical mechanics serves as the starting point for developing our concepts about quantum mechanics and quantum field theory, we shall take the classical results as a basis for characterizing the transformations in the quantum domain. However, it will be necessary to consider the fundamental structure of each theory in order to characterize the transformations that are of interest. Before doing so we shall examine some of the consequences of the assumption that the equations of motion of classical mechanics are invariant under time reversal.

2 | Time Reversal in Classical Mechanics

The Newtonian equations of motion of simple mechanical systems provide an opportunity to consider examples of the role time reversal arguments can play in the solution of specific problems. More important for our purposes is their use to develop the concept of "thought experiments" by means of which the effects of time reversal can be visualized. These imagined experiments lead to an understanding of the way actual experiments can be carried out to determine the behavior of systems under time reversal without doing the impossible—reversing the direction of flow of time.

For classical systems we will find that motion reversal acts as the surrogate for time reversal. To some extent the same will be true for the more fundamental quantum mechanical treatment, because, after all, the correspondence principle must be used to transfer our notions about time reversal to the quantum domain. However, there are additional implications of time reversal in quantum mechanics, making it a much richer source of information concerning the solution of problems and introducing important subtleties into the interpretation of experiments. This is a warning that some of the patently acceptable ideas developed in this chapter are due to be modified later.

2.1 Dynamic Symmetries

We introduce a simple test system to illustrate the invariance principles under consideration: a collection of N structureless mass points moving under the influence of mutual forces. The position of the αth mass point is denoted by $\vec{x}^{\alpha}$ with $\alpha = 1, 2, \ldots, N$. Given the existence of a set of kinematically admissible transformations of the $\vec{x}^{\alpha}$ and t, the question is: Do these principles relate to symmetry transformations of the real world? As pointed out in chapter 1, the way the definitions of our variables are related to measurement involves some assumptions about symmetries. The test of the validity of the assumptions rests directly on the effect of the transformations on the dynamics.

The dynamics are determined for our model system by the equations of motion

$$m_\alpha \frac{d^2\vec{x}^\alpha}{dt^2} = \vec{F}^\alpha(\vec{x}^1, \vec{x}^2, \ldots, \vec{x}^N; t), \tag{2.1}$$

where $\vec{F}^\alpha$ is the force acting on the αth mass point. The designation of the force as a three-vector implies that, under rotations of the coordinate axes, it transforms in the same way as the $\vec{x}^\alpha$. Thus if the *only* vectors entering into the determination of $\vec{F}^\alpha$ are $\vec{x}^1, \vec{x}^2, \ldots, \vec{x}^N$ it must be a vector function of these variables, and a given rotation applied to all of the $\vec{x}^\alpha$ will leave eq. (2.1) invariant. The equations are then consistent with the assumption of isotropy of the space. As we know, a particular consequence of this symmetry is conservation of total angular momentum.

Even if $\vec{F}^\alpha$ is a vector function of a set of internal vectors, $\vec{\sigma}^\beta$, that are associated with some of the mass points and independent of the corresponding $\vec{x}^\beta$, the rotational invariance will be maintained if the $\vec{\sigma}^\beta$ are subjected to the same rotation as the $\vec{x}^\beta$. However, if $\vec{F}^\alpha$ depends on any externally fixed vectors, as it will when external forces are acting, eq. (2.1) will no longer be invariant (because the fixed vectors are not rotated), and the total angular momentum will change with time in a manner determined by the torque associated with the external forces.

An isolated system is defined as one for which such external field vectors are absent; therefore we find that the Newtonian equations of motion for an isolated system reflect the (rotational) isotropy of space, that is, the absence of a fixed direction in empty space, in a natural way. The (translational) uniformity of space with the associated conservation of total linear momentum also is made evident in a natural way by remarking that, for an isolated system, the forces can depend only on the distance $(\vec{x}_\alpha - \vec{x}_\beta)$ between mass points. The assumption is that any specific dependence on the location of the origin of coordinates would imply the existence of an external agency fixed to the origin, contradicting the notion of isolation of the system.

As long as the internal properties of the mass points and the character of the three-dimensional configuration space are constant in time, there will be no explicit time dependences of the forces acting within the isolated system, and eq. (2.1) will be invariant under time translations, confirming the uniformity of the time variable. If the forces are "conservative," that is, expressible as the gradient of a potential that does not depend explicitly on the time, energy will be conserved.

Thus we find that the kinematically admissible *proper* transformations of space and time represent symmetries of Newton's equations of motion for isolated systems. The question is, What about the improper transformations P and T? Let us first consider the inversion P.

Since the $\vec{x}^\alpha$ are polar vectors, that is, vectors transforming in accordance with eq. (1.7), the Newtonian equations of motion will be invariant under P if the $\vec{F}^\alpha$ are also polar vectors; both sides of eq. (2.1) then change sign under P. However, even if the $\vec{F}^\alpha$ are vector functions only of the $\vec{x}^\beta - \vec{x}^\alpha$, there is no guarantee that they are purely polar vectors, because the vector products $\vec{x}^\alpha \times \vec{x}^\beta$ are axial vectors, which do not change sign under P:

(2.2) $$P: \quad \vec{A} \rightarrow \vec{A}$$

if $\vec{A}$ is any axial vector.

Furthermore, if there are internal vectors $\vec{\sigma}^\alpha$ that are axial vectors, for example, if the mass points are particles having a spin angular momentum, axial vector forces may be constructed from them. Finally, even the scalars that determine the size of the forces, such as the charge, could conceivably turn out to be pseudoscalars π, which have the property

(2.3) $$P: \quad \pi \rightarrow -\pi,$$

from which an axial vector force may be constructed, since a polar vector multiplied by a pseudoscalar is an axial vector. To turn this argument around, we note that a magnetic monopole moment would be expected to be just such a pseudoscalar because, when it is multiplied by the magnetic field, an axial vector, it yields a "natural" polar vector force.

This latter remark is an example of the conventional approach. We *assume* that the $\vec{F}^\alpha$ must be polar vectors and therefore restrict the functional dependence of the $\vec{F}^\alpha$ on the internal variables to stand in accord with this assumption. That is a severe restriction based on belief in invariance under P, and it can be justified as physics only by experiments.[1] Simple examples of the kinds of experiments to be used for this purpose will be considered later.

Turning now to time reversal, we see that the left-hand side of eq. (2.1) is independent of the sign of t because it involves the second time derivative:

(2.4) $$T: \quad \frac{d^2\vec{x}^\alpha}{dt^2} \rightarrow \frac{d^2\vec{x}^\alpha}{dt'^2} = \frac{d^2\vec{x}^\alpha}{dt^2}.$$

Therefore the requirement of invariance of the equations of motion under T

[1] The earliest reference to the need for such experiments that I am aware of is made by Purcell and Ramsey (1950). See also Ramsey (1957). I also recall earlier personal conversations with Ramsey during which he emphasized this point.

reduces to the requirement that the $\vec{F}^\alpha$ be independent of the sign of t. Even under the assumption that $\vec{F}^\alpha$ does not depend explicitly on the time or on mass point (particle) velocities, this is not a trivial requirement because of the possible involvement of internal parameters such as spin variables $\vec{\sigma}^\alpha$ and pseudoscalars. Spin, being an angular velocity or angular momentum, behaves like a velocity vector, eq. (1.3), under time reversal:

(2.5) $$T: \quad \vec{\sigma}^\alpha \rightarrow -\vec{\sigma}^\alpha.$$

Furthermore, a pseudoscalar defined within the constraints imposed by Lorentz covariance[2] will also be odd under T:

(2.6) $$T: \quad \pi \rightarrow -\pi.$$

Thus the only way to guarantee that the $\vec{F}^\alpha$ are even under T is to further restrict their possible dependence on the internal variables, and again, the only way to justify such restrictions is by resorting to experiments.

Evidently, invariance under the improper transformations is not as natural a characteristic of the laws of motion as we might have hoped. This distinct difference between the physics associated with proper and improper transformations arises because the former relate to transformations like translations and rotations of space-time that can be applied to real physical systems and are therefore "natural," whereas the discrete constraints imposed by the latter cannot be realized in any such direct physical manner.

The equations of motion, eq. (2.1), may be reformulated either as Lagrange equations or as Hamilton's canonical equations. The invariance properties are then properties of the Lagrangian or the Hamiltonian. In particular, invariance of the interaction term in the Lagrangian, referred to as the "interaction Lagrangian," or of the interaction term in the Hamiltonian, which is usually the negative of the interaction Lagrangian and is referred to as the "interaction energy" or just "the interaction," is needed to guarantee the invariance of the equations of motion. Therefore the requirement of invariance of the equations of motion under the improper transformations imposes specific restrictions in the form of invariance conditions on the way the interaction Lagrangian depends on the coordinates and the internal variables.

All these forms of the equations of motion are differential equations. In fact, not only for classical mechanics but also for electrodynamics and other field theories, differential equations are the elementary form of the equa-

[2] An antisymmetric tensor of rank four transforms as a scalar under proper Lorentz transformations. Since it has an odd number both of space and of time indices, three of the former and one of the latter, it is odd under P and under T.

tions of motion. I emphasized in chapter 1 that the invariance principles apply to the differential equations, but not necessarily to the solutions of these equations. The symmetries of the solutions can be no greater than the symmetries of the initial or boundary conditions that are imposed externally. This is the basis for the important distinction between the actual motion and the equations of motion.

Before delving into the general case of arbitrary initial conditions, let us consider an example (Painlevé 1904) for which the initial conditions share the time reversal symmetry of the dynamics. This example illustrates one way[3] T invariance may be used directly in the solution of a dynamic problem.

2.2 Painlevé's Theorem

In the introduction to his paper, Painlevé (1904) gives the impression that the dynamic question of how an arbitrarily oriented cat in free fall manages to land on its feet was a burning issue of the day. He raises the specific question whether the accomplishment would be possible if all the internal forces acting between the cat's moving parts were conservative forces, and he arrives at the conclusion that it would not be possible under that restriction.

Since a freely falling system may be treated as an isolated system in free space, a simplified version of the theorem, adequate for our purposes, may be stated as follows:

> If the relative motions of an isolated system of particles are governed by the actions of conservative forces acting between them, and if the initial velocities (time $= t_0$) of all particles vanish, the system cannot resume at a later time a configuration having the same relative positions as the initial one but with a different overall orientation in space.

Although there are some obvious counterexamples showing that this theorem as it stands is not exactly correct, I shall reproduce Painlevé's proof (in paraphrased form) because it appears to be a historic use of a time reversal argument. The weakness in the proof permitting the counterexamples will later be seen to be directly associated with an oversight in the use

[3] In fact, this is the only example in classical mechanics that I am aware of. I learned of it from Whittaker (1944, p. 70), where it is presented as an exercise. It is interesting that the invariance under time reversal of the equations of motion is not mentioned in Whittaker's book, and the only hint given for the solution of this problem is the reference to Painlevé, who made explicit use of the fact that, for conservative forces, the Newtonian equations of motion have this symmetry. For another type of classical application see note 5.

of T invariance that can be mitigated by stipulating an additional condition on the formulation of the theorem.

Painlevé's proof may be summarized as follows: Let $V(\vec{r}_{12}, \vec{r}_{13}, \ldots, \vec{r}_{\alpha\beta} \ldots)$ be the potential generating the conservative forces between particles, where

$$\vec{r}_{\alpha\beta} = \vec{x}^{\alpha} - \vec{x}^{\beta} \tag{2.7}$$

is the vector distance between the αth and β^{th} particles. As indicated, V is a scalar function assumed to depend only on the $\vec{r}_{\alpha\beta}$. The equations of motion, eq. (2.1), take the form

$$m_{\alpha} \frac{d^2 \vec{x}^{\alpha}}{dt^2} = -\vec{\nabla}_{\alpha} V. \tag{2.8}$$

It is clear that these equations are invariant under the transformation T.

The initial time t_0 will be taken to be $t_0 = 0$, at which time the particle positions are $\vec{x}_0^{\alpha}$ and, in accordance with the stated initial conditions,

$$\left(\frac{d\vec{x}^{\alpha}}{dt}\right)_{t=0} \equiv \dot{\vec{x}}_0^{\alpha} = 0; \tag{2.9}$$

that is, all velocities vanish. In this case the initial conditions are invariant under T, since

$$\left(\frac{d\vec{x}^{\alpha}}{dt}\right)_{t=0} = -\left(\frac{d\vec{x}^{\alpha}}{dt'}\right)_{t'=0} = 0, \tag{2.10}$$

and it follows that the *motion* is invariant under T; that is, the solutions $\vec{x}^{\alpha}(t)$ of eq. (2.8) and eq. (2.9) satisfy the relationships

$$\vec{x}^{\alpha}(-t) = \vec{x}^{\alpha}(t). \tag{2.11}$$

Let us suppose the contrary of the theorem, that at some time $t_1 > 0$ the system does return to its original relative configuration but is reoriented in space. Then, because the positions at $t = t_1$, denoted by $\vec{x}_1^{\alpha}$, differ from the $\vec{x}_0^{\alpha}$ by a common rotation R that is the same for all particles, the distance vectors $\vec{r}_{\alpha\beta}$ also are so related:

$$\vec{r}_{\alpha\beta}(t_1) = R[\vec{r}_{\alpha\beta}(0)], \tag{2.12}$$

where R is a three-dimensional orthogonal matrix (rotation) operating on the vector $\vec{r}_{\alpha\beta}$ and is the same for all α, β. Hence the potential energy, being a scalar function of the $\vec{r}_{\alpha\beta}$, has the same value at $t = t_1$ as at $t = 0$, and by conservation of energy, the kinetic energy must also have the same value. But the kinetic energy vanishes at $t = 0$; hence it must vanish again at

$t = t_1$. Since the kinetic energy is a positive definite function of the velocities, each velocity must vanish:

$$\left(\frac{d\vec{x}^\alpha}{dt}\right)_{t=t_1} = \dot{\vec{x}}_1^\alpha = 0. \tag{2.13}$$

Evidently the conditions at $t = t_1$ are identical with those at $t = 0$ except that the initial positions are rotated from $\vec{x}_0^\alpha$ to $\vec{x}_1^\alpha$. Therefore if $t = t_1$ rather than $t = 0$ is taken as the initial point, the foregoing arguments made relative to t may be repeated without change but starting from the new positions $\vec{x}_1^\alpha$. Hence, if we use a reference frame rotated by R with respect to the original frame, the motion from $t = t_1$ to $t = 2t_1$ in the new frame will be identical to that between $t = 0$ and $t = t_1$ in the old frame. Since the magnitude $r_{\alpha\beta}$ of the vector $\vec{r}_{\alpha\beta}$ is independent of the orientation of the reference frame, we find by n repetitions of this argument that

$$r_{\alpha\beta}(t + nt_1) = r_{\alpha\beta}(t); \tag{2.14}$$

the $r_{\alpha\beta}$ are periodic functions of t with period t_1.

We may also make use of the choice of t_1 as the initial time by introducing a translated time variable

$$\tau = t - t_1 \tag{2.15}$$

to rewrite the equations of motion and initial conditions. In the rotated frame, the equations written for position vectors $\vec{y}^\alpha = R^{-1}[\vec{x}^\alpha]$ in terms of τ will be identical with those given for $\vec{x}^\alpha$ in terms of t in the original frame. Therefore the motion will again be invariant under reversal of this translated time variable:

$$\vec{y}^\alpha(-\tau) = \vec{y}^\alpha(\tau). \tag{2.16}$$

It follows from eq. (2.15) that the solutions $\vec{x}^\alpha(t)$ satisfy the condition

$$\vec{x}^\alpha(t_1 + \tau) = \vec{x}^\alpha(t_1 - \tau) \tag{2.17}$$

as well as eq. (2.11) and, therefore,

$$r_{\alpha\beta}(t_1 + \tau) = r_{\alpha\beta}(t_1 - \tau). \tag{2.18}$$

Now let us consider a third time variable,

$$\tau' = t - \tfrac{1}{2}\, t_1 = \tau + \tfrac{1}{2}\, t_1. \tag{2.19}$$

Then from eq. (2.18) we have

$$r_{\alpha\beta}\left(t_1 + \tau' - \tfrac{1}{2}\, t_1\right) = r_{\alpha\beta}(t_1 - \tau' + \tfrac{1}{2}\, t_1) \tag{2.20}$$

or

$$r_{\alpha\beta}(\tfrac{1}{2}\, t_1 + \tau') = r_{\alpha\beta}(\tfrac{3}{2}\, t_1 - \tau'). \tag{2.21}$$

But then from the periodicity condition, eq. (2.14),

$$r_{\alpha\beta}(\tfrac{1}{2}\, t_1 + \tau') = r_{\alpha\beta}(\tfrac{1}{2}\, t_1 - \tau') \tag{2.22}$$

for arbitrary τ'. Thus $r_{\alpha\beta}(\tfrac{1}{2}\, t_1 + \tau')$ is an even function of τ', and its time derivative must be an odd function vanishing at $\tau' = 0$:

$$\left[\frac{dr_{\alpha\beta}}{dt}\right]_{t=t_1/2} = \left[\frac{d}{d\tau'}\, r_{\alpha\beta}(\tfrac{1}{2}t_1 + \tau')\right]_{\tau'=0} = 0. \tag{2.23}$$

By writing

$$r_{\alpha\beta} = (\vec{r}_{\alpha\beta} \cdot \vec{r}_{\alpha\beta})^{1/2} \tag{2.24}$$

we see that

$$\frac{dr_{\alpha\beta}}{dt} = \dot{\vec{r}}_{\alpha\beta} \cdot \hat{r}_{\alpha\beta}, \tag{2.25}$$

where $\hat{r}_{\alpha\beta}$ is the unit vector in the direction $\vec{r}_{\alpha\beta}$. Thus eq. (2.23) shows that the components eq. (2.25) of the relative velocities between the particles vanish at $t = \tfrac{1}{2}\, t_1$, so that at that instant the motion of the system may be described as a rigid body rotation. (Note that the center of mass is at rest by virtue of the initial conditions.) However, since total angular momentum is conserved and the total angular momentum vanished at $t = 0$, the system cannot be rotating at $t = \tfrac{1}{2}\, t_1$. Therefore

$$\left[\frac{d\vec{x}_\alpha}{dt}\right]_{t=t_1/2} = 0 \tag{2.26}$$

This result shows that at $t = \tfrac{1}{2}\, t_1$ the velocities satisfy the same initial conditions as those at $t = 0$, and all the arguments made using $t = t_1$ as an initial point may be repeated by using $t = \tfrac{1}{2}\, t_1$. In particular, the time reversal argument for the variable t' leads to the equivalent of eq. (2.17),

$$\vec{x}^\alpha(\tfrac{1}{2}\, t_1 + t') = \vec{x}^\alpha(\tfrac{1}{2}\, t_1 - \tau'). \tag{2.27}$$

Then for $\tau' = \tfrac{1}{2}\, t_1$,

$$\vec{x}^\alpha(t_1) = \vec{x}^\alpha(0), \tag{2.28}$$

the system has the same positions at $t = t_1$ as it did at $t = 0$. The only "rotated" set of positions $\vec{x}_1^\alpha$ that the system can attain are identical with the initial position $\vec{x}_0^{\alpha}$. This completes Painlevé's proof that there can be no rotation.

An obvious counterexample is offered by the two-particle harmonic oscillator, which comes to rest every half-period in a position rotated by 180° from the immediately preceding rest position.

The contradiction arises from an error in the application of the time reversal condition eq. (2.22) to obtain eq. (2.23). For the oscillators just described, the time t_1 is one-half the oscillator period. Therefore, in the particular case under consideration

$$r_{\alpha\beta}(\tfrac{1}{2}\, t_1) = 0. \tag{2.29}$$

However, this point is a branch point in the square root eq. (2.24) defining $r_{\alpha\beta}$. The requirement that $r_{\alpha\beta}$ be the positive square root implies a discontinuity in eq. (2.25) at $t = \frac{1}{2}\, t_1$, a point at which the unit vector $\hat{r}_{\alpha\beta}$ changes sign, yielding the required change in sign of $dr_{\alpha\beta}/dt$ for finite $\vec{r}_{\alpha\beta}$. Therefore one cannot conclude that the velocities vanish at $t = \frac{1}{2}\, t_1$; in fact, we know that for the example of the oscillator this is the point of maximum velocity.

The difficulty can be avoided by requiring that the particles be impenetrable infinitesimal spheres. It seems likely that Painlevé, in thinking about particles (or parts of the cat) in classical terms, would not have considered any other possibility.

2.3 The Role of Initial Conditions: Simple Systems

We now turn to the general situation in which the initial conditions are unrelated to the symmetry and consider how we might carry out experiments that are limited to observations of actual motions. The trick is to use the motion to reveal the underlying symmetry of the equations of motion in spite of the fact that the symmetries of the motions will be no greater than those of the initial conditions. Our interest concerns improper transformations for which the question is easily resolved for these classical experiments by making use of generalized "mirrors." We shall consider space reversal (P) and ordinary mirrors as a starting point to illustrate the method. It should be kept in mind here that reflections in a flat mirror, lying in, say, the y–z plane through the origin, are equivalent to the transformation P combined with a rotation by π about the x axis.

A mechanical system consisting of a structureless mass point ("particle") moving under the influence of attractive central forces exerted by a massive center will be taken as a model. A possible orbit of the particle is shown in figure 2.1*a*. Since space reversal is the interchange of right-handed and left-handed coordinate systems, its effect can be obtained by viewing the orbit in a mirror, shown as the image in figure 2.1*b*. Clearly the image

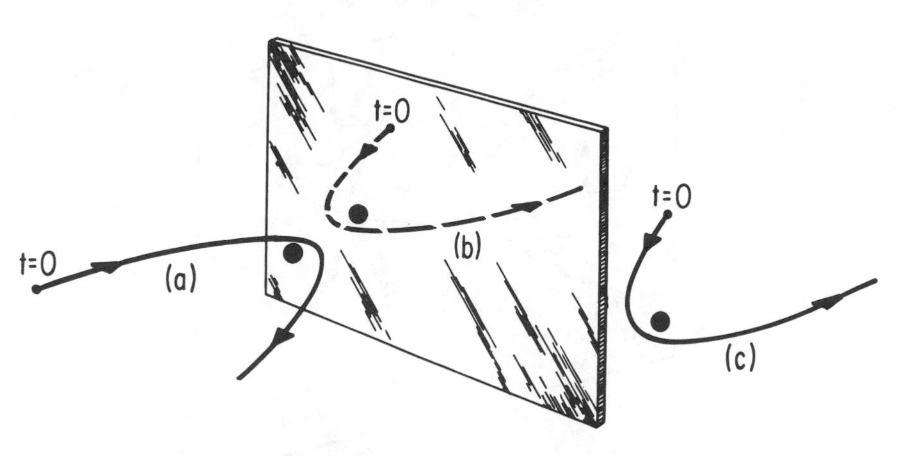

FIG. 2.1 (*a*) Orbit of a particle moving under the influence of an attractive central force. (*b*) Image of the orbit in the mirror—illustrates space reversal of the *motion*. (*c*) Actual orbit when initial conditions are same as in (*b*).

describes motion that is distinctly different from that of the original even if we move (translate) the image into a position bringing the two centers of force into coincidence. However, that does not mean that the equations of motion obtained from a point-by-point measurement of the image orbit will be different from those obtained by measuring the original orbit.

The point is that the initial conditions as seen in the mirror are different from the original ones. At $t = 0$, both the position and the velocity of the images are different from those of the original. To test the invariance of the equations of motion, we may repeat the experiment using initial conditions *as seen in the mirror*. This is illustrated in figure 2.1*c*. If this second experiment leads to an orbit that is point-by-point identical (after appropriate translations of coordinates) with the *image*, then we can say that the experiment is consistent with invariance of the equations of motion under the transformation P.

Of course, if the orbit in the second experiment, figure 2.1*c*, differed from that of the image, it would appear that P invariance was violated. However, the notion of P invariance is so deeply ingrained that we would be more likely to interpret such a "violation" as being due to an oversight on our part—to overlooking some hidden variables that were therefore not taken into account in our analysis of the experiment. Thus, although experiments viewed in mirrors appear to be utterly trivial, in principle they offer an opportunity to discover such hidden variables, which could represent profound information about the structure of the system being investigated.

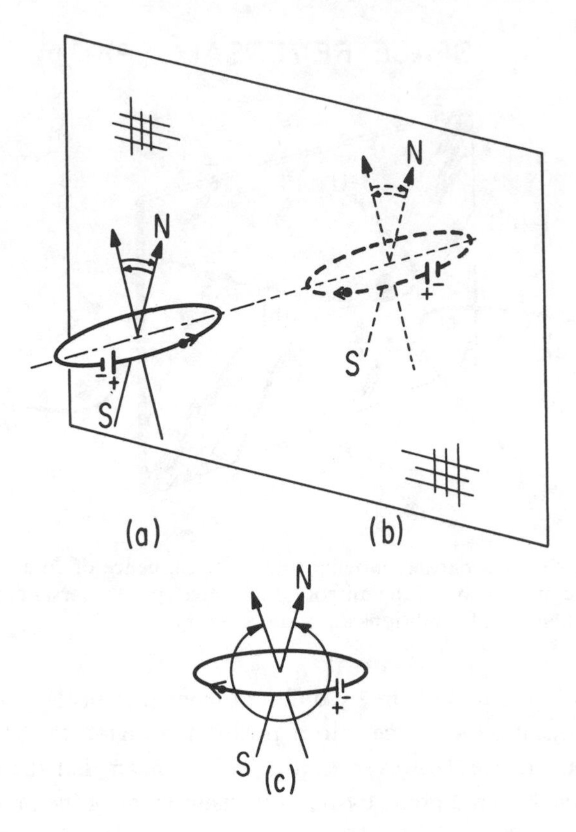

FIG. 2.2 (*a*) Small oscillations in a plane parallel to the mirror of an infinitely thin (one-dimensional) magnetic needle about an axis on the diameter of the loop. (*b*) The image in the mirror perpendicular to the axis of oscillation. (*c*) The motion in the laboratory when conditions are those seen in the mirror.

Let us illustrate this point by a second example, simple but quite realistic, illustrated in figure 2.2. A magnetic needle is pivoted at the center of a diameter of a current loop and set into small oscillations about its equilibrium position as shown in figure 2.2*a*. The motion is observed in a mirror whose plane is perpendicular to the diameter: figure 2.2*b*. This image motion is to be compared with actual motion when the conditions seen in the mirror are used as laboratory conditions in a second experiment, figure 2.2*c*.

Since the direction of current in the loop is reversed in the mirror and therefore in the second experiment, the magnetic field in the second experiment is opposite to that in the first. Thus we know that the motion in the latter case will be the large oscillations indicated in figure 2.2*c*; the motion is completely different from that seen in the mirror.

Before jumping to the conclusion that this is a violation of P invariance, we must ask whether all the properties of the magnetic needle have been taken into account. The needle, being a permanent magnet, has been treated as though it had no internal structure. If we assume that it consists of equal numbers of separated permanent magnetic north and south poles (static magnetic monopoles), the only way to preserve P invariance would be to conclude that magnetic monopoles are pseudoscalar,[4] that is, that under the transformation P a positive (north) magnetic pole changes into a negative (south) pole and vice versa. This pseudoscalar property suggests that the monopoles themselves have spatial internal structure in order to account for their dependence on space reflection.

On the other hand, if we ascribe the magnetism of the needle to (microscopic) electric currents, as we know we must, then the initial conditions on these currents are modified in the mirror in the same way as those of the large current loop. Because we had not observed the microscopic motions in the mirror, we did not realize that the mirror image of a north pole is a south pole. When the initial conditions used in the second experiment include this reversal of the needle, the oscillations in the reversed field will be identical with those seen in the mirror.

Such an approach to the analysis of permanent magnets could have been used before the discovery of the electron to argue that the microscopic magnetic dipole moments responsible for permanent magnetism must be axial vectors rather than polar vectors. This in turn implies that the microstructure is more complicated than a collection of simple (polar) vectors (the product of two polar vectors or one polar vector and a pseudoscalar is required to construct an axial vector). Imagine the models of magnetic structure that could have been invented on this basis!

A similar treatment of time reversal requires introducing a "mirror" of the time variable. A perfectly practical mirror exists for classical mechanical systems. We simply take a motion picture ("movie") of the system with high enough resolution (in both space and time) to make measurements of the motion in the projected film. Then by running the film backward we can observe and make measurements on the motion as a function of the reversed time variable, t'.

Let us return to consideration of the system consisting of a particle moving under the influence of attractive central forces. The original is shown in figure 2.3*a* and is the same as that of figure 2.1*a*. When the movie is reversed, the apparent motion is that shown as the dashed curve in figure 2.3*b*. It is clear that the "reflected" motion is just the reverse of the original motion.

[4] See note 2.

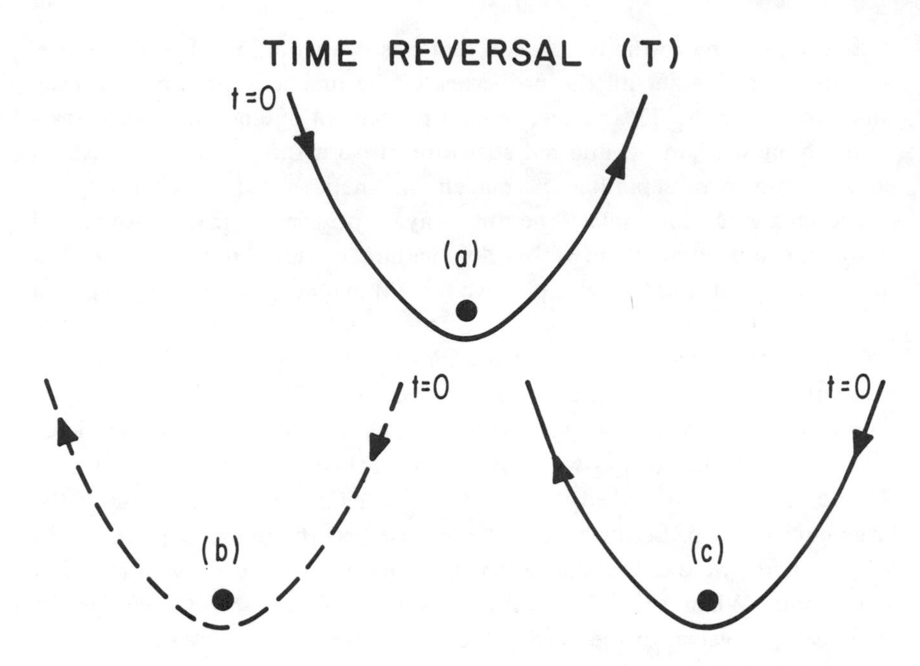

FIG. 2.3 (*a*) Orbit of a particle moving under the influence of an attractive central force. (*b*) The orbit as seen in a reversed movie. (*c*) Actual orbit when initial conditions are same as in (*b*).

Again, the reversal of the motion can be ascribed to the change in initial conditions. At the new initial point the position and the velocity differ markedly from the original; in fact these values depend on where we choose to terminate the original movie. We may proceed as before to determine what, if any, effect time reversal has on the equations of motion; the experiment is repeated with initial conditions as seen in the reversed movie. If the orbit for this second "real world" experiment, figure 2.3*c*, is exactly the same as the orbit seen in the reversed movie, then the results are consistent with T invariance of the equations of motion.

On the other hand, if the second experiment does not yield the orbit of figure 2.3*b*, we are again faced with the possibility either that T invariance is violated or that we have overlooked hidden variables. And again, our inclination would be to choose the second explanation.

As an illustration, we may pursue further the question of the dynamics of systems making use of permanent magnets. Consider an electrically charged particle moving through a static, uniform magnetic field as indicated in figure 2.4*a*. The magnetic field is assumed to be produced by a permanent magnet appearing as a static object in the movie. The orbit resulting from

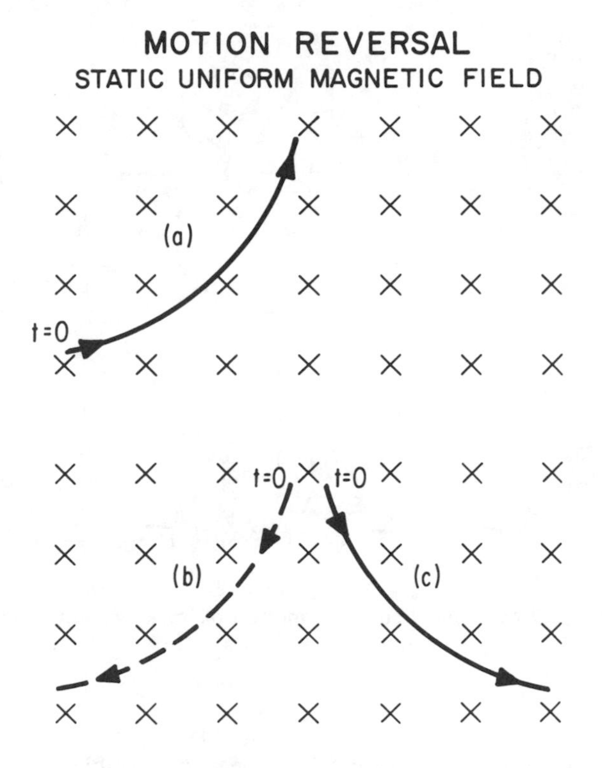

FIG. 2.4 (*a*) Orbit of a charged particle in a uniform magnetic field. (*b*) The same orbit as observed in a reversed movie. (*c*) The actual orbit when initial conditions are same as in (*b*), assuming a permanent magnet.

reversal of the movie is that shown in figure 2.4*b*. If we fix the conditions at the new initial time as those seen in figure 2.4*b*, then the actual motion will turn out to be that shown in figure 2.4*c*, which is utterly different from the orbit seen in the reversed movie.

If we are to explain this result without resorting to T violation, we must invoke some aspect of the structure of the magnet that is not manifest in the movie. This is, of course, the microscopic current structure, and if the resolving power of the camera had been great enough, these currents would appear to be reversed in the reversed movie. Therefore we would assign a magnetic field in the opposite direction from that shown in figure 2.4*b* and *c*, and the second experiment would not yield the orbit of figure 2.4*c* but would replicate the orbit of figure 2.4*b* in accordance with T invariance.

From the point of view of nineteenth-century physicists who were unaware of the existence of the electron and its contribution to the structure of matter, the importance of this set of experiments on the structure of permanent magnets as compared with the set investigating P invariance is that,

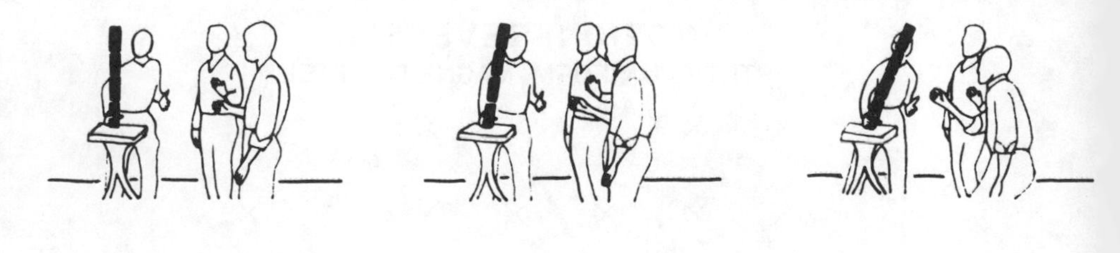

FIG. 2.5*a* Movie of experiment showing motion of bricks when the initial balanced stack is tipped.

while the results of the latter would have been susceptible to a static model, the former would have been needed to invoke *motion*, since the time dependence is essential for that purpose.[5] Although it is true that a permanent magnet constructed of magnetic monopoles would appear to be a static system, the fact that the pseudoscalar magnetic charges (monopole moments) must[6] change sign under time reversal suggests that, if such monopoles exist, they have a dynamic origin at some deep level.[7]

2.4. The Role of Initial Conditions: Complex Systems

The motion or "state" of an N-particle system is determined by the $3N$ initial positions and $3N$ initial velocities or, equivalently, by its initial position in the $6N$-dimensional phase space, when the forces acting on the

[5] Although this historical construct may seem very artificial, there is at least one example in which similar classical arguments have been used to obtain new results. Zocher and Török (1953) determine which crystal structures do and do not permit the various types of magnetomechanical effects by making use of the time reversal properties of the magnetic field.

[6] See note 2.

[7] The "natural" appearance of magnetic monopoles in versions of non-Abelian gauge theories is consistent with this statement. The required internal motions are provided, for example, by color currents of quantum chromodynamic origin. See Goddard and Olive (1978).

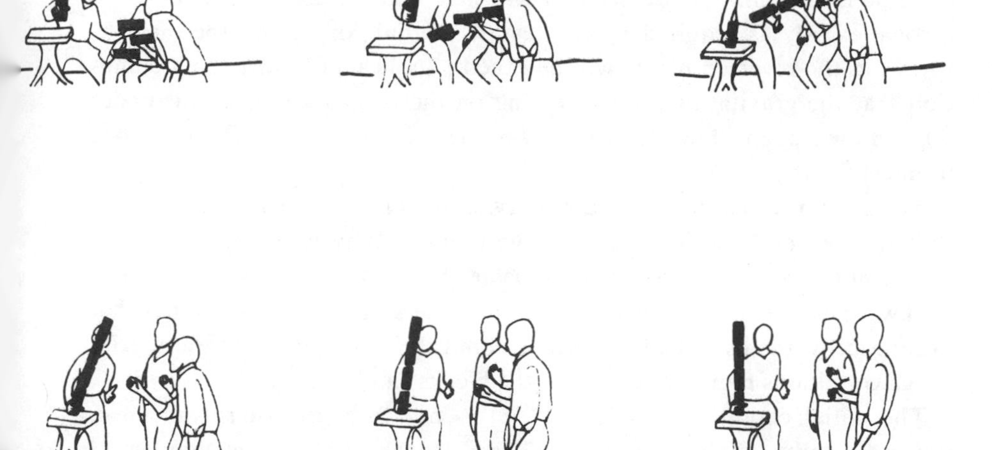

FIG. 2.5*b* Reversed movie using same film showing the bricks being thrown up into a stacked position.

particles are given. If one considers n possible one-particle initial states for each of the particles, the corresponding number of initial states for the system is n^N. Therefore, as the number of particles, N, increases, there is a very rapid increase in the number of available states, and hence in the amount of information required to specify the motion.

If the particles have structure, or if we consider dynamic systems requiring parameters other than the positions of a point in space to describe their motion, there is a corresponding further increase in the number of variables. As we shall see, this rapid increase in the complexity of the information required to specify a state of the system has a profound effect on its *apparent* behavior under time reversal whether or not the dynamics are T invariant.

As an illustration, consider the time-lapse sequence of events shown in figure 2.5*a*. The system under consideration consists of the four blocks, which constitute a moderately complex system. In the initial state the blocks are stacked vertically and aligned exactly. The initial velocities are prescribed by giving a common initial angular velocity about the bottom edge to the entire stack. Thus this initial condition describes a highly ordered system with sharply defined values of the variables.

To test the T invariance of the dynamics we now follow the procedure described in section 2.3; we run the movie represented by figure 2.5*a* backward and see the sequence shown in figure 2.5*b*. The test of T invariance is now to repeat the experiment as seen in the reversed movie with initial conditions matching those seen in the first frame 2.5*b*. If each block is thrown exactly as required by the reversed initial conditions, the motions shown in the reversed movie will be exactly duplicated (under the assumption that the gravitational forces acting on the blocks are invariant under T), and the stack will be reconstituted in the ordered state shown in the last frame of figure 2.5*b*.

There is no reason to doubt that gravitational forces are T invariant and that this reversed motion is a possible motion. Why is it, then, that any viewer of the reversed movie will immediately realize that it is a movie run backward rather than the actual motion of the system? The answer is that experience warns us to be incredulous about the possibility that the precise initial conditions required to restack the blocks can be attained.

The initial conditions of figure 2.5*b* do not have the sharply defined, ordered quality of those in figure 2.5*a*; because of the large number of variables in the system, there is a very large number of states having very nearly the same initial conditions, and it is highly improbable that we will be able to physically select from among this large number of possible initial states precisely the one that will lead to the exact motion of figure 2.5*b*.

Thus we find that the apparent irreversibility of motion from an ordered state to a disordered state of a complex system is associated with the irreversibility of the initial conditions; although in a T-invariant system the motion of each variable is exactly reversible, the probability of finding the precise initial conditions required to attain the exactly reversed motion is very small, exponentially smaller the greater the complexity of the system.

This observation makes possible an understanding of the apparently contradictory fact that the entropy of a system inevitably increases with time although the behavior of the atoms is determined by reversible (T-invariant) equations of motion. What may have been the most important application of T invariance in classical physics was made in this connection by Loschmidt (1877), who argued that Boltzmann's (1872) proof of the H theorem led to a paradox, since he used T-invariant equations of motion to derive a result that violated T invariance.

Boltzmann calculated the rate of change, owing to elastic collisions between molecules, of the distribution of velocities $f(\vec{v}, t)$ in a gas and, after making "plausible" assumptions about the separation of velocity and spatial distributions, found that

(2.30) $$\frac{\partial H}{\partial t} \leq 0,$$

where

$$H = \int d^3v f(\vec{v}, t) \ln f(\vec{v}, t). \tag{2.31}$$

Now if H_0 is defined by replacing $f(\vec{v}, t)$ by $f_0(\vec{v})$, the Maxwell-Boltzmann distribution for a perfect gas, H_0 turns out to be related to the equilibrium value of the entropy of the gas by

$$S_0 = -kH_0, \tag{2.32}$$

where k is the Boltzmann constant and S_0 the entropy per unit volume. This suggests that H be taken to be similarly related to the nonequilibrium value of the entropy and that the derivation of eq. (2.30), which implies the inevitable increase of entropy toward equilibrium, is a "proof" of the second law of thermodynamics for the perfect gas.

Loschmidt's observation was that in view of the T invariance of the dynamics underlying the motion of the molecules, with every distribution $f(\vec{v}, t)$ determined by the paths followed by all the particles there must be associated another possible distribution $f'(\vec{v}, t) = f(-\vec{v}, -t)$ obtained by time reversal of the motion (including the initial conditions). But the quantity H' defined as in eq. (2.31) with f replaced by f' then must satisfy.

$$\frac{\partial H'}{\partial t} \geq 0, \tag{2.33}$$

in contradiction to the second law of thermodynamics.

This paradox led to a very active debate concerning both the nature of the error in Boltzmann's proof and the way it could be rectified. A good review of the debate has been presented by the Ehrenfests (1912).[8] Let me remark here only that the central issue was Boltzmann's assumption (*Stosszahlansatz*), used in making an evaluation of the molecular collision rate, that the velocity distribution is independent of the spatial distribution of the molecules for every value of t.

Although resolution of the compelling questions raised by this issue goes to the very foundations of statistical mechanics, these issues are not our concern here. We are concerned with the more general question, How can the second law of thermodynamics, expressed as

$$\frac{\partial S}{\partial t} \geq 0,$$

[8] See especially the preface by Tatiana Ehrenfest in the translation of the original article. A thorough textbook discussion of the subject is to be found in the book by ter Haar (1954), especially app. 1, p. 331.

be consistent with time reversal invariance of the dynamics? Our answer is framed in terms of the way initial conditions are transformed under T.

The exact (microscopic) distribution function for the positions and velocities of each element of a complex system at a given instant t_1 may be considered a statement of initial conditions for the subsequent motion of the system. Let us assume that this initial function describes a highly ordered, nonequilibrium distribution. It can be expected that, as the result of collisions, the subsequent distribution at time t_2 will become increasingly disordered as time (t_2) goes on. If we now carry out the exercise that is the analogue of figure 2.5—that is, take as an initial distribution the exact distribution at t_2 with all velocities reversed—the original ordered distribution at t_1 (with all velocities reversed) will be restored in the time interval $t_2 - t_1$. It is assumed here that the equations of motion are T invariant.

The issue is not whether the motion is reversible but, rather, whether it is possible to prepare the required exactly reversed disordered distribution. The answer is no, as in the case of figure 2.5, because of the large number of parameters involved. Since the number of possible states grows exponentially with the number of degrees of freedom, the diffculty is greatly amplified as the complexity of the system increases.

For thermodynamic systems, the parameters that can be determined are gross averages over the multitude of microscopic parameters, so that an average over many possible microscopic distributions is used in describing a given thermodynamic state of the system. Thus, prescribing the thermodynamic state is not equivalent to prescribing the many initial conditions required to provide exact reversibility.

Gibbs (1931), in his formulation of statistical mechanics, introduced the concepts of "fine-grained" and "coarse-grained" averages to distinguish the methods of kinetic theory from the methods of statistical mechanics and thermodynamics. He suggested an example that serves nicely to illustrate how microscopic information required for exact reversibility is lost when the measured quantities consist only of coarse-grained averages (p. 144). Consider a fifty-fifty mixture of an opaque (opacity $\sigma = 1$) viscous liquid and a transparent ($\sigma = 0$) liquid in a beaker. Take as the initial state one in which the two components are completely separated, the opaque half at the bottom and the transparent half at the top. The average opacity is $\langle\sigma\rangle = \frac{1}{2}$, and the average of the square of the opacity is $\langle\sigma^2\rangle = \frac{1}{2}(0 + 1) = \frac{1}{2}$.

Now stir the liquid until it is thoroughly mixed. Its opacity σ' will then appear to be uniform and equal to the average, $\sigma' = \frac{1}{2}$, so that $\langle\sigma'^2\rangle = \frac{1}{4}$. However, σ' is a coarse-grained observation because we have measured only the gross opacity. To obtain the fine-grained structure, we examine the liquid with a microscope of sufficient resolution to see the intermingled threads of transparent and opaque liquid that have been created by the stirring. Then half of the volume will be found to be occupied by opaque threads and half by transparent threads, so the averages of the fine-grained opacity σ'' are $\langle\sigma''\rangle = \frac{1}{2}$ and $\langle\sigma''^2\rangle = \frac{1}{2}$, as in the initial situation.[9]

It will be noted here that the coarse-grained average of the square of the opacity is inevitably *decreased* by the mixing, whereas the fine-grained average is unchanged. Gibbs viewed the process of mixing in the fluid as an analogue to the motion of a set of points in a $6N$-dimensional phase space, each point representing the phase of a member of an ensemble of identical N-particle systems. The analogue of $\langle\sigma^2\rangle$ is $\int F \ln F$ over the phase space, where F is a coarse-grained distribution of the set of points in phase space. This integral is then the coarse-grained average of $\ln F$, which Gibbs showed to be a minimum for $F \equiv F_0$, the equilibrium (canonical) distribution, just as the coarse-grained average $\langle\sigma'^2\rangle$ attains its minimum value in the fully mixed case. The inevitable increase in entropy, which is defined as the coarse-grained average of $\ln(1/F)$, as the system approaches equilibrium, is thereby related to the increase in the variety of states that are consistent with the macroscopic measurements (of thermodynamic variables) on the system. The coarse-grained equilibrium state is the one having the greatest probability, that is, the greatest variety consistent with the imposed thermodynamic conditions.

Just as in the case of the stirred liquid, if f is the actual distribution of points in phase space for a system of particles, according to Liouville's theorem (Goldstein 1980, p. 426), it is constant in the neighborhood of a point in phase space that is moving with the system. Therefore the corresponding fine-grained average $\int f \ln f$ is unchanged by the motion and is invariant under time reversal. This result does not depend on the T invariance of the Hamiltonian. Only the form of the canonical equations underlying the proof of Liouville's theorem enters into it.

If in a reversible system f were measured by means of a supermicroscope at two instants of time t_1 and t_2, the observed motion between t_1 and t_2 would be completely reversible in the sense of figure 2.3. The irreversibility

[9] This may appear to be a contrived example, but Clyde A. Hutchison, Jr., has constructed an apparatus that demonstrates the behavior described here. This demonstration is shown on film no. 991135, "Time Reversal," produced by Bell Telephone Laboratories, Murray Hill, New Jersey. It is a beautiful demonstration of Gibbs's point because it shows that when two viscous liquids are mixed they can be unmixed and restored to the original state by reversing the mixing procedure, thus illustrating that all the information is stored in the threads of the liquid without requiring a microscopic examination to verify it. Unfortunately the title and narration of the film are misleading, because the dynamics of viscous fluids are not, in fact, T invariant. It is only because there is essentially no motion (the stirring must be done slowly) that the demonstration works.

of the macroscopic (coarse-grained) entropy results from the great variety of possible microscopic states that are consistent with a given macroscopic (thermodynamic) state. As in the case illustrated by figure 2.5, it would be incredibly difficult to select initial conditions from among this great variety of microscopic states with sufficient accuracy to take advantage of the reversibility of the motion.

The same line of reasoning leads to an understanding of the irreversibility of the physical and biological evolutionary processes of the universe, whether or not the dynamics of the evolutionary processes are invariant under time reversal. The number of parameters that must be controlled precisely to reverse the evolution of even the smallest organism is so large as to make the reverse motion incredible.

The irreversibility of the processes of the universe is often said to define an "Arrow of Time" that fixes the forward sense of time. From Gibbs's viewpoint this irreversibility is not specifically related to time but is characteristic of *any* change of the coarse-grained distribution of an isolated system in its approach to equilibrium. Even if the particles making up the system are subject to forces that violate T invariance, there is the same enormous statistical weight in favor of initial conditions leading to coarse-grained distributions that correspond to increasing entropy. Thus the Arrow of Time has little to do with the time variable as measured by physicists. In particular it has no bearing on the physics of time reversal.

3 | Time Reversal in Quantum Mechanics

The consequences of time reversal invariance for the solutions of the equations of quantum mechanics were formulated by Wigner (1932). He established the general notion of the time reversal transformation of quantum states, and his ideas will, from this point forward, provide the basis for all of our discussion. Although his development of the transformation was predicated on a belief in time reversal invariance, his results are more general.

To show that they are, I wish to avoid any a priori prejudice in favor of universal time reversal invariance. Therefore it is important to define the transformation in a manner independent of the T-invariance question.

As we have seen in dealing with classical theories, the time reversal transformation is to be defined by introducing a kinematically admissible transformation. This requires a clear distinction between kinematic and dynamic concepts and therefore leads us to give some attention to those formal aspects of the formulation of quantum mechanics that are essential to making the distinction.

This is, in fact, a good place to call attention to the three elements of the formal structure we have come to expect to encompass all dynamic theories of physics. They are:

A. The mathematical manifold within which the motions or states of the physical system are to be described.
B. A set of kinematic observables (measurable quantities whose definitions are independent of forces or interactions) and the kinematic (noncausal) relationships between them that determine the specific structure of the manifold for a given physical system.
C. The general structure of the dynamic equations giving the causal relationships between the kinematic variables.

Each of these elements has been invoked at some point in our discussion of T in Newtonian mechanics, and we will find that each is essential for our discussion of T and the implications of T invariance in both quantum mechanics and quantum field theories. To set the stage for treating both of these cases, I shall introduce the discussion of T invariance in quantum mechanics with a brief review of these three elements of its formal structure.

3.1 The Formal Structure of Quantum Mechanics

The foundations of quantum mechanics include three classes of assumptions, corresponding to each of the three requirements above:

A. Manifold

1. The (pure) physical state of a system is described by a ray $|\psi\rangle$ in a unitary function space (Hilbert space). The inner product between two such states $|\psi\rangle$ and $|\phi\rangle$ is denoted by $\langle\psi|\phi\rangle$.[1]
2. Each measurable physical quantity (observable) or measurable function of a physical quantity is represented by a Hermitian linear operator in this function space.
3. The only numerical values attainable in a measurement of an observable or a function of an observable are the eigenvalues of the associated operator, and any such eigenvalue is a possible result of such a measurement.
4. The expectation value for measurements of a given observable Q when the system is in state $|\psi\rangle$ is given by the Hermitian form[2]

(3.1) $$\langle Q\rangle = \langle\psi|Q|\psi\rangle$$

if the ray $|\psi\rangle$ is normalized so that

(3.2) $$\langle\psi|\psi\rangle = 1.$$

B. Kinematics

The kinematic relationships are associated with the algebra of the operators representing observables of the system and are usually formulated as commutation relations determining the algebra. For example, for a system consisting of structureless particles the kinematic observables

[1] I assume that the reader is familiar with quantum mechanics and the elementary mathematics of linear vector spaces in Dirac notation. My notation is such that if a member of the orthonormal set of basis vectors of the Hilbert space is denoted by $|\xi\rangle$, where ξ is a continuous or discrete index, or both, then $|\psi\rangle = \int d\xi\ |\xi\rangle\langle\xi|\psi\rangle$, where $\langle\xi|\psi\rangle$ is a complex number ("component" of $|\psi\rangle$) and $\langle\psi|\phi\rangle = \int d\xi\ \langle\psi|\xi\rangle\langle\xi|\phi\rangle$. Here $\langle\psi|\xi\rangle$ is the conjugate complex, $\langle\xi|\psi\rangle^*$, of the component so that $\langle\psi|$ is the Hermitian conjugate of $|\psi\rangle$. "Unitary function space" is defined in sec. 4, chap. 1 of Weyl (1931). See also p. 143 of Weyl.

[2] This is a bilinear form in the variables $\langle\xi|\psi\rangle$ and $\langle\psi|\xi\rangle$ in the notation of note 1:

$$\langle\psi|Q|\psi\rangle = \int d\xi \int d\xi'\langle\psi|\xi\rangle\langle\xi|Q|\xi'\rangle\langle\xi'|\psi\rangle.$$

Usually the basis vectors are chosen to be the eigenvectors $|q\rangle$ associated with the eigenvalue q of the observable Q.

may be taken to be the positions associated with the vector operators $\vec{X}^{\alpha}$ for the αth particle, and the momenta associated with the vector operators $\vec{P}^{\alpha}$. The algebraic structure is determined by the commutation relations

(3.3) $$[P_j^{\alpha}, X_k^{\beta}] = -i\hbar\delta_{jk}\delta_{\alpha\beta}.$$

If a particle has a spin $\frac{1}{2}\hbar$, there is associated with it a spin operator $\vec{\sigma}\hbar$ satisfying the kinematic commutation conditions

(3.4) $$[\vec{\sigma} \times \vec{\sigma}] = 2i\vec{\sigma}.$$

Other internal properties of the particles require the introduction of other operators and their commutation relations. A particular example is that of isotopic spin, for which an operator having commutation relations of the same form as eq. (3.4) is introduced. As we shall see later, the behavior of such internal variables under T must be included in establishing a kinematically admissible transformation.

C. Dynamics

The dynamics are determined by the operator H representing the Hamiltonian, which is an observable given as a function of the kinematic variables (observables). In the Schroedinger representation the time dependence of the state $|\psi(t)\rangle$ of a system is determined by the differential equation

(3.5) $$i\hbar\frac{\partial|\psi(t)\rangle}{\partial t} = H|\psi(t)\rangle.$$

An equivalent statement of the dynamics can be made in the Heisenberg representation, for which the explicit time dependence is assigned to the operators representing observables. The operators satisfy the dynamic equation

(3.6) $$-i\hbar\dot{Q}(t) = [H, Q(t)].$$

When the dependence of H on the kinematic variables is given, eq. (3.6) may be used along with the kinematic commutation relations to obtain equations of motion for each of those variables.

3.2 Wigner's Time Reversal Transformation

The transformation $t \to t' = -t$ of the time variable may be expected to transform the state vectors and observables of any system. The transformation of the state vector will be denoted by an operator T and the transformed state or operator representing an observable by the prime of the original state or observable. Thus the transformation in Schroedinger representation is written

$$|\psi'(t')\rangle = T|\psi(t)\rangle \tag{3.7}$$

and

$$Q' = TQT^{-1} \tag{3.8}$$

or, in Heisenberg representation,

$$|\psi'\rangle = T|\psi\rangle \tag{3.9}$$

and

$$Q'(t') = TQ(t)T^{-1}. \tag{3.10}$$

To establish the form of T we must impose two conditions: first, that it be a kinematically admissible transformation, and second, introducing the physical content, that it conform to the requirements of the correspondence principle—namely, operators representing classical kinematic observables must transform under T in a manner corresponding to classical motion reversal.

A requirement that must be satisfied by a kinematically admissible transformation is that it be consistent with the commutation relations, eqs. (3.3), (3.4), and so forth. Thus, if the transformations of the relevant observables (operators) are

$$\begin{aligned} X_k^{\beta'} &= TX_k^{\beta}T^{-1}, \\ P_j^{\alpha'} &= TP_j^{\alpha}T^{-1}, \\ \vec{\sigma}' &= T\vec{\sigma}T^{-1}, \end{aligned} \tag{3.11}$$

and so on, the commutation relations eqs. (3.3) and (3.4) must become

$$[P_j^{\alpha'}, X_k^{\beta'}] = T[P_j^{\alpha}, X_k^{\beta}]T^{-1} \tag{3.12}$$

and

$$[\vec{\sigma}' \times \vec{\sigma}'] = T[\vec{\sigma} \times \vec{\sigma}]T^{-1}. \tag{3.13}$$

At the same time, motion reversal imposes the requirements, in accordance with the classical conditions,

$$\vec{X}^{\beta'} = \vec{X}^{\beta}, \quad \vec{P}^{\alpha'} = -\vec{P}^{\alpha}, \quad \vec{\sigma}' = -\vec{\sigma}, \tag{3.14}$$

since momentum and angular momentum change sign on reversal. Thus we also arrive at the conditions

$$[P_j^{\alpha'}, X_k^{\beta'}] = -[P_j^{\alpha}, X_k^{\beta}] \tag{3.15}$$

and

(3.16) $$[\vec{\sigma}' \times \vec{\sigma}'] = [\vec{\sigma} \times \vec{\sigma}].$$

Comparing eq. (3.15) with eq. (3.3) and (3.12), or eq. (3.16) with eqs. (3.4) and (3.13) leads to

(3.17) $$TiT^{-1} = -i.$$

Therefore we arrive at the conclusion that T must include the operator K, which takes any complex number z into its conjugate complex:

(3.18) $$KzK^{-1} = z^*.$$

Thus, since the commutation relations are invariant under any linear similarity transformation, T can be written as

(3.19) $$T = UK,$$

where U is a linear transformation. Note that, since the meaning of K is just the conjugation of complex numbers,

(3.20) $$K^{-1} = K.$$

Under the assumption that $|\psi'\rangle$ must be an accessible state of the system if $|\psi\rangle$ is one, $|\psi'\rangle$ must belong to the same unitary function space. Therefore U is a *unitary* transformation. The explicit form of U is determined by the complete set of kinematic variables defining the applicable Hilbert space in any given case. Specific examples will be treated in the next section.

Another requirement on a kinematically admissible transformation is that, *in the absence of forces or interactions* (i.e., in the absence of causal effects), the dynamic equations must be left invariant. That is, if H_0 is the Hamiltonian operator in the absence of any force acting on or within the system, the dynamic equation

(3.21) $$i\hbar \frac{\partial |\psi\rangle}{\partial t} = H_0 |\psi\rangle$$

must be transformed under T to

(3.22) $$i\hbar \frac{\partial |\psi'\rangle}{\partial t'} = H_0 |\psi'\rangle,$$

where $t' = -t$. By applying $T = UK$ to both sides of eq. (3.21), we find

(3.23) $$-i\hbar \frac{\partial}{\partial t} |\psi'\rangle = TH_0 T^{-1} |\psi'\rangle,$$

which is consistent with eq. (3.22) if and only if

$$TH_0T^{-1} = H_0, \tag{3.24}$$

that is, if and only if H_0 is invariant under time reversal.

But H_0 includes only the kinetic energy terms in the Hamiltonian, which are quadratic functions of the momentum operators $\vec{P}^\alpha$. Therefore the conditions eqs. (3.11) and (3.14) guarantee the invariance eq. (3.24), and we conclude that eq. (3.14) represents the necessary and sufficient conditions on U for $T = UK$ to be a kinematically admissible time reversal transformation.

In quantum mechanics we deal with unitary function spaces[3] and, as a consequence, are accustomed to associating *unitary* transformations with physically meaningful operations such as symmetry operations. In that respect, time reversal symmetry is quite special because it does *not* lead to a unitary transformation but, rather, leads to the transformation $T = UK$, which is called "antiunitary."[4] Wigner (1959) has presented a comprehensive discussion of the special properties of antiunitary transformations in the English edition of his book on group theory and quantum mechanics. It is because of these special properties that the role played by time reversal in quantum mechanics is distinct from that of any of the other symmetry operations of physics.

T also belongs to a class of operators Wigner calls "involutional." The class includes those operators that, when repeated, restore the original state, a condition clearly satisfied by T. Since multiplication of a state $|\psi'\rangle$ by a phase factor is the only transformation that leaves the physical state unchanged, we have for any involutional operator, I,

$$I^2 = \eta_I \mathbf{1}, \tag{3.25}$$

where $\eta_I = e^{i\phi}$ is a constant phase factor and $\mathbf{1}$ is the identity operator. In particular,

$$T^2 = \eta \mathbf{1}, \tag{3.26}$$

[3] See note 1.

[4] In his original work Wigner (1932) assumed that the total Hamiltonian H is T invariant, that is, $THT^{-1} = H$, and applied this condition to the time-dependent Schroedinger equation to show that T must be antiunitary. The origin of the result when arrived at in this way is the appearance of $i\hbar(\partial/\partial t)$ in first order. Aside from questions concerning T violation, this approach would prove misleading in generalization to the one-particle relativistic case, as in chapter 6. The Klein-Gordon equation is of second order in $\partial/\partial t$, so that it is only through the kinematic conditions that we can argue for the antiunitarity of T.

and Wigner showed that

(3.27) $$\eta = \pm 1$$

are the only possibilities.

The proof is as follows:[5]

(3.28) $$T^2 = UKUK = UU^*.$$

Since U is unitary,

$$U^{-1} = U^\dagger = \tilde{U}^*$$

or

$$U^* = \tilde{U}^{-1},$$

whence, from eqs. (3.28) and (3.26),

$$T^2 = U\tilde{U}^{-1} = \eta \mathbf{1}$$

and

(3.29) $$U = \eta \tilde{U}.$$

The transpose of this equation is

$$\tilde{U} = \eta U,$$

which when inserted into eq. (3.29) gives

$$U = \eta(\eta U),$$

yielding

$$\eta^2 = 1,$$

from which eq. (3.27) follows. We may also write

(3.30) $$T^2 = \pm \mathbf{1}$$

or

(3.31) $$T^{-1} = \pm T.$$

The sign of T^2 is determined by the properties of U, which in turn are governed by the nature of the complete set of kinematic variables required by the definition of a given physical system. Therefore there are two classes

[5] The notation used here et seq. is as follows: For the matrix M representing an operator or a transformation, M^* is the conjugate complex, that is, the matrix obtained by taking the conjugate complex of each element of M, $\tilde{M}$ is the transposed matrix, and $M^\dagger = \tilde{M}^*$ is the Hermitian conjugate.

of quantum mechanical systems, "even" systems and "odd" systems. We shall find that for systems of particles these correspond to even and odd numbers of half-integral spin particles, respectively.

A fundamental property of an antiunitary operator that will be found to be most useful throughout our subsequent discussion is obtained by considering the inner product of two arbitrary states $|\psi\rangle$ and $|\phi\rangle$ of a system. If

$$|\psi'\rangle = T|\psi\rangle, \qquad |\phi'\rangle = T|\phi\rangle,$$

then we write[6]

$$\begin{aligned}\langle\psi'|\phi'\rangle &= \langle T\psi|T|\phi\rangle \\ &= \langle U\psi^*|U|\phi^*\rangle,\end{aligned}$$

where the star denotes conjugate complex. Hence

$$\begin{aligned}\langle\psi'|\phi'\rangle &= \langle\psi^*|U^\dagger U|\phi^*\rangle \\ &= \langle\psi|\phi\rangle^*,\end{aligned}$$

since U is unitary. Therefore the fundamental property of the inner product under antiunitary transformation is

(3.32) $$\langle\psi'|\phi'\rangle = \langle\phi|\psi\rangle.$$

This result clearly follows irrespective of the choice of U and is therefore valid for *any* antiunitary transformation.

The fundamental property of the inner product eq. (3.32) yields directly the important result that, for *odd* systems, the state $T|\psi\rangle$ is orthogonal to $|\psi\rangle$:

$$\langle T\psi|\psi\rangle = \langle T\psi|T^2\psi\rangle$$

from eq. (3.32) with $|\phi\rangle = T|\psi\rangle$, whence, for $T^2 = -1$

(3.33) $$\langle T\psi|\psi\rangle = -\langle T\psi|\psi\rangle = 0.$$

Thus, for "odd" systems, time reversal converts a state $|\psi\rangle$ into an independent state. This is the basis for the Kramers (1930) degeneracy.[7] The motivation for Wigner's initial work (1932) on time reversal invariance apparently arose in connection with the result owing to Kramers.

[6] For the sake of simplicity we write

$$[T|\psi\rangle]^\dagger = \langle T\psi|,\ [U|\psi\rangle]^\dagger = \langle U\psi|,\ K|\psi\rangle = |\psi^*\rangle,$$
$$[K|\psi\rangle]^\dagger = \langle\psi^*|, \text{ etc.}$$

[7] See section 4.1 and chapter 4, note 2. Kramers did not use T invariance to establish the result; see Griffith (1961, sec. 8.4, pp. 205 ff.).

3.3 One-Particle Systems

The simplest possible quantum mechanical system is that of a single, spinless particle having no internal structure. The complete set of kinematic operators defining the Hilbert space are the position $\vec{X}$ and the momentum $\vec{P}$ satisfying the commutation relations eq. (3.3):

(3.34)
$$[P_j, X_k] = -i\hbar\delta_{jk}$$

and transforming under T in accordance with eq. (3.14):

(3.35a)
$$T\vec{X}T^{-1} = \vec{X}$$

and

(3.35b)
$$T\vec{P}T^{-1} = -\vec{P}.$$

The eigenvalues of $\vec{X}$ are the position vectors $\vec{x}$ of the particle, whose vector components are real numbers. Therefore in the representation in which $\vec{X}$ is diagonal, eq. (3.35a) may be written as

(3.36a)
$$U\vec{X}U^{-1} = \vec{X}.$$

On the other hand, in this representation the commutation relations eq. (3.34) require that the nonvanishing matrix elements of $\vec{P}$ be purely imaginary. This is, of course, consistent with the usual representation:

(3.37)
$$\langle\vec{x}\,|\,\vec{P}\,|\,\vec{x}'\rangle = -i\hbar\vec{\nabla}\delta(\vec{x} - \vec{x}').$$

Therefore

$$K\vec{P}K^{-1} = -\vec{P}$$

and, from eq. (3.35b),

(3.36b)
$$U\vec{P}U^{-1} = \vec{P}.$$

Since $\vec{X}$ and $\vec{P}$ form a complete set of observables, eqs. (3.36a) and (3.36b) are the only requirements on U. It follows that, *for the structureless particle*,

(3.38)
$$U = \mathbf{1}e^{i\lambda},$$

where λ is an undetermined phase angle that may be chosen to be $\lambda = 0$, with the result

(3.39)
$$T = K$$

in the $|\,\vec{x}\rangle$ representation. If we carry through the analysis in the $|\,\vec{p}\rangle$ representation, in which the eigenvalue $\vec{p}$ is of course real, U is found to be the transformation (again setting $\lambda = 0$)

$$\langle \vec{p} \,|\, U \,|\, \vec{p}' \rangle = \delta(\vec{p} + \vec{p}'), \tag{3.40}$$

which transforms the state $|\vec{p}\rangle$ into the state $|-\vec{p}\rangle$, as it must to agree with eq. (3.35b). It is important to note here and for future reference the distinctive behavior of the matrix U on being subjected to a change in the representations of the states (in this case from the $|\vec{x}\rangle$ basis to the $|\vec{p}\rangle$ basis). Although a change of basis would ordinarily leave the unit matrix in eq. (3.38) unchanged, it is changed significantly to eq. (3.40) in this case. Such unfamiliar behavior is characteristic of antiunitary transformations.

For a particle having spin $\frac{1}{2}$, the Pauli spin matrices $\vec{\sigma}$, satisfying the commutation relations eq. (3.4), are introduced as additional kinematic variables. The spin condition eq. (3.14), which can be written as

$$T\vec{\sigma}T^{-1} = -\vec{\sigma}, \tag{3.41}$$

places a constraint on U in addition to those implied by eqs. (3.35). Again, the form of U will depend on the choice of representation. We use the $|\vec{x}, m_s\rangle$ representation where $m_s = \pm\frac{1}{2}$ is the eigenvalue of the z component of the spin and the three spin matrices are given in the standard representation:

$$\sigma_x = \begin{pmatrix} 0 & 1 \\ 1 & 0 \end{pmatrix}, \qquad \sigma_y = \begin{pmatrix} 0 & -i \\ i & 0 \end{pmatrix}, \qquad \sigma_z = \begin{pmatrix} 1 & 0 \\ 0 & -1 \end{pmatrix}. \tag{3.42}$$

Since σ_x and σ_z are real matrices whereas σ_y has purely imaginary elements, the operation K yields

$$K\sigma_x K^{-1} = \sigma_x, \qquad K\sigma_y K^{-1} = -\sigma_y, \qquad K\sigma_z K^{-1} = \sigma_z. \tag{3.43}$$

Therefore it follows from eq. (3.41) that

$$U\sigma_x U^{-1} = -\sigma_x, \qquad U\sigma_y U^{-1} = \sigma_y, \qquad U\sigma_z U^{-1} = -\sigma_z \tag{3.44}$$

or

$$U\sigma_x + \sigma_x U = 0, \qquad U\sigma_y - \sigma_y U = 0, \qquad U\sigma_z + \sigma_z U = 0. \tag{3.45}$$

It is well known that the matrix that commutes with σ_y and anticommutes with σ_x and σ_z is σ_y multiplied by any constant. Since σ_y is unitary, we write

$$T = \sigma_y K \tag{3.46}$$

for a single spin $\frac{1}{2}$ particle in the $|\vec{x}, m_s\rangle$ representation. Again the arbitrary phase has been set equal to zero.

At this point we note that, while for the *spinless* particle

$$T^2 = K^2 = \mathbf{1}, \tag{3.47}$$

for the spin $\frac{1}{2}$ particle

$$T^2 = \sigma_y K \sigma_y K = -\sigma_y^2 = -\mathbf{1}. \tag{3.48}$$

Therefore the spinless particle is to be identified as an "even" system and the spinning particle as an "odd" system by the definition established on the basis of eq. (3.31). Then, in accordance with eq. (3.33), we find that a state of a spin $\frac{1}{2}$ particle is always transformed into an orthogonal state by time reversal.

It should be emphasized that the properties described here are kinematic in character. The behavior of the dynamics under time reversal, in particular the notion of time reversal invariance, has *not* been introduced and is *not* relevant to these results.

3.4 Many-Particle Systems

The kinematic manifold for a system consisting of many structureless particles of the types treated in section 3.3 consists of the outer product of the single-particle manifolds, since the kinematic variables may be taken to be the independent (commuting) one-particle operators $\vec{X}^\alpha$, $\vec{P}^\alpha$, $\vec{\sigma}^\alpha$ for the αth particle satisfying the commutation relations eqs. (3.3) and (3.4).

Let us assume that, among the N particles, n have spin $\frac{1}{2}$ and $N-n$ are spinless. Since the transformations under T of the variables associated with each particle are those described in the one-particle case, the unitary factors U in the transformation $T = UK$ of the N-particle manifold will be obtained simply as the outer product of one-particle unitary transformations. Therefore, in the representation $|\vec{x}^1, \vec{x}^2, \ldots, \vec{x}^N; m_s^1, m_s^2, \ldots, m_s^n\rangle$, U takes the form of the product

$$U = \sigma_y^1 \sigma_y^2 \cdots \sigma_y^n. \tag{3.49}$$

We notice immediately that

$$T^2 = \sigma_y^1 \sigma_y^2 \cdots \sigma_y^n K \sigma_y^1 \sigma_y^2 \cdots \sigma_y^n K = (-1)^n \mathbf{1}. \tag{3.50}$$

Therefore, if there are an even number of spin $\frac{1}{2}$ particles we are dealing with an "even" system, and if there are an odd number we are dealing with an "odd" system as defined on the basis of eq. (3.31). It follows as before that, for an odd number of spin $\frac{1}{2}$ particles, a state $|\psi\rangle$ is transformed into an orthogonal state under time reversal.

In general, many relative phases of the coefficients in an expansion of the state $|\psi\rangle$ of a many-particle system in terms of products of one-particle states are arbitrary because of the arbitrariness of the phases of the basis vectors in the Hilbert space. However, because of the antiunitary character of T, the behavior of $|\psi\rangle$ under T depends on the way these arbitrary phases are chosen. We shall find in the next section that one consequence of this is that invariance of the dynamics under time reversal imposes conditions on the choice of phases.

In the special case of an isolated system of particles in a state of specified angular momentum $j\hbar$, where j is integral for even n and half-integral for odd n, the arbitrariness in phases may be used to obtain another useful representation of U. For this purpose, we denote the total angular momentum operator by $\vec{J}\hbar$. The eigenvalues of $(\vec{J}\cdot\vec{J})$ are then $j(j+1)$, and those of J_z are denoted by m. As usual, the raising and lowering operators are defined as

(3.51) $$J_{\pm} = J_x \pm iJ_y .$$

Also,

(3.52) $$J_{-} = J_{+}^{\dagger} .$$

The matrix elements of the operators are determined by the commutation relations .

(3.53) $$[\vec{J} \times \vec{J}] = i\vec{J}$$

and are known to be of the form

(3.54a) $$\langle m | J_z | m' \rangle = m\delta_{mm'},$$

(3.54b) $$\langle m' | J_{+} | m \rangle = c_{jm}\,\delta_{m',\,m+1},$$

(3.54c) $$\langle m | J_{-} | m' \rangle = c_{jm}\,\delta_{m',\,m+1},$$

where arbitrary phases may be and have been chosen in such a way that the c_{jm} are real numbers.

Useful relationships among the different c_{jm} may be obtained by considering the quantization of the angular momentum with respect to a new set of axes rotated about the x axis by an angle π relative to the original set of axes. This rotation reverses the directions of the z axis and y axis so that the operators $\vec{J}''$ in the new system are related to those in the old by

(3.55) $$J''_x = J_x, \qquad J''_y = -J_y, \qquad J''_z = -J_z,$$

whence

(3.56) $$J''_{\pm} = J_{\mp} .$$

But the components of $\vec{J}''$ satisfy the same commutation relations as the components of $\vec{J}$; therefore the matrix elements take the same form as eqs. (3.54) but with m, m' replaced by $-m$, $-m'$.

From eq. (3.56), we find then

(3.57) $$\langle m' | J_+ | m \rangle = \langle -m' | J_- | -m \rangle,$$

whence from eqs. (3.54b) and (3.54c)

(3.58) $$c_{jm}\delta_{m', m+1} = c_{j, -m'}\delta_{-m, -m'+1}.$$

Therefore

(3.59) $$c_{j, -(m+1)} = c_{jm}.$$

The transformation of the total angular momentum operator $\vec{J}$ under T that follows from eq. (3.14) is

(3.60) $$T\vec{J}T^{-1} = -\vec{J},$$

as is to be expected for motion reversal. Therefore

(3.61) $$TJ_zT^{-1} = -J_z, \qquad TJ_\pm T^{-1} = -J_\mp,$$

the switch of signs in the subscripts of the second equation being due to the operator K in T. Since the matrix elements of J_z and $J_\pm$ are real, T can be replaced by U in eq. (3.61). Thus

(3.62) $$UJ_zU^{-1} = -J_z, \; UJ_\pm U^{-1} = -J_\mp$$

or

(3.63a) $$UJ_z + J_zU = 0$$

and

(3.63b) $$UJ_\pm + J_\mp U = 0.$$

Also,

(3.63c) $$UJ^2 - J^2U = 0.$$

Eq. (3.63a) yields for the matrix elements of U, which is diagonal in j by virtue of eq. (3.63c),

(3.64) $$(m + m')\langle m | U | m' \rangle = 0$$

Therefore

(3.65) $$\langle m | U | m' \rangle = 0 \quad \text{if } m' \neq -m;$$

the only nonvanishing matrix elements of U in this representation are those on the line orthogonal to the diagonal of the submatrix for given j.

Eq. (3.63b) yields

(3.66) $$\sum_{m''} [\langle m | U | m'' \rangle \langle m'' | J_+ | m' \rangle + \langle m | J_- | m'' \rangle \langle m'' | U | m' \rangle] = 0,$$

or, from eqs. (3.54),

$$\langle m | U | m' + 1\rangle c_{jm'} + c_{jm}\langle m + 1 | U | m'\rangle = 0. \tag{3.67}$$

However, according to eq. (3.65) the matrix elements of U appearing in this last equation vanish unless

$$m' + 1 = -m. \tag{3.68}$$

Therefore the nonvanishing elements satisfy the conditions

$$\langle m | U | -m\rangle c_{j,-(m+1)} + \langle m + 1 | U | -(m + 1)\rangle c_{jm} = 0$$

or

$$\frac{\langle m + 1 | U | -(m + 1)\rangle}{\langle m | U | -m\rangle} = -\frac{c_{j,-(m+1)}}{c_{jm}}. \tag{3.69}$$

It follows from eq. (3.59) that

$$\frac{\langle m + 1 | U | -(m + 1)\rangle}{\langle m | U | -m\rangle} = -1. \tag{3.70}$$

A stepwise application of eq. (3.70) leads immediately to the general result determining the ratios of all nonvanishing matrix elements of U,

$$\frac{\langle m' | U | -m'\rangle}{\langle m | U | -m\rangle} = (-1)^{m'-m} = i^{2(m'-m)}. \tag{3.71}$$

By choosing the arbitrary phase of U appropriately, we then find that the only nonvanishing matrix elements of U are

$$\langle m | U | -m\rangle = i^{2m}. \tag{3.72}$$

Therefore the phase of a state $|j, m\rangle$ of given angular momentum may always be chosen so that[8]

$$T | j, m\rangle = i^{2m} | j, -m\rangle. \tag{3.73}$$

[8] To make use of this result, one must always make certain of the choice of phases. For example, two different choices of phase are found in the literature for the spherical harmonics. Clearly, the one that must be used to be consistent with eq. (3.73) is $Y_l^{m*} = (-1)^m Y_l^{-m}$, since in the coordinate representation, to which Y_l^m refers, $T = K$. This is the convention used, for example, in the standard work on atomic spectra by Condon and Shortley (1951). In combining angular momentum states, the phases of the so-called Clebsch-Gordan coefficients must also be chosen in a particular way. See Sachs (1953, app. 3, pp. 356–57). One should recognize that this is another example of the way the form of U depends on the choice of the basis for representation of the states.

Note that

$$T^2 |j, m\rangle = Ti^{2m} |j, -m\rangle = i^{-2m} T |j, -m\rangle,$$

whence

$$T^2 |j, m\rangle = i^{-4m} |j, m\rangle. \tag{3.74}$$

Since systems with even numbers of spin $\frac{1}{2}$ particles have integral values of m and those with odd numbers have half-integral values, eq. (3.74) is consistent with eq. (3.50) and our conclusion that these are "even" and "odd" systems respectively.

Again the kinematic nature of these results must be emphasized. I shall now turn to a discussion of some of the dynamic consequences in the *special* case when we are dealing with a system whose dynamics are invariant under time reversal.

3.5. Dynamic Considerations

The dynamics of a system are determined by the Hamiltonian operator H by means of either eq. (3.5) or its equivalent, eq. (3.6). The operator H is a function of the kinematic variables $\vec{X}^\alpha$, $\vec{P}^\alpha$, $\vec{\sigma}^\alpha$, etc. Therefore the way it transforms under time reversal is governed by the general operator transformation eq. (3.8):

$$H'(\vec{X}^\alpha, \vec{P}^\alpha, \vec{\sigma}^\alpha \cdots) = TH(\vec{X}^\alpha, \vec{P}^\alpha, \vec{\sigma}^\alpha \cdots)T^{-1}. \tag{3.75}$$

Then if T is applied to eq. (3.5) it is transformed to

$$i\hbar \frac{\partial |\psi'(t')\rangle}{\partial t'} = H' |\psi'(t')\rangle, \tag{3.76}$$

and eq. (3.6) is transformed to

$$-i\hbar \dot{Q}'(t') = [H', Q'(t')]. \tag{3.77}$$

where the dot denotes the derivative with respect to t' and the *form* of dynamic equations eqs. (3.5) and (3.6), with t' replacing t and H' replacing H, is unchanged. Note that the operator K appearing in $T = UK$ is essential in order to compensate for the change in sign of the first time derivative, since $t' = -t$.

That the form of the dynamics is unchanged does not imply *invariance* under T, because if there is a difference between H' and H the equations of motion are different. The condition for time reversal invariance is that

$$H'(\vec{X}^\alpha, \vec{P}^\alpha, \vec{\sigma}^\alpha \cdots) = H(\vec{X}^\alpha, \vec{P}^\alpha, \vec{\sigma}^\alpha \cdots), \tag{3.78}$$

that is, the *Hamiltonian* function of the kinematic variables *must be invariant under T*:

$$TH(\vec{X}^\alpha, \vec{P}^\alpha, \vec{\sigma}^\alpha, \cdots)T^{-1} = H(\vec{X}^\alpha, \vec{P}^\alpha, \vec{\sigma}^\alpha, \cdots). \quad (3.79)$$

The question of T invariance therefore reduces to the determination of the functional form of H. In this connection it is important to recognize that although it is Hermitian, H may involve complex functions of the Hermitian operators $\vec{X}^\alpha$, $\vec{P}^\alpha$, etc., and these complex functions will be conjugated by the operator K appearing in T.

For one class of physical systems, including atoms and molecules, to which nonrelativistic quantum mechanics applies, the interactions are electromagnetic in origin, and the Hamiltonian is therefore given in a time reversal invariant form. Therefore time reversal considerations are useful in dealing with these systems and, as we shall see, restrictions are imposed on the states of the system by the invariance condition. Of course, any well-established violation of these restrictions would imply that the Hamiltonian must contain an additional interaction term violating T invariance. Experiments designed to validate the restrictions therefore provide a means of testing for the existence of such a term. No violation has been established for systems in this category on the basis of experimental tests that have been performed, some of which will be discussed later.

Another class of nonrelativistic systems is characterized by interactions that are not fully known or understood, such as nuclear systems for which our knowledge of the strong interactions between nucleons is incomplete. The trial-and-error process of determining appropriate phenomenological forms for the interactions entails starting from the simplest forms and adding complications only as they are required by experimental evidence. Time reversal invariance, as well as other symmetry arguments, is used as a "natural" guiding principle in formulating model interactions. This assumption places severe restrictions on the form of the interactions to be considered and thereby reduces the ambiguities that must be faced in exploring them. Of course it also places restrictions characteristic of T invariance on the states of the system. Again, any well-established violation of these restrictions would require the introduction of T-violating interactions into the model Hamiltonian.

The following chapter is devoted to determining restrictions on the states and dynamics of T-invariant nonrelativistic quantum systems and to describing the interpretation of experiments designed to validate them. I conclude this chapter by considering examples of the way T invariance, as well as the other usual assumptions of invariance under translations, rotations, and inversion, leads to restrictions on the form of interaction. In particular, the results will be applied to the determination of allowable forms of spin-spin and spin-orbit couplings between nucleons in nuclei (Sachs 1953, pp. 213 ff.).

Let us again consider an isolated system of spin $\frac{1}{2}$ particles and assume that there are only two-body interactions among them. In keeping with the nonrelativistic approximation, dependence of the interactions on velocity will be limited to terms that are at most quadratic. Then the requirement of translational and Galilean[9] invariance will limit the interactions to the form

$$H_{\text{int}} = \sum_{\alpha<\beta} V_{\alpha\beta}(\vec{X}^{\alpha\beta}, \dot{\vec{X}}^{\alpha\beta}, \vec{\sigma}^{\alpha}, \vec{\sigma}^{\beta}), \tag{3.80}$$

where

$$\vec{X}^{\alpha\beta} = \vec{X}^{\alpha} - \vec{X}^{\beta}. \tag{3.81}$$

The requirement of rotational invariance limits the possible forms of $V_{\alpha\beta}$ to functions of scalar products of its vector arguments. Therefore if $X^{\alpha\beta} = |\vec{X}^{\alpha\beta}|$, the potential $V_{\alpha\beta}$ might be expected to be the product of a function $V(X^{\alpha\beta})$, giving the shape of the potential, and a factor $S_{\alpha\beta}$ that is a scalar product or a product of scalar products of the arguments:

$$V_{\alpha\beta} = V(X^{\alpha\beta})S_{\alpha\beta}. \tag{3.82}$$

Since $S_{\alpha\beta}$ must be a scalar rather than a pseudoscalar if invariance under inversion is to be preserved, it cannot include odd powers of a scalar product of the axial vector $\vec{\sigma}$ with one of the polar vectors $\vec{X}^{\alpha\beta}$ or $\dot{\vec{X}}^{\alpha\beta}$. Time reversal invariance imposes a similar requirement; products of odd powers of $\dot{\vec{X}}^{\alpha\beta}$ and $\vec{\sigma}$ with $\vec{X}^{\alpha\beta}$ are forbidden by eqs. (3.35) and (3.41). Thus the allowed forms of $S_{\alpha\beta}$ include only

$$S_{\alpha\beta} = 1, \tag{3.83a}$$

$$S_{\alpha\beta} = \vec{\sigma}^{\alpha} \cdot \vec{\sigma}^{\beta}, \tag{3.83b}$$

$$S_{\alpha\beta} = ([\vec{X}^{\alpha\beta} \times \dot{\vec{X}}^{\alpha\beta}] \cdot [\vec{\sigma}^{\alpha} + \vec{\sigma}^{\beta}]), \tag{3.83c}$$

$$S_{\alpha\beta} = ([\vec{X}^{\alpha\beta} \times \dot{\vec{X}}^{\alpha\beta}] \cdot [\vec{\sigma}^{\alpha} - \vec{\sigma}^{\beta}]), \tag{3.83d}$$

$$S_{\alpha\beta} = (\vec{\sigma}^{\alpha} \cdot \vec{X}^{\alpha\beta})(\vec{\sigma}^{\beta} \cdot \vec{X}^{\alpha\beta}), \tag{3.83e}$$

$$S_{\alpha\beta} = (\vec{\sigma}^{\alpha} \cdot \dot{\vec{X}}^{\alpha\beta})(\vec{\sigma}^{\beta} \cdot \dot{\vec{X}}^{\alpha\beta}). \tag{3.83f}$$

Although these six forms admit a considerable ambiguity in the form of the interaction, the requirement of invariance under P and T has reduced

[9] That is, invariance under the transformation to a reference frame moving at a constant velocity with respect to the original frame: $\vec{X}^{\alpha} \to \vec{X}^{\alpha} + \vec{c}t$. See eq. (1.6).

the ambiguity by eliminating forms such as $(\vec{X}^{\alpha\beta} \cdot \dot{\vec{X}}^{\alpha\beta})^n$, where n is an odd integer, $(\vec{X}^{\alpha\beta} \cdot [\vec{\sigma}^\alpha \pm \vec{\sigma}^\beta])$, $(\dot{\vec{X}}^{\alpha\beta} \cdot [\vec{\sigma}^\alpha \pm \vec{\sigma}^\beta])$, and $([\vec{X}^{\alpha\beta} \times \dot{\vec{X}}^{\alpha\beta}] \cdot [\vec{\sigma}^\alpha \times \vec{\sigma}^\beta])$. The limitation to, at most, bilinear expressions in $\vec{\sigma}^\alpha$ and $\vec{\sigma}^\beta$ results from the fact that the algebra of the $\vec{\sigma}$ matrices is closed with the unit matrix and the three $\vec{\sigma}$ matrices.

Another one of the forms is eliminated if all the particles are identical, for then $V_{\alpha\beta}$ must be symmetrical in the labels α and β. Thus the spin-orbit coupling eq. (3.83d) must be excluded for identical particles. Since in nuclear physics the proton and the neutron are treated as two states of the same particle, called the "nucleon," the possible forms of the strong interactions between nucleons may be treated as interactions between identical particles if a new internal variable is introduced to distinguish the two states. This variable is the isotopic spin $\tau_3 = \pm 1$, corresponding to proton and neutron states, respectively.

The introduction of the isotopic spin vector $\vec{\tau}$ with components τ_1, τ_2, τ_3 satisfying the same commutation relations as the Pauli spin matrices $\sigma_x, \sigma_y, \sigma_z$ makes it possible to generalize the forms eq. (3.83) to the case of identical nucleons by including factors depending on $\vec{\tau}^\alpha$, $\vec{\tau}^\beta$, which are now additional kinematic variables.

The commutation relations for $\vec{\tau}$ are of the same form as those of $\vec{\sigma}$, eq. (3.4). Thus

(3.84) $$[\vec{\tau} \times \vec{\tau}] = 2i\vec{\tau},$$

and the kinematically admissible transformations in addition to the space-time transformations are three-dimensional rotations of $\vec{\tau}$, the underlying group being the $SU(2)$ transformations of the isotopic spin wave functions. The behavior of $\vec{\tau}$ under time reversal is of particular interest. Since τ_3 determines the charge of the nucleon, it is unchanged by T:

(3.85a) $$T\tau_3 T^{-1} = \tau_3.$$

Furthermore, the matrix elements of τ_3 are real, from which it follows that

(3.86) $$U\tau_3 U^{-1} = \tau_3.$$

Therefore U may be an arbitrary rotation about the three-axis in isotopic spin space and the three-component of any isotopic spin vector is left invariant by U, whence

(3.85b) $$T[\tau_3^\alpha \pm \tau_3^\beta]T^{-1} = \tau_3^\alpha \pm \tau_3^\beta,$$

(3.85c) $$T\tau_3^\alpha \tau_3^\beta T^{-1} = \tau_3^\alpha \tau_3^\beta.$$

But, although

(3.87) $$U[\vec{\tau}^{\alpha} \times \vec{\tau}^{\beta}]_3\, U^{-1} = [\vec{\tau}^{\alpha} \times \vec{\tau}^{\beta}]_3,$$

we have

(3.85d) $$T[\vec{\tau}^{\alpha} \times \vec{\tau}^{\beta}]_3\, T^{-1} = -[\tau^{\alpha} \times \tau^{\beta}]_3,$$

because the commutation relation eq. (3.84) requires that the third component of the vector product be imaginary when τ_3 is real as in eq. (3.42).

Interactions that conserve electric charge, that are symmetric in α and β, and that are invariant under T include the fourfold possibilities

(3.88a) $$S_{\alpha\beta},$$

(3.88b) $$(\tau_3^{\alpha} + \tau_3^{\beta})S_{\alpha\beta},$$

(3.88c) $$\tau_3^{\alpha}\tau_3^{\beta}S_{\alpha\beta},$$

(3.88d) $$(\vec{\tau}^{\alpha} \cdot \vec{\tau}^{\beta})S_{\alpha\beta},$$

where $S_{\alpha\beta}$ is any one of the expressions eq. (3.83) *except* (3.83d).[10] In the excepted case we have in addition

(3.88e) $$(\tau_3^{\alpha} - \tau_3^{\beta})([\vec{X}^{\alpha\beta} \times \dot{\vec{X}}^{\alpha\beta}] \cdot [\vec{\sigma}^{\alpha} - \vec{\sigma}^{\beta}]).$$

Finally, the form

(3.88f) $$[\vec{\tau}^{\alpha} \times \vec{\tau}^{\beta}]_3\, ([\vec{X}^{\alpha\beta} \times \dot{\vec{X}}^{\alpha\beta}] \cdot [\vec{\sigma}^{\alpha} \times \vec{\sigma}^{\beta}])$$

is also permitted. Note that eqs. (3.88e) and (3.88f) will contribute only to the interaction between *unlike* particles (a proton and a neutron).

Although the electromagnetic interactions depend on τ_3, $SU(2)$ invariance is a good approximation for the strong interactions between nucleons. Thus the dominant nuclear interactions are limited to the $SU(2)$ invariant forms (3.88a) and (3.88d). However, any analysis of deviations from $SU(2)$ symmetry should include the other four possible forms given by eqs. (3.88).

[10] Note that eq. (3.88d) leads to what is known in nuclear physics as a "charge exchange force." See Sachs (1953, pp. 161, 215).

4 | Applications to Atomic and Nuclear Systems

As far as is known at present, the electromagnetic and strong interactions responsible for the structure and general dynamic behavior of atoms and atomic nuclei are invariant under time reversal. This invariance has important consequences for the properties of stationary states, scattering and reaction amplitudes, and (electromagnetic) radiative transitions of such systems. It is the purpose of this chapter to illustrate many of these consequences and to show how they may be used to test the assumption that these or other interactions are in fact T invariant.

Usually in quantum mechanics there are associated with an invariance of the Hamiltonian a conservation law and some degree of degeneracy of the energy states. Invariance of the Hamiltonian under T has different implications. Because T is antiunitary rather than unitary, it is not directly related to a Hermitian observable, and the invariance does not lead to a conservation law. There is an implication of twofold degeneracy for "odd" systems, as defined in section 3.2, and there are additional implications for the stationary states of any multiparticle system. The latter may be expressed as reality conditions on the wave functions, as will be demonstrated in the following section.

The consequence of T invariance for electromagnetic, scattering, and reaction processes will be taken up in subsequent sections. It should be kept in mind that the quantitative formulation of these consequences provides the means both for testing the validity of the assumption of T invariance and for measuring the degree of T violation if the assumption turns out to be incorrect.

4.1 Conditions Imposed on Stationary States by *T* Invariance

Since the stationary states of energy E are the state vectors $|\psi_E\rangle$ determined by

(4.1) $$H|\psi_E\rangle = E|\psi_E\rangle,$$

where H is the Hamiltonian of the system under consideration, the transformed state

(4.2) $$|\psi'_E\rangle = T|\psi_E\rangle$$

satisfies the equation

(4.3) $$THT^{-1}|\psi'_E\rangle = E|\psi'_E\rangle.$$

Therefore, if H satisfies the condition for T invariance

(4.4) $$THT^{-1} = H,$$

$|\psi'_E\rangle$ satisfies the same equation as $|\psi_E\rangle$:

(4.5) $$H|\psi'_E\rangle = E|\psi'_E\rangle,$$

and either the state $|\psi'_E\rangle$ is equivalent to $|\psi_E\rangle$ or there is (at least) a twofold degeneracy of the state of energy E.

Eq. (3.33) shows that for "odd" systems of particles, for example, systems including an odd number of spin 1/2 particles, $|\psi'_E\rangle$ is *always* orthogonal to $|\psi_E\rangle$. Therefore for such systems each energy state will be at least twofold degenerate if the Hamiltonian is T invariant.

The Hamiltonian of a system having T-invariant internal interactions will be T invariant in the presence of external electric fields, which are independent of the sense of time. External magnetic fields destroy this symmetry, as was illustrated in section 2.3. Consider then an *odd* system of particles subjected to no external forces or to external forces arising *only* from external electrical fields. If the internal interactions of the system are T invariant, every stationary state of such a system is at least doubly degenerate. This is Wigner's (1932) proof of the "Kramers degeneracy" (Kramers 1930). Because static magnetic fields change sign under T, they will in general split this degeneracy.

As an example of the application of this result, consider the rotational levels of a diatomic molecule. The quantum number representing the absolute value of the projection of the electron orbital angular momentum on the molecular axis is usually denoted by Λ and that of the total electron angular momentum by $\Omega = |\Lambda + \Sigma|$, where Σ is the projection of the total electron spin.[1] If the molecule is in the total angular momentum state $J = 0$, every such level except that with $\Lambda = 0$ is doubly degenerate because the projection of the orbital angular momentum on the axis may point either way. However, for $J \neq 0$, this degeneracy is split ("Λ-doubling"). Nevertheless, if all the interactions of the molecule are T invariant, there

[1] For these and other details of diatomic molecular states see Herzberg (1950, chap. 5, pp. 212 ff.).

must be a residual twofold degeneracy when the molecule contains an odd number of electrons, if hyperfine structure (i.e., nuclear spin) effects are neglected, although $J \neq 0$. Henley (1969, p. 374) has pointed out that an analogous statement can be made about the degeneracies of states of deformed nuclei containing an odd number of nucleons when such nuclei are treated as having a body-fixed axis of symmetry.

These twofold degeneracies will survive in the presence of an external electric field, but they will be split by an external magnetic field, a splitting that is significant for the case of diatomic molecules but, ordinarily, not for nuclei because the magnitude of the magnetic fields required to yield significant splitting is so large.

Other important applications of the Kramers degeneracy occur in the treatment of the dynamics of paramagnetic ions in crystals.[2] The crystal fields destroy most of the degeneracies of the electronic energy levels of the free ions, but as long as they are electric fields they do not destroy the Kramers degeneracy, which occurs if the ions comprise an odd number of electrons. The resulting twofold degeneracy can be removed by applying an external magnetic field, and transitions between the split pair of levels may be induced by applying an external oscillatory electromagnetic (microwave) field. The resulting paramagnetic resonance transitions provide a very sensitive measure of the local magnetic field at the ion because the splitting depends directly on that field.

If the nucleus of the ion has a nonvanishing spin, there is always an "external" magnetic field acting on the electrons of the ion, the field produced by the magnetic moment of the nucleus. Therefore the electronic Kramers degeneracy may be expected to be removed by this field, the splitting being of the order of a hyperfine structure effect. However, when the nucleus is treated as a part of the complete ionic system, its magnetic field is internal to the ion rather than external. Under the assumption that the Hamiltonian of the entire system, electrons plus nucleus, is invariant under time reversal, there is a Kramers degeneracy whenever the total number of spin $\frac{1}{2}$ particles, that is electrons, protons, and neutrons, is odd. Therefore this degeneracy persists for an ion having an odd number of electrons and an even number of nucleons (protons and neutrons)—that is, *if the nucleus has integral spin.* Furthermore, there *is* a Kramers degeneracy for ions having an even number of electrons and *a nucleus with half-integral spin.* This latter degeneracy is at the hyperfine level, of course, and will be split appreciably only by external magnetic fields strong enough to influence the

[2] For an excellent discussion of the role of Kramers degeneracy in the understanding of electronic states of paramagnetic crystals, see Abragam and Bleaney (1970, chap. 15).

orientation of the nuclear spin either directly or through stronger magnetic fields induced within the electronic structure of the atom.

Let us now consider systems of particles having spherical symmetry (atoms, nuclei) so that, in the absence of external fields, their energy states may be characterized by a total angular momentum j and a magnetic quantum number $m = j, j-1, \ldots, -j+1, -j$. The spherical symmetry introduces the degeneracy associated with the $2j+1$ values of m. If the convention for the phases of the angular momentum states is that of section 3.4 then, in accordance with eq. (3.73), the transformation under time reversal of the state $|\psi_{E,j,m}\rangle$ is

$$T|\psi_{E,j,m}\rangle = i^{2m}|\psi_{E,j,-m}\rangle, \tag{4.6}$$

under the assumption that the Hamiltonian is T invariant and that the $(2j+1)$-fold degeneracy is the only degeneracy (no "accidental" degeneracy). The associated degeneracy between states of $\pm m$ is not new, of course, since it is also a consequence of spherical symmetry.

In the presence of a static uniform external electric field along the z axis, the spherical symmetry is reduced to cylindrical symmetry about that axis, and the degeneracy between states of different m is destroyed in part. However, since the electric field does not remove the T invariance, states of $\pm m$ will remain degenerate with one another. This twofold degeneracy clearly applies to both even and odd systems so that it need not be an example of the Kramers degeneracy. In fact, the degeneracy also follows from the fact that a plane parallel to the z axis is a plane of symmetry. Reflection in this plane reverses the direction of the z component of angular momentum so that states of $\pm m$ are degenerate even in the absence of T invariance.

On the other hand, in the presence of an inhomogeneous static electric field for which there is no plane of symmetry, the Kramers degeneracy will be maintained for odd systems as long as the internal Hamiltonian is T invariant. The degeneracy is not associated with states having definite magnetic quantum numbers in this case unless the inhomogeneous electric field has an axis of symmetry (as in the case of Stark broadening by collision between atoms and atomic ions). Both the Kramers degeneracy and the degeneracy for $\pm m$ are destroyed if the electric field is replaced by a magnetic field, the former because the magnetic field changes sign under T and the latter because it changes sign under reflection in a plane that includes the axis of symmetry of the feld.

Although T invariance does not introduce any new degeneracies in the case of central symmetry, it does impose reality conditions on the wave functions of a complex system. For example, the energy eigenstate $|\psi_{E,j,m}\rangle$ may be a linear combination of states that are not themselves eigenstates of the energy. The mixing of different electron configurations in an atom and different nucleon configurations in a nucleus to form the energy eigenstates are typical examples. Thus consider an energy state having the form

$$|\psi_{E,j,m}\rangle = \sum_{\iota} a_{\iota} |\psi^{\iota}_{j,m}\rangle, \tag{4.7}$$

where ι is a collective index enumerating the set of states of total angular momentum j and magnetic quantum number m that are being combined.

Application of T to eq. (4.7) yields

$$T|\psi_{E,j,m}\rangle = \sum_{\iota} a^{*}_{\iota} T |\psi^{\iota}_{j,m}\rangle. \tag{4.8}$$

Now we may assume that the phases have been chosen so that the kinematic condition

$$T|\psi^{\iota}_{j,\,m}\rangle = i^{2m} |\psi^{\iota}_{j,\,-m}\rangle \tag{4.9}$$

is fulfilled for each value of ι. Then comparison with eq. (4.6) shows that

$$|\psi_{E,\,j,\,-m}\rangle = \sum_{\iota} a^{*}_{\iota} |\psi^{\iota}_{j,\,-m}\rangle \tag{4.10}$$

is an eigenstate of energy E as a result of the T invariance. But by rotational invariance, eq. (4.7) can be transformed to

$$|\psi_{E,\,j,\,-m}\rangle = \sum_{\iota} a^{*}_{\iota} |\psi^{\iota}_{j,\,-m}\rangle \tag{4.11}$$

which is also an eigenstate. Therefore, unless there is an "accidental" degeneracy of the states $|\psi_{E,\,j,\,m}\rangle$ other than the $(2j + 1)$-fold degeneracy associated with the m values, the ratio a^{*}_{ι}/a_{ι} must be independent of ι. For two different values ι and ι', it follows that

$$\frac{a_{\iota'}}{a_{\iota}} = \left(\frac{a_{\iota'}}{a_{\iota}}\right)^{*}, \tag{4.12}$$

or $a_{\iota'}/a_{\iota}$ is a real number; if the phases of the states $|\psi^{\iota}_{j,\,m}\rangle$ are chosen in accordance with eq. (4.9), *the coefficients in the expansion eq. (4.7) are relatively real* as a result of the T invariance of the Hamiltonian (if there is no accidental degeneracy).

This result is useful, for example, in writing a trial function for the ground state of a system in terms of a linear combination of some specified set of simple functions. A judicious choice of the phase conventions makes it possible to conclude that the number of real parameters that must be determined (for example, by a variational method) is only about half as large as might be expected.[3]

The reality conditions on the wave functions of T-invariant systems having central symmetry have physical consequences that are susceptible to measurement. Such measurements provide an important means for testing the validity of our assumptions. Generally speaking, the physical consequences of T invariance take the form of a statement that some measurable quantity must vanish, and the test consists either of placing an upper limit on this quantity, if there is no evidence for T violation, or of determining the value of this quantity if there is T violation.

The quantity to be measured can be expressed, directly or indirectly, in terms of the phase difference between terms in a wave function—that is, in terms of the deviations from the reality condition eq. (4.12). This use of a phase difference as a measure of T violation will be found to be quite general.

The condition that a physically measurable quantity must vanish as a consequence of a symmetry principle is usually transparent and does not require the intricacies of an analysis of the phases of the wave functions. However, in order to interpret the results of measurements in terms of the limitations placed on T-violating interactions, such a detailed analysis is needed. To help clarify the relationship between the general conditions imposed by T invariance and T violating interactions, let us consider as an example the implications of one such general condition for a simple dynamic system.

4.2 Electric Dipole Moment as a Measure of *T* Violation

Our example makes use of the result that a system having spherical symmetry cannot have a static electric dipole moment in the absence of an accidental degeneracy if the dynamics are T invariant. The result follows immediately from the fact that the electric dipole moment operator $\vec{\mathbf{D}}$, which is $\Sigma_\alpha e_\alpha \vec{x}^\alpha$ where e_α is the electric charge of the particle at the point $\vec{x}^\alpha$, is even under T,

$$T\vec{\mathbf{D}}T^{-1} = \vec{\mathbf{D}}, \tag{4.13}$$

while the angular momentum operator, or "spin" $\vec{\mathbf{J}}$ of the system is odd,

$$T\vec{\mathbf{J}}T^{-1} = -\vec{\mathbf{J}} \tag{4.14}$$

[3] I first became aware of the importance of time reversal arguments in quantum mechanics when Wigner pointed out to me that in my work on the structure of the nuclei ^3H and ^3He the number of real parameters in the wave function I had used could be reduced by almost a factor of two. See Sachs (1953, p. 187).

in accordance with eq. (3.60). Since $\vec{\mathbf{J}}$ is the only available internal vector observable of the isolated system, rotational invariance requires that $\vec{\mathbf{D}} \sim \vec{\mathbf{J}}$, but the conditions eqs. (4.13) and (4.14) then show that the coefficient of proportionality must vanish. Therefore $\vec{\mathbf{D}} = 0$, as stated. Note that here and henceforth I revert to the more conventional notation; the lowercase $x(p)$ in $\vec{x}^{\alpha}$ ($\vec{p}^{\alpha}$) denotes either the position (momentum) operator or its eigenvalue, the choice being made clear by the context.

It is not necessary, in fact it is not usual, to use T invariance to establish this result. The more usual argument is to make use of invariance of the system under space inversion, the transformation P. Since $\vec{\mathbf{D}}$ is a polar vector and $\vec{\mathbf{J}}$ is an axial vector, they have opposite signatures under P ($\vec{\mathbf{D}}$ is odd and $\vec{\mathbf{J}}$ is even), whence it follows as above that $\vec{\mathbf{D}} = 0$.

The experimental demonstration that a system with spherical symmetry, such as a particle, atom, or nucleus, has a permanent electric dipole moment (i.e., $\vec{\mathbf{D}} \sim \vec{\mathbf{J}}$) would therefore be an unequivocal demonstration that *both* T invariance *and* P invariance are violated (Landau 1957). And even if it is known that P invariance is violated, as in the case of weak interactions, the existence of a permanent electric dipole moment would be unequivocal evidence for T violation.

The way the dynamics of the system relate to this general result may be demonstrated by a simple hypothetical example. Consider a system consisting of an electron moving in a potential having central symmetry. The Hamiltonian is assumed to be made up of two terms

(4.15) $$H = H^e + H^0,$$

where H^e is even under T:

(4.16) $$TH^eT^{-1} = H^e,$$

and H^0 is odd:

(4.17) $$TH^0T^{-1} = -H^0.$$

H^e is then T invariant, and H^0 may be said to be "maximally" T violating. There are two (doublet) eigenstates of H^e that are taken here for the sake of simplicity to be the only ones of interest, a $^2S_{1/2}$ ground state ψ_1 and a $^2P_{1/2}$ excited state ψ_2, and it is assumed again for the sake of simplicity that there are no other multiplets having the same energies. The eigenfunctions of H^e having magnetic quantum number $m = \pm\frac{1}{2}$ may then be written in terms of the real functions $f_1(x)$ and $f_2(x)$ as

(4.18) $$\psi_1^m = f_1(x)\chi^m$$

and

$$\psi_2^m = if_2(x)(\vec{\sigma} \cdot \vec{x})\chi^m, \tag{4.19}$$

where χ^m is the electron spin function defined so as to satisfy the phase condition eq. (4.9). The vector $\vec{x}$ of magnitude x is the position of the electron, and $\vec{\sigma}$ is the Pauli spin matrix of the electron. Since $\vec{\sigma}$ is odd under T, the factor i is required in eq. (4.19) to establish the phase condition eq. (4.9) for ψ_2^m.

To be specific, we take

$$H^0 = (\vec{\sigma} \cdot \vec{x})U(x), \tag{4.20}$$

where $U(x)$ is a real "potential" function. This expression clearly violates T invariance maximally. Note that it also violates parity conservation, since $\vec{x}$ is a polar and $\vec{\sigma}$ an axial vector. Then the condition eq. (4.12) should not be fulfilled by the eigenstates of H; they must be linear combinations of ψ_1^m and ψ_2^m having complex coefficients differing in phase. To demonstrate this result and to determine the relative phase, we calculate the matrix elements of H^0, which are easily found to be

$$\begin{aligned} H_{11}^0 &= H_{22}^0 = 0, \\ H_{12}^0 &= -H_{21}^0 = i \int d^3x\, f_1(x)x^2U(x)f_2(x). \end{aligned} \tag{4.21}$$

The eigenstates of the 2×2 matrix of H are then given by

$$\begin{aligned} \varphi_1^m &= (1 + c^2)^{-1/2}(\psi_1^m - ic\psi_2^m), \\ \varphi_2^m &= (1 + c^2)^{-1/2}(\psi_2^m - ic\psi_1^m), \end{aligned} \tag{4.22}$$

with

$$c = \frac{iH_{12}^0}{E_2^e - E_1}, \tag{4.23}$$

and E_i are the eigenvalues of H while E_i^e are the eigenvalues of H^e. Since c is a real quantity, the ratio of the coefficients of ψ_1^m and ψ_2^m in eq. (4.22) is purely imaginary rather than real as required by T invariance.

We now show by direct calculation that the coefficient in $\vec{\mathbf{D}} \sim \vec{\mathbf{J}}$ does not vanish for this hypothetical case. Since the expectation value of $\vec{\mathbf{J}}$ is in the z direction for a state of given magnetic quantum number, only the expectation value of D_z need be calculated. It is given by

$$D_z = -e\langle \varphi_1^m \mid z\varphi_1^m \rangle \tag{4.24}$$

for the state φ_1^m, which is taken here to be the ground state. Thus

(4.25) $$D_z = -e(1+c^2)^{-1}\{\langle\psi_1^m|z|\psi_1^m\rangle - ic\langle\psi_1^m|z|\psi_2^m\rangle + ic\langle\psi_2^m|z|\psi_1^m\rangle + c^2\langle\psi_2^m|z|\psi_2^m\rangle\}.$$

From the form of ψ_1^m and of ψ_2^m, eqs. (4.18) and (4.19), it is clear that the first and last terms vanish. The other two terms are given by

(4.26) $$\langle\psi_1^m|z|\psi_2^m\rangle = \langle\psi_2^m|z|\psi_1^m\rangle^* = im\int d^3x\, f_1(x)z^2 f_2(x).$$

Therefore

(4.27) $$D_z = -2ec(1+c^2)^{-1}m\int d^3x\, z^2 f_1(x)f_2(x),$$

and D_z is proportional to $\mathbf{j}_z = m$, the "spin" (expectation value of the z-component of the total angular momentum) of the system.

It should be clear that the phase difference between the two terms in eq. (4.22) is essential to the result. If the coefficient of ψ_2^m in eq. (4.22) had been real, as required by T invariance, the second and third terms in eq. (4.25) would have made equal and opposite (imaginary) contributions to D_z, and the projection of the electric dipole moment on the angular momentum would be zero, as it must be.

Our model also provides an opportunity to understand the way the Kramers degeneracy is broken by T violation. As we have already seen, for a rotationally invariant system, the Kramers degeneracy is indistinguishable from the degeneracy associated with $m = \pm|m|$, where m is the magnetic quantum number of the total angular momentum. Since the rotational symmetry is retained in our model, there is no splitting of this degeneracy by the T-violating interaction H^0, as is clear from the fact that the matrix elements of H^0, given by eq. (4.21) are independent of m.

The Kramers degeneracy of a T-invariant system is maintained even when the spherical symmetry is destroyed by an electric field. However, in our model an applied electric field $\vec{E}$ will couple to the electric dipole moment with an interaction energy

(4.28) $$\Delta E = -(\vec{D}\cdot\vec{E}),$$

which from eq. (4.27) is seen to change sign with m and therefore to break the Kramers degeneracy.

A hypothetical case has been treated here in order to expose the salient physical aspects of the relationship between electric dipole moments and T invariance. However, that both P invariance and CP invariance, which is presumably equivalent to T invariance according to the CPT theorem (see sec. 1.3 and chap. 8), are violated by weak interactions makes the subject one of considerable physical import. Since the interaction of an electric dipole moment $\vec{\mathbf{D}} \sim \vec{\mathbf{J}}$ with an electric field, eq. (4.28) has the same form as the interaction of a magnetic moment with a magnetic field, it is possible to measure D for nuclei and atoms by comparing spin precession resonance frequencies in electric fields with those in magnetic fields. This was the basis of the original method that Purcell and Ramsey (1950) used to place an upper limit on the electric dipole moment of the neutron.[4] Although the precision of their work was remarkable enough at the time, a series of experiments of even greater precision have been done and continue to be carried out by Ramsey and his collaborators as well as by others.[5] At this time the most precise experiment is that of Alterev and others (1981), giving an upper limit

$$|D_n| < 6 \times 10^{-25} e\ cm, \tag{4.29}$$

where e is the magnitude of the charge of the electron.[6]

The interpretation of this result in terms of the limits on a T-violating weak interaction or in terms of the limits on a phase difference between interaction terms is quite model dependent (see Ramsey 1981), but eq. (4.29) does not provide a low enough limit to discriminate among current models of CP violation and T violation in weak interactions, although it has ruled out many models proposed in the past.

Sandars (1967) has shown that a highly sensitive measure of the electric dipole moment of the proton (in a nucleus) can be obtained by a molecular beam resonance method that makes use of the electric field induced at the nucleus by applying an external electric field to polarize the molecule. Then

[4] As I noted in chapter 2, note 1, Purcell and Ramsey (1950) first called attention to the importance of a measurement of the electric dipole moment of the neutron, D_n. At that time, which was before the discovery of parity violation, they emphasized that their measurement was a test of parity conservation. After the discovery of P violation, Landau (1957) pointed out that the neutron can be expected to have an electric dipole moment only if T as well as P is violated. Since T invariance and CP invariance (C is charge conjugation; see sec. 7.4) are believed to be related by the CPT theorem and CP invariance appeared to hold for many weak interaction processes, the measurement of D_n was expected to yield a null result. When CP violation was discovered, interest was again aroused in obtaining as precise a measurement of D_n as possible in the hope that its determination would shed light on the origins of the CP violation.

[5] Summaries of the status of this work are given by Ramsey (1978, 1981). The latest published result of Ramsey and collaborators is presented by Dress et al. (1977), with a promise of more to come.

[6] A method for measuring the electric dipole moment of the neutron and for testing P and T invariance of its gravitational interaction by means of slow neutron interferometry has been suggested by Anandan (1982).

the frequency of the nuclear spin resonance in a magnetic field parallel to the external electric field would change slightly when the relative directions of the two fields are reversed.

This method has been applied by Harrison, Sandars, and Wright (1969) and by Hinds and Sandars (1980) to place a value of

$$D_p = (-1.4 \pm 6) \times 10^{-21} e\ cm \tag{4.30}$$

on the electric dipole moment of the proton. It is, of course, consistent with zero.

Player and Sandars (1970) also have made a measurement of the electric dipole moment of the $^3P_2$ metastable state of the xenon atom and have used it to obtain a value (again consistent with zero) of the magnitude of the electric dipole moment of the *electron*:[7]

$$|D_e| = (0.7 \pm 2.2) \times 10^{-24} e\ cm. \tag{4.31}$$

Again the interpretation of these results bears on the nature of the weak interaction, but the limits are too high to discriminate among current models of *CP* violation that will be discussed later. In this regard, I should mention the current description of these measurements of electric dipole moments as "tests of *CP* invariance," a usage that has arisen from reference to models constructed to account for the known violation of *CP* invariance. The electric dipole moment is a direct measure of *T* violation, not *CP* violation. To arrive at an *experimental* conclusion concerning *CP* violation from these measurements requires an experimental test of *CPT* invariance *on the same physical system at the same level of precision.* Compare the discussion of the converse problem of arriving at a conclusion about *T* violation from measurements on the known *CP* violation in the *K* meson system in chapter 9.

The fundamental character of the electric dipole moment measurement as a direct test of *T* invariance is attested by its connection with the Kramers degeneracy. Other direct tests rely on effects relating to motion reversal, and as we shall discover later, their interpretation is often clouded by the effects of interactions between particles ("final state interactions"; see sec. 5.2). Therefore the electric dipole measurements are, in principle, the most unambiguous test of *T* invariance available at present.

4.3 Radiative Transitions of Atoms and Nuclei: Lloyd's Theorem

The equations of motion for the free electromagnetic field, that is, the classical Maxwell equations in the absence of electric charges, are manifest-

[7] A recent measurement of the electric dipole moment of ^{129}Xe in the $^1S_0$ state yields $D_{Xe} = (-0.3 \pm 1.1) \times 10^{-26}$ e cm as reported by Vold et al. (1984).

ly T invariant. Even in the presence of charged sources, the invariance condition is manifest *if* the charge density is invariant (even) under T and the current density is odd (changes sign), as would normally be expected. If the source of the electromagnetic field is a system whose internal dynamics violate T invariance, however, then the effective charge and current densities (i.e., the matrix elements of their operators) may not satisfy these conditions. In fact, we have just seen how T violation can lead to a permanent electric dipole moment for an atomic or a nuclear system, in which case the effective charge density is *not* T invariant.

Because of a possible effect of T violation on matrix elements, the electromagnetic radiation emitted or absorbed by a system of electric charges can provide information concerning the T invariance of the internal dynamics of the system. As it turns out, radiation emitted or absorbed in an allowed transition does not depend on the behavior of the system under T, but the interference that can occur between various types of forbidden transitions does provide a direct test of T invariance in the system undergoing the transition.

Transitions between stationary states of atoms and nuclei leading to emission of photons are called "allowed" or "forbidden" depending on whether the change in angular momentum and parity in the transition satisfies the electric dipole ($\Delta j = 0, \pm 1$, parity change) or higher electric multipole selection rules.[8] The distinction between "allowed" and "forbidden" is made on the basis of an expansion in powers of $(\vec{k} \cdot \vec{x})$, where $|\vec{k}|$ is 2π times the reciprocal of the photon wavelength and $\vec{x}$ is the position vector of a charged particle in the atom or nucleus. In effect, this is an expansion in powers of the small quantity ak, where a is the size of atom or nucleus and k^{-1} is the wavelength of the photon (divided by 2π). The allowed, electric dipole ($E1$) transition arises from the leading term in this expansion, and if the selection rule is satisfied, the $E1$ term will dominate by itself. However, for a forbidden transition—that is, when the $E1$ selection rule is not satisfied—there are circumstances such that two multipole moments, one the electric $2^{\lambda+1}$ pole, denoted by $E(\lambda + 1)$, the other the magnetic 2^{λ} pole, denoted by $M\lambda$, can make nonvanishing contributions of the same order of magnitude. The result is that one might expect to see interference between the competing electric and magnetic multipoles, which would make it possible to determine relative phases of the matrix elements of the multipole moments.

Lloyd (1951) was the first to show that these phases are determined by the requirement of T invariance. In fact, the relative phase of the competing

[8] For a general treatment of multipole transitions by the method used here see Sachs and Austern (1951) and Sachs (1953, pp. 232 ff.).

electric and magnetic terms is then either 0 or π. At the time, the importance of Lloyd's work was that it dashed the hopes of those who were expecting to obtain information about phases of nuclear energy states by measuring such interference terms. Later, when interest in the possibility of T violation was aroused by the discovery of P violation, it was realized that a measurement of such an interference term could be used as a test of T invariance and that a determination of the phase difference between $E(\lambda + 1)$ and $M\lambda$ would serve as a quantitative measure of T violation. The generic theoretical basis for such measurements was presented by Jacobsohn and Henley (1959).[9]

It should be recognized that though the same principles apply to atomic and nuclear transitions, the latter have been given more emphasis because the detection of higher-order multipole transitions is much more likely in nuclei than in atoms. The ratio of size, a, to photon wave length, k^{-1}, is normally much smaller for atoms than for nuclei, so that higher multipole matrix elements, which depend on higher powers of ak, are relatively much smaller in atoms. However, since the same principles apply to both atomic and nuclear systems, no distinction need be made between them at this point.

Interference between the $E(\lambda + 1)$ and the $M\lambda$ transitions may occur if the selection rules for both are satisfied simultaneously—that is, when $|\Delta j| \leq \lambda$ and the relative parity for the transitions is $(-1)^{\lambda+1}$. Thus for $|\Delta j| = 1$ and no parity change, the electric quadrupole ($\lambda + 1 = 2$) and magnetic dipole transitions can compete. On the other hand, in the converse situation for which both the $E\lambda$ and $M(\lambda + 1)$ selection rules are satisfied, no appreciable interference is expected. For example, $|\Delta j| \leq 1$ with a change of parity allows both the $E1$ and $M2$ transitions to occur, but the $M2$ amplitude is suppressed by an extra factor of $(ak)^2$. Therefore, for $|\Delta j| \leq 1$ strong interference will occur only when there is no parity change.

In general, for $|\Delta j| = \lambda$ and parity change $(-1)^\lambda$ the competing transitions are $E(\lambda)$ and $M(\lambda + 1)$ and the magnetic amplitude is suppressed by a factor of $(ak)^2$, whereas for parity change $(-1)^{\lambda+1}$ the $E(\lambda + 1)$ and $M\lambda$ transitions are of the same order. Then the total amplitude is proportional to the matrix element of the linear combination of multipole moments[10]

$$\mathbf{A}_\lambda = |\vec{k}|(\hat{e} \cdot \vec{\mathbf{D}}_{\lambda+1}) + i(\hat{e} \times \hat{k} \cdot \vec{\mathbf{M}}_\lambda), \tag{4.32}$$

where $\hat{e}$ is the polarization of the photon and $\hat{k}$ is the unit vector in the direction of propagation of the photon so that $\hat{e} \cdot \hat{k} = 0$. $\vec{\mathbf{D}}_\lambda$ is the 2^λ pole

[9] Henley (1969, pp. 404 ff.) provides an excellent review of time reversal tests in nuclear physics.

[10] See note 8.

electric moment

$$\vec{\mathbf{D}}_\lambda = (\lambda!)^{-1} \sum_\alpha e_\alpha \vec{x}^\alpha (\hat{k} \cdot \vec{x}^\alpha)^{\lambda-1} \tag{4.33}$$

of a system of particles having charges e_α, while $\vec{\mathbf{M}}_\lambda$ is the magnetic 2^λ pole magnetic moment. The usual[11] form of the Hermitian operator $\vec{\mathbf{M}}_\lambda$ is

$$\vec{\mathbf{M}}_\lambda = \frac{1}{(\lambda-1)!}\frac{\hbar}{2c}\sum_\alpha \frac{e_\alpha}{m_\alpha}\left\{\frac{1}{\lambda+1}[\vec{l}^\alpha(\hat{k}\cdot\vec{x}^\alpha)^{\lambda-1} + (\hat{k}\cdot\vec{x}^\alpha)^{\lambda-1}\vec{l}^\alpha] + g_\alpha \vec{\sigma}^\alpha(\hat{k}\cdot\vec{x}^\alpha)^{\lambda-1}\right\}, \tag{4.34}$$

where $\vec{l}^\alpha\hbar$ is the orbital angular momentum of the αth particle with respect to the center of mass and m_α and g_α are the mass and gyromagnetic ratio of that particle, respectively.

The amplitude of the radiation is determined by the matrix element of $\mathbf{A}_\lambda$ for the transitions from the initial state $|\psi_{E,j,m}\rangle$ to the final state $|\psi_{E',j',m'}\rangle$:

$$A_\lambda = \langle E', j', m', |\mathbf{A}_\lambda| E, j, m\rangle, \tag{4.35}$$

where[12] $\Delta j = j' - j = \pm\lambda$, $\Delta m = |m' - m| \leq \lambda$ and it is understood that the relative parity of the two states is $(-1)^{\lambda+1}$. Then if the electric and magnetic multipole matrix elements are defined by

$$A_\lambda^D = \langle E', j', m' |(\hat{e}\cdot\vec{\mathbf{D}}_\lambda)| E, j, m\rangle, \tag{4.36}$$

$$A_\lambda^M = \langle E', j', m' |(\hat{e}\times\hat{k}\cdot\vec{\mathbf{M}}_\lambda)| E, j, m\rangle, \tag{4.37}$$

we have

$$A_\lambda = |\vec{k}| A_{\lambda+1}^D + iA_\lambda^M. \tag{4.38}$$

The phases of $A_{\lambda+1}^D$ and A_λ^M may now be determined by making use of the fundamental property of the inner product, eq. (3.32):

$$A_{\lambda+1}^{D*} = \langle T(E', j', m')| T(\hat{e}\cdot\vec{\mathbf{D}}_{\lambda+1})| E, j, m\rangle \tag{4.39}$$

[11] Although the form of the electric multipole moments is independent of dynamic details, the magnetic multipole moments depend on these details. See note 8. For specific examples of the effects of spin-orbit, velocity-dependent, and exchange forces, see Austern and Sachs (1951). My purpose here is to consider only the time reversal and some other simple symmetry properties of the moments, and these are independent of the particular form in T-invariant systems. Therefore I consider only the usual forms, for the sake of simplicity. The word "usual" is used here to mean that $\vec{\mathbf{M}}_\lambda$ includes only orbital and spin moments. It is more common to see the moments expressed in terms of vector spherical harmonics, but the form used here serves just as well and is simpler.

[12] For the special case $\lambda = 1$, $\Delta j = 0$ must also be included because there is no electric monopole transition to overwhelm the electric quadrupole and magnetic dipole transitions.

or

$$A^{D*}_{\lambda+1} = i^{2(m-m')}\langle E', j', -m' | (\hat{e} \cdot \vec{\mathbf{D}}_{\lambda+1}) | E, j, -m\rangle, \tag{4.40}$$

in accordance with the T-invariance conditions eq. (4.6), and

$$T\vec{\mathbf{D}}_{\lambda+1}T^{-1} = \vec{\mathbf{D}}_{\lambda+1}, \tag{4.41}$$

the latter following from the fact that $\vec{\mathbf{D}}_{\lambda+1}$ depends only on particle coordinates.

In order to relate the right-hand side of eq. (4.40) to the original matrix element, we choose the z axis, the axis of quantization, to be in the direction $\hat{k}$ and the x axis in the direction $\hat{e}$. Then rotating these axes 180° about the y axis changes the signs of $(\hat{k} \cdot \vec{x}_j)$ and $(\hat{e} \cdot \vec{x}_j)$, multiplying $(\hat{e} \cdot \vec{\mathbf{D}}_{\lambda+1})$ by $(-1)^{\lambda+1}$. But since the axis of quantization has been reversed, the corresponding transformed state is $| E, j, -m\rangle e^{i\alpha(j, m)}$, where $\alpha(j, m)$ is the phase change induced by the rotation. Therefore

$$\begin{aligned} &\langle E', j', -m' | (\hat{e} \cdot \vec{\mathbf{D}}_{\lambda+1}) | E, j, -m\rangle \\ &\qquad = (-1)^{\lambda+1} e^{i\beta}\langle E', j', m' | (\hat{e} \cdot \vec{\mathbf{D}}_{\lambda+1}) | E, j, m\rangle, \end{aligned} \tag{4.42a}$$

with

$$\beta = \alpha(j, m) - \alpha(j', m') \tag{4.42b}$$

and

$$A^{D*}_{\lambda+1} = (-1)^{\lambda+1} i^{2(m-m')} e^{i\beta} A^{D}_{\lambda+1}. \tag{4.43}$$

This analysis may be repeated for the magnetic term, taking into account that angular momenta change sign under T, so that eq. (4.41) is replaced by

$$T\vec{\mathbf{M}}_{\lambda}T^{-1} = -\vec{\mathbf{M}}_{\lambda}. \tag{4.44}$$

Also, in considering the rotation about the y axis it must be noted that $(\hat{e} \times \hat{k} \cdot \vec{l}^{\alpha})$ and $(\hat{e} \times \hat{k} \cdot \vec{\sigma}^{\alpha})$ remain unchanged because they are the y components of the vectors. Thus eq. (4.42a) is replaced by

$$\begin{aligned} &\langle E', j', -m' | (\hat{e} \times \hat{k} \cdot \vec{\mathbf{M}}_{\lambda}) | E, j, -m\rangle \\ &\qquad = (-1)^{\lambda-1} e^{i\beta}\langle E', j', m' | (\hat{e} \times \hat{k} \cdot \vec{\mathbf{M}}_{\lambda}) | E, j, m\rangle, \end{aligned} \tag{4.45}$$

and the final result is:

$$A^{M*}_{\lambda} = (-1)^{\lambda} i^{2(m-m')} e^{i\beta} A^{M}_{\lambda}. \tag{4.46}$$

Comparison with eq. (4.43) shows that the two terms in A_{λ}, iA^{M}_{λ} and $A^{D}_{\lambda+1}$, have the same phase (within a sign); that is, their ratio is a real number if the dynamics of the system are T invariant. This is *Lloyd's theorem* (1951).

Lloyd's theorem has proved very useful in pointing the way to tests of T invariance for nuclear systems by measuring the interference between electric and magnetic transitions. However, since all the evidence indicates that any violation of the theorem for such systems is very small, these tests require precise measurements. Therefore it is necessary to take into account the small corrections to the results of Lloyd's theorem that occur because the proof of the theorem is based on an approximation.

The approximation is that the amplitude of emitted radiation $\mathbf{A}$ is assumed to be given by eq. (4.32), obtained from a first-order perturbation treatment of the electromagnetic interactions. Although this is an excellent approximation, the higher-order corrections, even in a T-invariant system, introduce a small phase difference ξ between the electric and magnetic terms in eq. (4.32). Any effect of T violation would appear as an additional phase difference η, and the two phases must be determined separately to arrive at the limits on T invariance.

The occurrence of the phase ξ is associated with the fact that emission of radiation by an atom or nucleus is a decay process. The generic treatment of decay processes in chapter 5 shows that decay amplitudes can be expected to have a phase, such as ξ, that arises from interactions between the decay products, called "final state interactions." In the case of radiative transitions the magnitude of these higher-order effects can be estimated with reasonable accuracy for systems on which measurements have been made, and the associated emendation of Lloyd's theorem can be taken into account.

The next section will analyze the interference occurring when the amplitude is given by eq. (4.32). The results will be used as a simple illustration of how interference may be used in principle to test T invariance. For actual tests at a high level of precision, the corrections owing to final state interactions must be taken into account. We shall do so at the end of section 4.4, where some of the results of precision measurements will be summarized.

4.4 Radiative Interference as a Measure of T Violation

The experimental methods for measuring the interference between $A^D_{\lambda+1}$ and A^M_λ that were considered by Jacobsohn and Henley (1959) involve the determination of various angular correlations among the spins of the nuclear states, the photon polarizations, and the directions of propagation. There is no need to repeat the details of their analysis here. Our principal interest is in the physical consequences of a deviation from Lloyd's theorem; that is, how a relative phase factor $e^{i\eta}$ between $A^D_{\lambda+1}$ and iA^M_λ owing to T violation would manifest itself in such experiments. It is clear from

Lloyd's theorem that

(4.47) $$\sin \eta \neq 0$$

would be a direct indication that the dynamics of the system being observed are *not* invariant under time reversal. In fact, $\sin \eta$ would be a measure of the T violation.

To show how measurements of the interference terms in the emitted radiation may be used as a measure of T violation, we shall determine the way certain interference terms in the emitted intensity depend on η. It will suffice to limit attention to the particular case of interference between electric quadrupole and magnetic dipole radiation as an illustration of the general principles. The radiation is then emitted in a transition between states of the same parity, the initial state having angular momentum j and the final state,[13] $j' = j \pm 1$ or j.

If the initial magnetic quantum states with $m = j, j - 1, \ldots, -j$ are uniformly occupied ($\vec{\mathbf{J}}$ randomly oriented) and the final state magnetic quantum numbers m' are not specified, the intensity of the emitted radiation is proportional to

(4.48) $$(2j+1)^{-1} \sum_{m,\, m'} |A_1|^2 = (2j+1)^{-1} \sum_{m,\, m'} [|\vec{k}|^2 |A_2^D|^2 - 2|\vec{k}| \operatorname{Im} A_2^{D*} A_1^M + |A_1^M|^2],$$

according to eq. (4.38). A_2^D and A_1^M are given by eqs. (4.36) and (4.37).

(4.49a) $$A_2^D = \langle E', j', m' | \hat{e} \cdot \vec{\mathbf{D}}_2 | E, j, m \rangle,$$

(4.49b) $$A_1^M = \langle E', j', m' | (\hat{e} \times \hat{k} \cdot \vec{\mathbf{M}}_1) | E, j, m \rangle.$$

In terms of the quadrupole moment tensor,

(4.50) $$\mathbf{Q}_{ii'} = \frac{1}{2} \sum_{\alpha} e_\alpha x_i^\alpha x_{i'}^\alpha,$$

the definition of $\vec{\mathbf{D}}_2$, eq. (4.33) combined with eq. (4.49a), then yields

(4.49c) $$A_2^D = \sum_{i,\, i'} \langle E', j', m' | \hat{e}_i \mathbf{Q}_{ii'} \hat{k}_{i'} | E, j, m \rangle.$$

According to the Wigner-Eckart theorem[14] the matrix elements of the vector and tensor operators may be expressed in terms of the matrix elements of a general vector Hermitian operator $\vec{\mu}$ as

(4.51a) $$\langle E', j', m' | \vec{\mathbf{M}} | E, j, m \rangle = \kappa_M \langle j', m' | \vec{\mu} | j, m \rangle,$$

[13] See note 11.

[14] See Messiah (1962, pp. 573 ff.).

and for $i' \neq i$,

$$\textbf{(4.51b)} \qquad \langle E', j', m' | \mathbf{Q}_{ii'} | E, j, m \rangle = \kappa_Q \langle j', m' | \{\mu_i \mu_{i'}\} | j, m \rangle,$$

where κ_M and κ_Q are independent of m and m' and the curly brackets, $\{\ \}$, denote the irreducible symmetric product of noncommuting operators: $\{\mu_i \mu_{i'}\} = \frac{1}{2}\{\mu_i \mu_{i'} + \mu_{i'} \mu_i\} - \frac{1}{3} \sum_j \mu_j^2$.

Introduction of the projection operator $\rho_0(j)$ defined by

$$\textbf{(4.52)} \qquad \langle J, M | \rho_0(j) | J', M' \rangle = \frac{1}{2j + 1} \delta_{Jj} \delta_{JJ'} \delta_{MM'}$$

makes it possible to write the double sum in the expression for the intensity eq. (4.48) as a trace over all angular momentum states J, M. Then the interference term in eq. (4.48) takes the form

$$\begin{aligned} \textbf{(4.53)} \qquad & 2|\vec{k}|(2j + 1)^{-1} \operatorname{Im} \sum_{m,\, m'} A_2^{D*} A_1^M \\ & = 2(2j' + 1) \operatorname{Im} \kappa_Q^* \kappa_M \\ & \cdot \operatorname{Tr}[\rho_0(j)\{(\hat{e} \cdot \vec{\mu})(\vec{\mu} \cdot \vec{k})\} \rho_0(j')(\hat{e} \times \vec{k} \cdot \vec{\mu})]. \end{aligned}$$

Again, we may choose the z axis to be along $\vec{k}$ and the x axis along $\hat{e}$. An alternative choice would be to rotate these axes 90° about the y axis, changing x to z and z to $-x$, thereby changing the sign of the matrix product in eq. (4.53). But because the trace of the product includes all values of m and m' with equal weight, the expression is invariant under the transformation of the quantum states $|m\rangle$ and $|m'\rangle$ induced by the rotation. Therefore the quantity eq. (4.53) vanishes; there is no interference term in a transition between unpolarized states. The same argument will clearly apply to any of the interference terms between $E(\lambda + 1)$ and $M\lambda$ radiation.

To obtain interference experimentally we must observe radiation either from an initially polarized state of the system or by selective detection of polarized final states, or both. Then there is a coherent mixture of magnetic substates, and the interference term may not vanish. Since there are many possibilities for such coherence of states, we first consider the specific case of a coherent mixture of initial states but a detection system that does not distinguish between final states of polarization.

The coherence of the initial states is best described by introducing in place of the density matrix for random orientation, $\rho_0(j)$, the statistical density matrix[15] $\rho(j)$, whose J, J' matrix elements relate the amplitudes and phases of the $2j + 1$ initial magnetic substates of angular momentum j but vanish for $J \neq j$ and $J' \neq j$. Then the intensity of the emitted radiation is

[15] The density matrix is treated in the textbook by Messiah (1961, pp. 331 ff.). For a thorough authoritative treatment see Fano (1957).

given by

$$I = (2j' + 1) \text{ Tr } \rho(j)\mathbf{A}^{\dagger}\rho_0(j')\mathbf{A}, \tag{4.54}$$

where the matrix emission amplitude

$$\mathbf{A} = \kappa_Q\{(\hat{e}\cdot\vec{\mu})(\vec{\mu}\cdot\hat{k})\} + i\kappa_M(\hat{e}\times\hat{k}\cdot\vec{\mu}) \tag{4.55}$$

has been introduced in such a way as to be in accord with eq. (4.38). Note that for emission from a pure state $|j, m\rangle$ (into all subsets $|j', m'\rangle$ of $|E', j'\rangle$) the intensity is given by the diagonal element $\langle j, m|\mathbf{A}^{\dagger}\rho_0(j')\mathbf{A}|j, m\rangle$.

Since the density matrix is Hermitian, the interference terms in the emission are given by

$$\Delta I = -2(2j' + 1) \text{ Im } \{\kappa_Q^* \kappa_M \text{ Tr } [\rho(j)\{(\hat{e}\cdot\vec{\mu})(\hat{k}\cdot\vec{\mu})\} \cdot \rho_0(j')(\hat{e}\times\hat{k}\cdot\vec{\mu})]\}, \tag{4.56}$$

in place of eq. (4.53).

The coefficient $\kappa_Q^* \kappa_M$ is, in general, a complex constant depending on the initial and final states of the system. Since $\vec{\mu}$ may be defined as a magnetic moment operator so that $T\vec{\mu}T^{-1} = -\vec{\mu}$, the phase of its matrix element will be the same as the phase of the corresponding matrix element of $\vec{\mathbf{M}}_1$ (see eqs. 4.44 to 4.46) in the case of a T-invariant system. Also the phase of a matrix element of $\mathbf{Q}_{ii'}$ will, in that case, be the same as that of the corresponding matrix element of the tensor $\{\mu_i \mu_{i'}\}$. Therefore $\kappa_Q^* \kappa_M$ will be real *if the dynamics of the system are T invariant.* In general, we may write

$$\kappa_Q^* \kappa_M = |\kappa_Q \kappa_M| e^{i\eta}, \tag{4.57}$$

where η, the relative phase of the electric and magnetic terms, will differ from 0 or π only when there is a violation of T invariance. The interference term eq. (4.56) then takes the form

$$\Delta I = -(2j' + 1)|\kappa_Q \kappa_M|[T_- \cos\eta + T_+ \sin\eta], \tag{4.58a}$$

where

$$T_{\pm} = \text{Tr } \{\rho(j)[\{(\hat{e}\cdot\vec{\mu})(\vec{\mu}\cdot\hat{k})\}\rho_0(j')(\hat{e}\times\hat{k}\cdot\vec{\mu}) \pm (\hat{e}\times\hat{k}\cdot\vec{\mu})\rho_0(j')\{(\hat{e}\cdot\vec{\mu})(\vec{\mu}\cdot\hat{k})\}]\}. \tag{4.58b}$$

The existence of a term having the form T_+ would therefore be evidence of T violation, and measurement of its magnitude would make it possible to determine $\sin\eta$.

An experimental test of the existence of the term T_+ requires that the correlation of the initial magnetic substates be capable of producing a nonvanishing contribution T_+ to the intensity. The required correlations may be described in terms of the $(2j+1) \times (2j+1)$ submatrix of $\rho(j)$, which will also be designated by $\rho(j)$ and takes the form of a polynomial in the components of the total angular momentum matrix $\vec{\mathbf{J}}$:

$$\rho(j) = \frac{1}{2j+1}\left[\mathbf{1} + \frac{3}{j(j+1)}(\vec{\mathbf{j}} \cdot \vec{\mathbf{J}}) + \sum_{i,k} \theta_{ik} \mathbf{J}_i \mathbf{J}_k + \sum_{i,k,l} \theta_{ikl} \mathbf{J}_i \mathbf{J}_k \mathbf{J}_l + \cdots\right], \tag{4.59}$$

where $\vec{\mathbf{j}}$ is the expectation value of the total angular momentum operator $\vec{\mathbf{J}}$ with components $\mathbf{J}_i$ and θ_{ik}, θ_{ikl}, etc., are irreducible symmetric tensors. They are real constants determined by experimental conditions on the initial state. Each of the terms in $\rho(j)$ has a simple physical interpretation: the first describes the unpolarized system; in the second $\vec{\mathbf{j}}/j$, where j is the angular momentum quantum number, is the "polarization" of the system, θ_{ik} measures the "alignment" with respect to $\vec{\mathbf{j}}$, and θ_{ikl} is a measure of the "orientation of the third degree" with respect to its three components. For $j = 1/2$ only the first two terms occur, for $j = 1$ only the first three terms, and so forth, as is easily established by counting the number of independent matrices that can be constructed from products of the $\mathbf{J}_x$, $\mathbf{J}_y$, $\mathbf{J}_z$. Clearly, terms higher than the fourth may be written as irreducible tensor products of the matrices $\mathbf{J}_x$, $\mathbf{J}_y$, $\mathbf{J}_z$ multiplied by irreducible tensors $\theta_{ikl\cdots}$.

Both of the interference terms T_+ and T_- may be calculated by inserting this expression for $\rho(j)$ into eq. (4.58b). Since our principal purpose is to obtain physical insight into methods for testing T invariance, we consider only the T-violating term T_+.

Because of the invariance of the trace under rotations of the coordinate axes, the only terms that will survive in T_+ or T_- are those that can be expressed as the invariant products of tensor operators constructed from products of $\mathbf{J}_x$, $\mathbf{J}_y$, and $\mathbf{J}_z$ and the tensor formed by permutation of the operator product $\mu_x \mu_y \mu_z$ appearing in eq. (4.58b). Such an invariant can be formed from the contraction of the irreducible symmetric tensors $\{\mathbf{J}_i \mathbf{J}_k \mathbf{J}_l\}$ and $\{\mu_i \mu_k \mu_l\}$. Therefore there is a contribution to T_+ from the θ_{ikl} term in ρ. Since the product of lower-rank tensors in the $\mathbf{J}_i$ with the irreducible symmetric tensor $\{\mu_i \mu_k \mu_l\}$ contains no such invariant, the third-rank term is the lowest in ρ that will contribute to T_+. This result shows that the initial state must be assigned at least a third-order orientation in order to produce a term in the intensity proportional to $\sin \eta$ if the orientation of the final state is undefined.

The initial state may be assumed to have a prescribed polarization so that the vector $\vec{\mathbf{j}}$ is given. In the absence of any other information about the orientation of the initial system, the symmetric tensor θ_{ikl} is proportional to

$\mathbf{j}_i\mathbf{j}_k\mathbf{j}_l - \frac{1}{3}\mathbf{j}^2\delta_{ik}\mathbf{j}_l$, where

$$\vec{\mathbf{j}} = \mathbf{j}_1\hat{e} + \mathbf{j}_2(\hat{e} \times \hat{k}) + \mathbf{j}_3\hat{k}, \quad (4.60)$$

when the coordinate axes are defined by the polarization and propagation vectors of the photon. Then the contribution of this term in $\rho(j)$ to ΔI, eq. (4.58a), is proportional to

$$\mathbf{j}_1\mathbf{j}_2\mathbf{j}_3 \sin\eta = (\hat{k}\cdot\vec{\mathbf{j}})(\hat{e}\cdot\vec{\mathbf{j}})(\hat{e}\times\hat{k}\cdot\vec{\mathbf{j}}) \sin\eta. \quad (4.61)$$

That the existence of a term in the intensity depending on the product $(\hat{k}\cdot\vec{\mathbf{j}})(\hat{e}\cdot\vec{\mathbf{j}})(\hat{e}\times\hat{k}\cdot\vec{\mathbf{j}})$ would be an apparent violation of time reversal invariance is evident on qualitative grounds; the expectation value of the angular momentum, $\vec{\mathbf{j}}$, changes sign under T, or under "motion reversal," as does the photon momentum $\hat{k}$, while $\hat{e}$, which determines the plane of polarization, is invariant under motion (or time) reversal. Therefore the product changes sign. But in a T-invariant system the *intensity* of the radiation must be invariant under T, and it cannot contain an odd term. Therefore the appearance of the odd term appears to imply T violation. However, as indicated in section 1.2, a distinction must be made between motion reversal and time reversal. This distinction arises here as a consequence of the corrections to the perturbation approximation mentioned at the end of section 4.3. We shall find in section 5.3 that final (or initial) state interaction effects, although completely consistent with T invariance, lead to terms that are odd under *motion reversal* like eq. (4.61) and therefore mimic T-violation effects. These terms are consistent with T invariance because they include additional factors (arising from phases of the scattering matrix) that are not affected by motion reversal but do change sign under *time reversal.* Their overall sign is therefore left unchanged by T.

Inclusion of the final state interaction in the interference term eq. (4.61) introduces the additional phase ξ; therefore the factor $\sin\eta$ is simply replaced by a factor $\sin(\eta + \xi)$. Although it might appear that the experimental demonstration of the mere existence of such a motion reversal violating term would demonstrate the failure of T invariance, we now see that that is not so, because ξ is usually different from zero. But accurate measurement of the term does provide a test of T invariance if ξ can be determined independently, since any significant deviation of the measured factor $\sin(\eta + \xi)$ from $\sin\xi$ provides a measure of the T violation, η. Extracting the factor $\sin(\eta + \xi)$ requires a determination of the coefficient of the interference term for the given transition. These coefficients are given, for example, by Dehn, Marzolf, and Salmon (1964) and by Boehm (1968, p. 279).

Measurements of this particular interference effect in nuclear transitions

TABLE 4.1 Experimental Results on T Violation from Nuclear Radiative Transitions

Odd Term	Nucleus	$\eta \times 10^3$(rad)	Reference
$(\hat{k}\cdot\mathbf{j})(\hat{e}\cdot\mathbf{j})(\hat{e}\times\hat{k}\cdot\mathbf{j})$	^{99}Ru	0 ± 1.7	Kistner (1967)
	^{193}Ir	0.2 ± 1.7	Atac et al. (1968)
	^{191}Ir	-0.4 ± 0.5	Gimlett et al. (1981)
	^{57}Fe	-0.31 ± 0.65	Cheung et al. (1976) Cheung, Henrikson, and Boehm (1977)
	^{131}Xe	-1.1 ± 1.1	Gimlett, Henrikson, and Boehm (1982)
	^{197}Au	-0.33 ± 0.66	Tsinoev et al. (1982)
$(\hat{k}\cdot\mathbf{j}')(\hat{k}\cdot\mathbf{j}\times\mathbf{j}')$	^{56}Mn	26 ± 14	Garrell et al. (1969)
	^{106}Pd	-45 ± 27	Perkins and Ritter (1968)
	^{49}Ti	17 ± 25	Kajfosz, Kopecky, and Honzatko (1965, 1968)
	^{36}Cl	0.8 ± 2.3	Eichler (1968)
$(\hat{k}\cdot\hat{k}')(\mathbf{j}\cdot\hat{k}\times\hat{k}')$	^{49}Ti	-10 ± 6	Sharman et al. (1978)
	^{180}Hf	48 ± 87	Krane, Murdoch, and Steyert (1974)
	^{169}Tm, ^{75}Lu	-170 ± 500	Murdoch, Olsen, and Steyert (1974)
	^{110}Ag	1.5 ± 2.2	Wang et al. (1978)
$(\mathbf{j}\cdot\hat{k})(\mathbf{j}\cdot\hat{k}\times\hat{k}')$	^{180}Hf	<10 (?)	Murdoch et al. (1974)

were first made by Kistner (1967), Blume and Kistner (1968), and Atac and others (1968). Recent measurements are presented by Gimlett and others (1981), Gimlett, Henrikson, and Boehm (1982), and Tsinoev and others (1982). Theoretical estimates of the final state interaction effects owing to radiative corrections and interaction of the gamma rays with atomic electrons have been given for the measured transitions by Hannon and Trammell (1968), Henley (1969), Goldwire and Hannon (1977), and Davis, Koonin, and Vogel (1980). The results of these and other experiments are given in table 4.1 and may be summarized for our purposes by the statement that they place an upper limit on T violation of the order of

$$|\eta| \lesssim 10^{-3} \text{ radians.} \tag{4.62}$$

Thus they are consistent with T invariance of nuclear (and electromagnetic) interactions.

The upper limit that can be placed on $|\eta|$ is governed by both the limits of precision of the experiments and the uncertainties in the calculations of ξ. There appears to be only one experiment (Gimlett et al. 1981) leading to an interference term of the form eq. (4.61) that is significantly different from zero. The result is consistent with the calculated value of ξ (Davis, Koonin, and Vogel 1980).

The experimental methods that may be used to determine the nuclear spin orientation have been described by Jacobsohn and Henley (1959). For the case we have been considering, only the initial state is oriented, and the direction of $\vec{\mathbf{j}}$ for these experiments has been established by applying an external magnetic field. Other methods have been used to carry out experiments in which the interference terms involve the orientations of both the initial and final states. They make use of correlations between the directions of two gamma rays in cascade from a polarized initial state or between the directions of gamma rays and of an electron from beta decay. The capture of polarized neutrons has also been used to produce polarized nuclei for this purpose.

The analysis of some of the experiments requires a generalization of our treatment of the $M1$-$E2$ interference to include polarization of both the initial state $|j, m\rangle$ and the final state $|j', m'\rangle$. For this purpose the density matrix $\rho_0(j')$ describing the unpolarized final state in eq. (4.54) is replaced by a density matrix $\rho'(j')$ expressible as a polynomial in the matrices $\mathbf{J}_i'$ having the same form as eq. (4.59). Then the general expression for the intensity is

(4.63) $$I = \mathrm{Tr}\ \rho(j)\mathbf{A}^\dagger \rho'(j')\mathbf{A},$$

and the interference terms include various products of $(\hat{e} \cdot \vec{\mathbf{j}})$, $(\hat{k} \cdot \vec{\mathbf{j}})$, $(\hat{e} \times \hat{k} \cdot \vec{\mathbf{j}})$ with $(\hat{e} \cdot \vec{\mathbf{j}}')$, $(\hat{k} \cdot \vec{\mathbf{j}}')$, $(\hat{e} \times \hat{k} \cdot \vec{\mathbf{j}}')$, etc. We can immediately identify those products that are odd under motion reversal as the ones that can be used to measure $\sin \eta$. Tests for T violation are therefore provided by measurements of the emitted intensity for correlated systems capable of yielding such terms in principle.

It is a straightforward matter to identify the possible forms of terms odd under motion reversal that could appear in I. Typical forms are displayed in table 4.2, classified in terms of the observed polarization of the emitted photon. The unit vector $\hat{\sigma}$ is used here to give the direction of circular polarization, and it is to be noted that σ, which is an angular momentum, is odd under T. The direction $\hat{k}'$ is that of a second photon following the $M1 + E2$ transition in cascade, and the displayed expression represents the correlation between the two photons, which serves as a substitute for the correlation of the first photon with the orientation of the final state of the nucleus from which the photon $\hat{k}'$ originates. Note that the term $(\vec{\mathbf{j}} \cdot \hat{k})$

TABLE 4.2 Typical Forms of Interference Terms That Are Odd under Motion Reversal

Photon Polarization	Interference Term
None	$(\hat{k}\cdot\mathbf{j}_1)(\hat{k}\cdot\mathbf{j}_1\times\mathbf{j}_2)$
	$(\hat{k}\cdot\hat{k}')(\mathbf{j}_1\cdot\hat{k}\times\hat{k}')$
	$(\mathbf{j}_1\cdot\hat{k})(\mathbf{j}_2\cdot\hat{k}\times\hat{k}')$
Linear	$(\hat{k}\cdot\mathbf{j}_1)(\hat{e}\cdot\mathbf{j}_2)(\hat{e}\times\hat{k}\cdot\mathbf{j}_3)$
Circular ($\hat{\sigma}$ = spin of photon)	$(\hat{k}\cdot\hat{\boldsymbol{\sigma}})(\hat{k}\cdot\mathbf{j}\times\mathbf{j}')$ $(\hat{k}\cdot\boldsymbol{\sigma})(\hat{k}\cdot\mathbf{j}\times\mathbf{j}')(\mathbf{j}\cdot\mathbf{j}')$

$\mathbf{j}_1$, $\mathbf{j}_2$, or $\mathbf{j}_3 = \mathbf{j}$ or $\mathbf{j}'$ in all combinations

Note: Table 4.1 presents the results of experiments designed to measure the indicated correlation terms.

$(\mathbf{j}\cdot\hat{k}\times\hat{k}')$ provides a direct measure of parity violation as well as motion reversal violation, since $\hat{k}$ and $\hat{k}'$ are polar vectors.

We shall find that the form of argument used here may be applied not only to intensities of radiative transitions but also to any other quantity determined by the modulus of an amplitude, such as a decay rate or a cross section. Opportunities to determine $\sin\eta$ are provided by those interference terms that are odd under motion reversal. In each such case the determination of $\sin\eta$ requires independent information concerning the final state phase ξ, which is small for electromagnetic and leptonic (electron, muon, neutrino) effects but may be quite large for effects involving particles that interact strongly (hadrons) in the final state.

An important question remaining to be answered concerns how failure of T invariance in the dynamics of the radiating system introduces a phase η between electric and magnetic terms in the amplitude. This abrogation of Lloyd's theorem occurs through failure of the fundamental phase relation eq. (4.6) for the energy eigenstates, which was essential for the proof of the theorem. The way the phase relation is altered by a T violation in the dynamics—that is, when $THT^{-1}\neq H$—has been illustrated in a simple example in section 4.2. Specifically, eq. (4.22) shows that the phase condition is not satisfied by the φ_i^m of that example because the linear combination violates the condition eq. (4.12).

In a more general case, if we assume that the T-violating term in the Hamiltonian H^0, satisfying eq. (4.17), is small enough, the effect of T violation on the energy states of H can be obtained by treating H^0 as a small perturbation on the states of the T-invariant Hamiltonian H^e.

When the zero-order states are denoted by $|\psi^{\iota}_{j,m}\rangle$, the eigenstates $|\psi_{E,j,m}\rangle$ of H may be written as a linear combination of the $|\psi^{\iota}_{j,m}\rangle$, as in eq. (4.7), with the small coefficients $a^{\iota}_{j,m}$ proportional to matrix elements of H^0. Application of the fundamental eq. (3.32) and use of eq. (4.17) show that these matrix elements are imaginary relative to the zero-order term. Therefore the coefficients no longer satisfy the phase conditions eq. (4.12). Although the phases of the $|\psi^{\iota}_{j,m}\rangle$ have been chosen to satisfy the kinematic condition eq. (4.9), the corresponding condition eq. (4.6) on the energy states $|\psi_{E,j,m}\rangle$, which is essential to the proof of Lloyd's theorem, is no longer satisfied.

The additional phase parameter (or parameters) introduced by the perturbation will usually manifest itself (or themselves) in the matrix elements of the electric and magnetic multipoles in different ways, with the result that there is a phase difference η between the two multipole terms, leading to the introduction of correlation terms such as eq. (4.61) in the radiated intensity.

The relationship between the magnitude of η and that of H^0 is determined from the amplitudes $a^{\iota}_{j,m}$ of the states introduced by the perturbation. These amplitudes are imaginary and of the order of (compare eq. 4.23)

(4.64)
$$|a^{\iota}_{j,m}| = \left|\frac{H^0_{0\iota}}{\Delta_\iota E}\right|,$$

where $\iota = 0$ corresponds to the unperturbed state and $\Delta_\iota E$ is the energy difference $E_\iota - E_0$. When the matrix elements of the magnetic dipole and electric quadrupole moments are calculated with respect to the perturbed wave functions, the (small) added contributions are again imaginary and provide the required measure of η. An upper limit on the value of $|\eta|$ corresponding to a given perturbation may be found by assuming that

(4.65)
$$\langle\psi^0_{j,m}|\vec{\mathbf{M}}_1|\psi^{\iota}_{j',m'}\rangle \approx 0,$$

as would be the case even if $\psi^{\iota}_{j',m'}$ and $\psi^0_{j,m}$ belong to the same spin-orbit configuration, because $\vec{\mathbf{M}}_1$ depends only on spin and angular momentum operators (if exchange magnetic moments are neglected), and the radial wave functions of the states between which the transition takes place are orthogonal.

Then the relative phases of magnetic dipole and electric quadrupole matrix elements will depend only on the perturbation of the quadrupole transition. If all matrix elements of the quadrupole moment are of the same order of magnitude, we can conclude that

(4.66)
$$|\eta| \lesssim 2\sum_{\iota\neq 0}\left|\frac{H^0_{0\iota}}{\Delta_\iota E}\right|.$$

Thus a measurement of η would provide a lower limit on the size of the T-violating term in the Hamiltonian relative to the level spacing. On the basis of the result eq. (4.62), however, one can only say that, for nuclei, there is *no* evidence for a T-violating interaction greater than $10^{-3}\Delta E$, where ΔE is some average level spacing. It *cannot* be said, as we might be tempted to say, that the T-violating interaction is less than this, the reason being the possibility of cancellations, a possibility expressed by the inequality in eq. (4.66). Of course a more definite statement can be made about a specific nucleus for a specific form of H^0, but eq. (4.66) serves as an adequate illustration of the kind of information to be obtained by interference experiments.

4.5 Collision Processes: Formalism

The time reversal properties of strongly interacting systems of particles manifest themselves most directly in scattering and reaction processes. The basis for analysis of these processes is generally referred to as "collision theory," where the collision may be elastic or inelastic and may involve more or less than the two particles ("fragments," as defined below) involved in the usual visualization of a collision.

The general quantum mechanical treatment of collisions between complex bodies (atoms, molecules, nuclei) leading to the production of the same or different complex bodies (fragments) may be formulated in a variety of equivalent ways.[16] For our purposes in this chapter, the formal stationary state methods[17] based on the reaction amplitude matrix (T matrix) introduced by Møller (1945, 1946) are most useful, and they have the advantage that their formal structure is directly applicable to relativistic quantum field theory.

We consider an initial state comprising a set of fragments (usually two) in one channel being transformed by their interaction into a set (two or more) in another channel. A "channel" is defined here as a specified set of separated fragments, each in a specified quantum state and having no mutual

[16] The classic textbook over the years has been that of Mott and Massey (1933, 1949, 1965). The textbook of Goldberger and Watson (1964) is more in keeping with the approach taken here. Neither of these texts treats the collision theory of Wigner and Eisenbud (1947), whose formulation of the general problem of the reactions of complex bodies subject to short-range forces probably has the most rigorous foundation. See Wigner (1948, 1949), Teichmann (1950), Teichmann and Wigner (1952), and especially Wigner and von Neumann (1954) and Wigner (1955). This formulation is not presented in the textbooks mentioned above, but it is given in Sachs (1953, pp. 284 ff.), where the consequences of T invariance are also treated (p. 289). See also Blin-Stoyle (1952). The results of the Wigner-Eisenbud treatment differ from those used here because it is carried out entirely in an angular momentum representation, whereas a mixed representation of linear momentum and spin is more useful for our purposes. Of course the results are equivalent.

[17] These methods are treated in detail by Goldberger and Watson (1964).

interaction (with the possible exception of a Coulomb potential) when the distance of separation is large. We denote the identity of the channel and all the internal quantum numbers associated with each of the fragments collectively by a lowercase letter from the beginning of the alphabet: *a*, *b*, or *c*.

Since we shall be dealing with stationary states, it is appropriate to define the total energy of the system, which will be denoted by E. One important point of Møller's work was to emphasize the advantage of treating E as a complex variable extending throughout the complex plane. Of course, eigenvalues of the Hamiltonian occur only on the real energy axis, and for the situation we are considering here, the spectrum is continuous for energies above the threshold for formation of a channel. When the energy is below that threshold the channel is said to be "closed"; when it is above, the channel is "open." For values of E below the lowest threshold of the complete system (which *might* be defined as the zero of energy), all channels are closed and the spectrum of eigenvalues is discrete. As E increases, the eigenstates associated with E will correspond to an increasing number of different open channels. For example, the collision of two complex nuclei at high energy may lead to excitation or breakup of either or both of the nuclei in a great variety of ways. Each excitation state defines a new channel in our usage of the term, but a given channel may occur with the spins or momenta having different values, denoted by a_1, a_2, and so forth.

Under these conditions it is clear that many different channels *a*, *b*, *c*, and so on, may have the same energy eigenvalue. Since a channel is defined in terms of noninteracting fragments, we may associate with any given channel an eigenstate of a Hamiltonian H_a that includes the (internal) Hamiltonian of each fragment and the total kinetic energy of the fragments. Such an eigenstate is then the solution of

(4.67a) $$H_a|\chi_a\rangle = E_a|\chi_a\rangle,$$

where the label "a" describes the channel and, where needed, includes all other relevant quantum numbers except the energy.

The total Hamiltonian of the system may then be written as

(4.68a) $$H = H_a + V_a,$$

where V_a is the operator describing all the (short-range) interactions among the fragments making up this particular channel. If we consider a different channel, denoted by "b," made up of different fragments or the same fragments in different states of excitation, the "free particle" Hamiltonian will be H_b, different from H_a, and

(4.67b) $$H_b|\chi_b\rangle = E_b|\chi_b\rangle.$$

For this second channel, the interaction V_b is also necessarily different because we still must have

(4.68b)
$$H = H_b + V_b$$

for the same complete Hamiltonian. It is useful to note that when E is above threshold for both channels, we may have $E_a = E_b$, although $|\chi_a\rangle$ and $|\chi_b\rangle$ describe entirely different states of the system.

The stationary states of the system are the solutions of

(4.69)
$$H|\psi\rangle = E|\psi\rangle,$$

and when the initial state is $|\chi_a\rangle$, that is, when the collision takes place in channel a, the solutions may be expressed by the formal integral equation of Lippmann and Schwinger (1950):

(4.70)
$$|\psi_a^\pm\rangle = |\chi_a\rangle + \frac{1}{E_a - H_a \pm i\varepsilon} V_a|\psi_a^\pm\rangle.$$

The two choices of Green's functions correspond to the outgoing (+) and incoming (−) solutions of the inhomogeneous equation, while $|\chi_a\rangle$ is the solution of the "homogeneous" equation, eq. (4.67a).

Each of the sets of states $|\psi_a^+\rangle$ and $|\psi_a^-\rangle$ forms a complete orthonormal set when a runs over all channels and all quantum numbers. The "S matrix" or "characteristic matrix" of Heisenberg and Møller is defined in terms of these latter two complete sets by

(4.71)
$$S_{ba} = \langle\psi_b^- | \psi_a^+\rangle.$$

It is the probability amplitude for producing the system in channel b, with the associated quantum numbers, when the collision takes place in channel a, with its associated quantum numbers. The matrix S_{ba} is unitary, since it represents the transformation from one complete orthonormal set to another.

To obtain reaction cross sections from the S matrix, it is convenient to separate the states representing the products of the collision from the initial state by noting that S can be written in either of the forms (Goldberger and Watson 1964, p. 194)

(4.72a)
$$S_{ba} = \delta(b - a)\delta(E_b - E_a) - 2\pi i\, \delta(E_b - E_a)T_{ba}^+$$

(4.72b)
$$= \delta(b - a)\delta(E_b - E_a) - 2\pi i\, \delta(E_b - E_a)T_{ab}^{-*},$$

where

(4.73)
$$T_{ba}^\pm = \langle\chi_b| V_b |\psi_a^\pm\rangle_{E_b = E_a}.$$

Comparing eqs. (4.72a) and (4.72b) leads to the important general result

(4.74) $$T^{+}_{ba} = T^{-*}_{ab}.$$

For our purposes, we need know only that measurable quantities such as cross sections and transition probabilities for the process $a \rightarrow b$ are proportional to $|T^{+}_{ba}|^2$, the coefficients of proportionality being determined by the specific process and the way the wave functions are normalized as well as by density of states and degeneracy factors (Goldberger and Watson 1964, pp. 90 ff.).

4.6 Collision Processes: Consequences of *T* Invariance

The consequences of time reversal invariance for reaction processes may now be determined by considering the way the Lippmann-Schwinger equation is transformed. Since the states are labeled with the eigenvalues a_1, a_2, etc., of operators A_1, A_2, etc., representing the kinematic variables associated with channel a, the behavior of the state $|\chi_a\rangle$ is determined by the behavior of the operators A_i under T. Now

(4.75) $$TA_iT^{-1} = \pm A_i,$$

the sign depending on whether A_i is even or odd under motion reversal; that is, the sign is determined by the behavior of the classical physical quantity represented by the operator A_i. Therefore the behavior of the eigenstates may be obtained by applying T to the equation

(4.76) $$A_i|\chi_a\rangle = a_i|\chi_a\rangle,$$

with the result

$$TA_iT^{-1}T|\chi_a\rangle = a_i T|\chi_a\rangle$$

or, by eq. (4.75),

(4.77) $$\pm A_i T|\chi_a\rangle = a_i T|\chi_a\rangle.$$

Therefore $T|\chi_a\rangle$ is an eigenstate of A_i with eigenvalue

(4.78a) $$a_i' = \pm a_i,$$

and we can write

(4.78b) $$T|\chi_a\rangle = e^{i\phi_a}|\chi_{a'}\rangle,$$

the sign in eq. (4.78a) again depending on the behavior of A_i under motion reversal. The phase factor $e^{i\phi_a}$ depends on the choice of phases in defining the wave functions. Usually the relative motion of fragments is described by plane waves and the associated a_i are momenta $\vec{k}_a$, so that $\vec{k}_a' = -\vec{k}_a$ and the associated phase may be taken to be zero. The variable associated with the spin of each fragment is the angular momentum with quantum numbers (j_a, m_a) for which the motion-reversed values are $j'_a = j_a$ and $m'_a = -m_a$. The conventional choice of phases leads to a factor $e^{i\phi_a} = i^{2m_a}$.

The consequences of T invariance for the exact solutions $\psi_a^{\pm}$ may be found by making use of eq. (4.70) and the requirements

(4.79) $$THT^{-1} = H, \quad TH_aT^{-1} = H_a, \quad TV_aT^{-1} = V_a, \quad \text{etc.}$$

Then, since T is antiunitary,

(4.80) $$T|\psi_a^{\pm}\rangle = e^{i\phi_a}|\chi_{a'}\rangle + \frac{1}{E_a - H_a \mp i\varepsilon} V_a T|\psi_a^{\pm}\rangle,$$

for which it follows that

(4.81) $$T|\psi_a^{\pm}\rangle = e^{i\phi_a}|\psi_{a'}^{\mp}\rangle,$$

the outgoing state is converted into the incoming state with the quantum numbers (momenta, spins) reversed, and vice versa.

The consequences of T invariance for the S matrix then follow immediately from the fundamental property of the inner product, eq. (3.32):

$$S_{ba} = \langle\psi_b^- | \psi_a^+\rangle = \langle T\psi_a^+ | T\psi_b^-\rangle = e^{i(\phi_b - \phi_a)}\langle\psi_{a'}^- | \psi_{b'}^+\rangle.$$

Thus

(4.82) $$S_{ba} = S_{a'b'}\, e^{i(\phi_b - \phi_a)}.$$

Because of the direct connection between S_{ba} and T_{ba} given by eq. (4.72a), it follows from eq. (4.82) that, also,

(4.83) $$T_{ba}^+ = T_{a'b'}^+\, e^{i(\phi_b - \phi_a)}$$

if the dynamics are T invariant.

Another statement of the consequences of T invariance may be obtained by using the explicit expression, eq. (4.73), for $T_{ba}^{\pm}$. Again the fundamental property of the inner product may be used to write

(4.84) $$T_{ba}^{-*} = \langle T\chi_b | TV_b|\psi_a^-\rangle = e^{-i(\phi_b - \phi_a)}\langle\chi_{b'} | V_b|\psi_{a'}^+\rangle$$

by virtue of eqs. (4.78b) and (4.81). Therefore

(4.85) $$T_{ba}^{-*} = T_{b'a'}^+\, e^{-i(\phi_b - \phi_a)}.$$

This equation also could have been obtained by applying the general result[18] eq. (4.74) to eq. (4.83). Therefore, *under the assumption of time*

[18] The implications of this result, including those to be discussed here later, are treated thoroughly in Goldberger and Watson (1964, pp. 202 ff.).

reversal invariance, the rate of the transition $a \rightarrow b$, which depends only on $|T^{+}_{ba}|^2$, will always be equal to the rate of the transition $b' \rightarrow a'$ when degeneracy and density of states factors are taken into account. Note that all quantum numbers are subjected to motion reversal *and* the initial and final channels (states) are interchanged.

We have here an example of the distinction between motion reversal and time reversal. The requirement that initial and final states be interchanged relates to the important issue of the role of initial conditions, discussed at length for classical mechanics in section 2.3. In the treatment of a collision process in quantum mechanics, or any other form of wave mechanics, the choice of the incoming (or initial) state sets the boundary conditions. Just as in classical mechanics, in order that the motion—that is, the collision amplitude—be invariant, the boundary conditions as well as the motion must be reversed; thus initial and final states must be interchanged.

Eq. (4.82) or eq. (4.83) is the general basis of the *principle of detailed balance*,[19] which is of considerable importance in the study of the foundations of statistical mechanics, kinetic theory, transport theory, and so forth. The equation also provides opportunities for experimental tests of T invariance, especially in nuclear reactions (Henley and Jacobsohn 1959). However, in considering tests based on eq. (4.82) it is very important to recognize that T invariance is only a *sufficient* condition; it is *not* a necessary condition. There are a variety of processes for which detailed balance will hold in some form even when time reversal is violated (Henley and Jacobsohn 1957, 1959). They include reactions caused by small perturbations, for which only the fact that the matrix of the perturbation is Hermitian is needed to establish detailed balance on the average over degenerate states (Pauli 1928). Also, as noted by Henley and Jacobsohn, for a two-channel process the unitarity of the S matrix guarantees that the off-diagonal elements differ only by a phase factor and therefore the reactions $a \rightarrow b$ and $b \rightarrow a$ have equal rates. There is, in addition, the case of the

[19] Although Wigner and Eisenbud (1947) used T invariance to establish the symmetry of the collision matrix equivalent to eq. (4.82), it was Coester (1951, 1953) who derived the result in essentially the form given here and called attention to the connection between T invariance and the principle of detailed balance. Since the usual early applications of detailed balance are statistical in nature, they involved averages of the rates over all possible spin states of the colliding objects and their products. These averages do not distinguish between states a,b and a',b'. Therefore Blatt and Weisskopf (1952, p. 530) distinguish between detailed balance, which they define as $S_{ba} = S_{ab}$, and "reciprocity," defined by eq. (4.82). Although $S_{ba} = S_{ab}$ also leads to detailed balance in the average over spins, it does not follow from general principles except in certain special cases. Since $|S_{ba}| = |S_{a'b'}|$ does follow from T invariance in general and leads to the same result upon averaging, it will be my preferred meaning for the phrase "detailed balance."

scattering of a spin $\frac{1}{2}$ particle from a spinless target, where, as we shall see, the combination of angular momentum and parity conservation is sufficient to guarantee detailed balance. Biedenharn (1959) has identified various other conditions under which T invariance is irrelevant to detailed balance.

The ambiguous nature of detailed balance implies that great care must be taken in using it as a test of time reversal. In such tests, as in every other test of T invariance, the safest criterion is that the experiment be designed to provide a direct quantitative measure of T violation.

Experimental tests of detailed balance for processes due to strong interactions have been carried out for a variety of reactions between unpolarized nuclei, following especially the early suggestions of Henley and Jacobsohn (1957, 1959). Some of these experiments involved determining the relative angular distributions of "forward" and "backward" reactions at a given energy in the center-of-mass system, others used the determinations of the relative energy dependence of the differential cross sections at a fixed angle, and some made use of a comparison of absolute cross-section measurements. Table 4.3 summarizes methods and results for a variety of reactions.

The interpretation of a violation of detailed balance in terms of a T-violating term in the reaction amplitude usually requires a detailed analysis involving a model of the reaction process. An even more detailed model is required to make an interpretation in terms of a T-violating (T-odd) term in the Hamiltonian of the system.[20] Since the relationship between the cross section and the amplitude is straightforward, it is a simple matter to show by repeating the argument leading to eq. (4.85) and using the general eq. (4.74) that, if the T-odd term $T^{+}_{ba}(o)$ is small compared with the T-even term $T^{+}_{ba}(e)$, and the relative deviation from detailed balance is defined by

(4.86a) $$\Delta = [|T^{+}_{ba}|^2 - |T^{+}_{a'b'}|^2]/|T^{+}_{ba}|^2,$$

then

(4.86b) $$\Delta = 4\,\mathrm{Re}\{T^{+}_{ba}(o)/T^{+}_{ba}(e)\}.$$

A model is required to estimate the relative phase of $T^{+}_{ba}(o)$ and $T^{+}_{ba}(e)$. If V_b in eq. (4.73) is weak enough to justify an estimate based on first-order perturbation theory, the relative phase would be $\pi/2$, as in eq. (4.21). However, strong interaction effects introduce phases that depend on their details, as shown in chapter 5. Therefore it may be assumed that on the average the relative phase would turn out to be about $\pi/4$, leading to a Δ of the order of twice the relative size of the T-violating amplitude. Thus we may conclude from table 4.3 that an upper limit on the T-violating amplitude for strong

[20] See, for example, Henley and Jacobsohn (1959), Ericson (1966), Mahaux and Weidenmüller (1966), Robson (1968), Moldauer (1968*a*,*b*), and Henley and Huffman (1968).

TABLE 4.3 Tests of Detailed Balance in Reactions between Unpolarized Nuclei

Reactions	Lab Energy (Range) MeV $\longrightarrow$	$\longleftarrow$	Cross-Section Measurement	Violation of Detailed Balance	Reference
$p + {}^3H \rightleftarrows d + d$	8.3	(4.4)	$\sigma(\theta)$ absolute	$< 10\%(?)$	Rosen and Brolley (1959)
$\alpha + {}^{12}C \rightleftarrows d + {}^{14}N$	41.7	20.0	$\sigma(\theta)$ absolute	$< 6\%$	Bodansky et al. (1959)
$d + {}^{24}Mg \rightleftarrows p + {}^{25}Mg$	~ 10	~ 15	$\sigma(\theta)$ relative	$< 0.3\%$	Weitkamp et al. (1968)
$\alpha + {}^{24}Mg \rightleftarrows p + {}^{27}Al$	$(13.2 \rightarrow 13.8)$	$(10.0 \rightarrow 10.7)$	$\sigma(E)$ relative	$< 0.63\%$	von Witsch, Richter, and von Brentano (1968)
$p + {}^{27}Al \rightleftarrows \alpha + {}^{24}Mg$	$(1.35 \rightarrow 1.46)$	$(3.38 \rightarrow 3.52)$	$\sigma(E)$ relative*	$< 2\%$	Driller et al. (1979)
$d + {}^{16}O \rightleftarrows \alpha + {}^{14}N$	$(3.6 \rightarrow 5.3)$	$(8.5 \rightarrow 10.0)$	$\sigma(E)$ absolute	$< 0.5\%$	Thornton et al. (1971)

Note: Columns 2 and 3 give the energy of the reaction in the indicated direction, column 4 indicates the nature of the measurement on the two reactions, and column 5 gives the relative difference in cross sections.

*Relative values of differential cross sections at two resonances. Relative phase of reduced widths (essentially equivalent to relative T-violation phase of resonance states) is estimated to be $(0.3 \pm 3)^\circ$.

interactions is of the order of a few tenths of a percent of the T-invariant amplitude.

The limits on the T-violating Hamiltonian are much more obscure, because such an effect may arise in at least two ways. For example, if the reaction goes through a resonance (compound nucleus), the origin of $T^{+}_{ba}(o)$ may arise from T-violating terms in the wave functions of the resonant states like the term described by eq. (4.66). On the other hand, it may be due to a T-violating term in the interaction V appearing in eq. (4.73) that is directly responsible for the reaction process. Most likely, if there is an effect, it would be due to both these causes. The one example of a phase "determination" cited in the footnote in table 4.3 indicates the level at which the limits on a T-violating phase may be set by these methods. They do not compare in sensitivity to the interference effects in radiative transitions, as can be seen by examining table 4.1.

Barshay (1966) suggested that detailed balance in the processes

$$\gamma + d \rightleftarrows n + p$$

at energies capable of exciting the first nucleon resonance at a mass (mc^2) of 1232 MeV, $\Delta(1232)$, is sensitive to T-violating interactions. This suggestion was explored by Sober and others (1969) and Anderson, Prepost, and Wiik (1969), who carried out the photodisintegration experiment, and by Bartlett and others (1969), who measured the cross section for the inverse process. The conclusion of Bartlett and others is that detailed balance is well satisfied within the estimated experimental errors.

Abegg and others (1982) have measured the cross section for $p + d \rightarrow {}^3\text{He} + \gamma$ as a function of energy and have compared the results with measurements by others of the photo cross section of ^{3}He, that is, to the inverse reaction. Although they do not find close agreement with detailed balance, they conclude that there may be a discrepancy in cross-section normalizations among the various experiments and are therefore unable to draw a conclusion concerning T invariance.

Neither of these experiments places a quantitative limit on T violation, but again, the experiments do not compare in sensitivity to the low-energy nuclear radiative interference effects.

One reason for the lack of sensitivity is that the full power of detailed balance is not exploited by a comparison of average cross sections. The information that may be obtained by comparing the rates of transition between substates (e.g., spin states) has been lost in the averaging process. For example, there is information about T invariance to be obtained from detailed cross-section measurements even in the case of elastic scattering.

Let us consider the simple case of the elastic scattering of a spin $\frac{1}{2}$ particle by a spin 0 particle, which is the simplest case for which eq. (4.83) offers some interesting rewards. The kinematic variables for the system are taken to be the spin operator of the one particle and the operators of relative momentum and total momentum. Since total momentum $\vec{P}$ is conserved, we label the scattering amplitude with the initial and final values of the spin magnetic quantum number $m = \pm\frac{1}{2}$ and the relative momentum $\vec{k}$ and write:

$$T^{+}_{fi} = t(m_f, \vec{k}_f; m_i, \vec{k}_i)\delta(\vec{P}_f - \vec{P}_i). \tag{4.87}$$

It is convenient to introduce an amplitude that is a 2×2 matrix (the "T matrix") in the spin labels m_i and m_f denoted by $\mathbf{t}(\vec{k}_f, \vec{k}_i)$, so that

$$T^{+}_{fi} = \langle m_f | \mathbf{t}(\vec{k}_f, \vec{k}_i) | m_i \rangle \delta(\vec{P}_f - \vec{P}_i). \tag{4.88}$$

Since any 2×2 matrix may be written as a linear combination of Pauli spin matrices σ_x, σ_y, σ_z given by eq. (3.42) and the unit matrix, **1**, we can write

$$\mathbf{t} = g\mathbf{1} + (\vec{\sigma} \cdot \vec{h}). \tag{4.89}$$

Like **t**, g and $\vec{h}$ are functions of $\vec{k}_i$ and $\vec{k}_f$. Rotational invariance of the scattering amplitude requires that g be a scalar function and $\vec{h}$ a vector function of these variables. Thus g and $|\vec{h}| = h$ can be functions of the scalar variables $|\vec{k}_i| = k_i$, $|\vec{k}_f| = k_f$ and $(\vec{k}_i \cdot \vec{k}_f)$ only.

By making use of the orthogonal set of unit vectors

$$\hat{e}_1 = \frac{\vec{k}_f \times \vec{k}_i}{|\vec{k}_f \times \vec{k}_i|}, \quad \hat{e}_2 = \frac{\vec{k}_f - \vec{k}_i}{|\vec{k}_f - \vec{k}_i|}, \quad \hat{e}_3 = \frac{\vec{k}_f + \vec{k}_i}{|\vec{k}_f + \vec{k}_i|} \equiv \hat{e}_1 \times \hat{e}_2 \tag{4.90}$$

that can be constructed from $\vec{k}_f$ and $\vec{k}_i$, we find that the most general rotationally invariant form of **t** is

$$\mathbf{t} = g\mathbf{1} + \sum_{l=1}^{3} (\vec{\sigma} \cdot \hat{e}_l) h_l, \tag{4.91}$$

where g and h_l are functions of the scalar variables. Since energy is conserved in the collision, there are just two scalar variables $k = k_f = k_i$ and $\cos\theta = (\vec{k}_f \cdot \vec{k}_i)/k^2$.

If P invariance is assumed, then $h_2 = h_3 = 0$, because $\vec{\sigma}$ is an axial vector and $\hat{e}_2$ and $\hat{e}_3$ are polar vectors, while $\hat{e}_1$ is an axial vector. It is usually assumed that P invariance holds, in which case **t** takes the simple form

$$\mathbf{t} = g(k, \cos\theta)\mathbf{1} + g'(k, \cos\theta) \sin\theta(\vec{\sigma} \cdot \hat{e}_1), \tag{4.92}$$

where $g'(k, \cos\theta) \sin\theta = h_1$ and $\hat{e}_1$ is the unit vector normal to the scattering plane. However, our interest is in the consequences of T invariance, that is, the condition eq. (4.83), which may now be written

$$\langle m_f|\mathbf{t}(\vec{k}_f, \vec{k}_i)|m_i\rangle = \langle -m_i|\mathbf{t}(-\vec{k}_i, -\vec{k}_f)|-m_f\rangle i^{2(m_f-m_i)} \tag{4.93}$$

if the phase conventions are chosen in accordance with eq. (3.73). From the definition of the $\vec{\sigma}$ matrices eq. (3.42) it is evident that

$$\langle -m_i|\vec{\sigma}|-m_f\rangle i^{2(m_f-m_i)} = -\langle m_f|\vec{\sigma}|m_i\rangle, \tag{4.94}$$

and the requirement eq. (4.93) is equivalent to the condition that the matrix $\mathbf{t}$ given by eq. (4.91) be invariant under the substitutions

$$\vec{\sigma}\to -\vec{\sigma}, \quad \vec{k}_i \leftrightarrow -\vec{k}_f \tag{4.95a}$$

or

$$\vec{\sigma}\to -\vec{\sigma}, \quad \hat{e}_1 \to -\hat{e}_1, \quad \hat{e}_2 \to \hat{e}_2, \quad \hat{e}_3 \to -\hat{e}_3. \tag{4.95b}$$

But eqs. (4.95) describe motion reversal combined with the interchange of initial and final momentum states. Therefore the *invariance of the matrix* $\mathbf{t}$ *under the combination of motion reversal and the interchange of initial and final momentum states follows* from the T invariance of the dynamics.

In eq. (4.91) the first term and the $l = 1$ and $l = 3$ terms are invariant under the transformation eq. (4.95b), but the $l = 2$ term changes sign. Therefore this last term is forbidden by T invariance or

$$h_2 \equiv 0 \tag{4.96}$$

if the interaction between the spin 1/2 and spin 0 particles is T invariant.

A measurement establishing that $h_2 \neq 0$ would be evidence for both T violation and P violation of the interaction. It is clear that $h_3 \neq 0$ implies P violation. Since for strong interactions it is generally assumed that both invariance principles are valid, the form of the scattering amplitude for this case of spin 1/2 on spin 0 is usually taken to be of the form eq. (4.92).

Since the collision cross section is proportional to $|\langle m_f|\mathbf{t}|m_i\rangle|^2$, it is clear that eq. (4.91) with $h_2 \neq 0$ will lead to a violation of detailed balance because the cross section includes an interference term that changes sign when the indices f, i are replaced by i', f'. Therefore h_2 can be measured, at least in principle, by comparing the rate $i \to f$ with the rate $f' \to i'$, if the states of spin polarization are specified in initial and final states. For this elastic process, the required information is provided by a measurement of the spin and angular dependence of the differential cross section. Thus we are led to consideration of the measurements of spin polarization in elastic scattering.

4.7 Polarization in the Scattering of Spin $\frac{1}{2}$ Particles from Unpolarized Targets

Although the ideal method for testing detailed balance is to work with specified spin quantum numbers in both the initial and the final state, it is possible to obtain reliable information by measuring certain spin averages. For example, one can measure the average polarization $\vec{P}$ of a beam of spin $\frac{1}{2}$ particles. Even if such a beam is initially unpolarized, when it is scattered from an unpolarized target it will usually become polarized. And if the beam is initially polarized, the scattering from such a target will show a dependence of the angular distribution of scattering on the initial polarization, which may be used to measure polarization of the beam.

The amount of polarization produced by an unpolarized beam scattered from an unpolarized target is measured by a quantity $\mathscr{P}$ called "polarizing power" that will be defined later. The quantitative measure of the angular dependence of scattering on the polarization of an incident beam is called the "analyzing power," $\mathscr{A}$, also to be defined later. The necessary analysis for determining and interpreting these quantities, and the consequences for them of T invariance, will be treated for the scattering of spin $\frac{1}{2}$ particles in the remainder of this chapter, which will also include a summary of the related experimental tests of T invariance. First, to illustrate the method, let us consider the example of the scattering from spin 0 targets.

For the purpose of dealing with measurements of averages, the spin states are described in terms of the 2×2 density matrices ρ_i and ρ_f for the initial and final states. They can always be written in the form (compare eq. 4.59)

$$\rho = \tfrac{1}{2}[1 + (\vec{P} \cdot \vec{\sigma})]. \tag{4.97}$$

Thus $\vec{P} = \vec{\mathbf{j}}/j$ is "the polarization," where $\mathbf{j}$ is the expectation value of the spin operator $\vec{\mathbf{S}} = \frac{1}{2}\vec{\sigma}$, whence

$$\vec{P} = \mathrm{Tr}(\rho\vec{\sigma}). \tag{4.98}$$

The component of the polarization $\vec{P}$ in a given direction for an ensemble can be measured by determining the number N_+ and N_- of particles having spin $+\frac{1}{2}$ and $-\frac{1}{2}$, respectively, along the direction. Then for, say, the z direction,

$$P_z = \frac{N_+^z - N_-^z}{N_+^z + N_-^z}, \tag{4.99}$$

since the density matrix in that case is

$$\rho^z = \begin{pmatrix} N_+^z & 0 \\ 0 & N_-^z \end{pmatrix} (N_+^z + N_-^z)^{-1}. \tag{4.100}$$

Similar expressions can be written for the x and y components, in which cases the x and y axes, respectively, would be taken as the axes of quantization of the spin.

When collision measurements are carried out with an ensemble of spin $\frac{1}{2}$ particles on spin 0 targets for which the scattering amplitude is **t** given by eq. (4.89), the initial polarization is $\vec{P}_i$ and the corresponding density matrix is ρ_i, the differential cross section for production of a final state having spin magnetic quantum number m_f is given by the diagonal element $\langle m_f | \mathbf{t}\rho_i \mathbf{t}^\dagger | m_f \rangle$. The average of the differential cross section over final spin states is then

(4.101) $$\sigma = \mathrm{Tr}(\mathbf{t}\rho_i \mathbf{t}^\dagger),$$

if **t** is normalized appropriately.

The effect of the collision is to transform the density matrix from its initial form ρ_i to

(4.102) $$\rho_f = \mathbf{t}\rho_i \mathbf{t}^\dagger/\sigma.$$

Then the polarization in the final state is

(4.103) $$\vec{P}_f = \mathrm{Tr}(\rho_f \vec{\sigma}) = \mathrm{Tr}(\mathbf{t}\rho_i \mathbf{t}^\dagger \vec{\sigma})/\sigma.$$

We may now generalize these results to the case of the scattering of a spin $\frac{1}{2}$ particle from an *unpolarized* target of spin j.[21] The scattering amplitude is then a $2(2j + 1) \times 2(2j + 1)$ matrix in the space formed by the outer product of the space of the spin $\frac{1}{2}$ projectile with the target spin space. The amplitude may again be written in the form of eq. (4.89):

(4.104) $$\mathbf{t} = \mathbf{g} + (\vec{\sigma} \cdot \vec{\mathbf{h}}),$$

where, now, $\mathbf{g}$ and $\vec{\mathbf{h}}$ are $(2j + 1) \times (2j + 1)$ matrix functions of the target angular momentum matrix $\vec{\mathbf{J}}$ as well as being functions of $\vec{k}_i$ and $\vec{k}_f$.

The density matrix $\boldsymbol{\rho}$ also is associated with the $2(2j + 1)$ dimensional space, and $\boldsymbol{\rho}_i$ can be written as the outer product of the target density matrix with the 2×2 matrix ρ_i given by eq. (4.97) with $\vec{P} = \vec{P}_i$, because the spins are initially uncorrelated. In fact, since our considerations will be limited to the case of an initially unpolarized target, the target density matrix is just $(2j + 1)^{-1}$ times the unit matrix. Therefore eqs. (4.101) and (4.103) are replaced by

(4.105) $$\sigma = \mathrm{Tr}(\mathbf{t}\rho_i \mathbf{t}^\dagger)/(2j + 1)$$

[21] Methods for treating the general case are given by Wolfenstein and Ashkin (1952), Dalitz (1952), Blin-Stoyle (1952), Wolfenstein (1956), Henley and Jacobsohn (1959), and Thorndike (1965).

and

(4.106) $$\vec{P}_f = \mathrm{Tr}(\mathbf{t}\rho_i \mathbf{t}^\dagger)/(2j+1)\sigma,$$

where the trace is taken over the $2(2j+1)$ dimensions of the product space.

When eqs. (4.97) and (4.104) are used for ρ_i and $\mathbf{t}$ respectively, we find[22]

(4.107a) $$\sigma = \sigma_0 + \frac{1}{2j+1}(\mathrm{Tr}_j\,\{\mathbf{g}\vec{\mathbf{h}}^\dagger + \vec{\mathbf{h}}\mathbf{g}^\dagger - i\vec{\mathbf{h}} \times \vec{\mathbf{h}}^\dagger\} \cdot \vec{P}_i),$$

with

(4.107b) $$\sigma_0 = \frac{1}{2j+1}\,\mathrm{Tr}_j\,\{\mathbf{g}\mathbf{g}^\dagger + (\vec{\mathbf{h}} \cdot \vec{\mathbf{h}}^\dagger)\}$$

and

(4.108) $$\sigma\vec{P}_f = \frac{1}{2j+1}\,\mathrm{Tr}_j\{\mathbf{g}\vec{\mathbf{h}}^\dagger + \mathbf{g}^\dagger\vec{\mathbf{h}} + i\vec{\mathbf{h}} \times \vec{\mathbf{h}}^\dagger + (\mathbf{g}\mathbf{g}^\dagger - \vec{\mathbf{h}} \cdot \vec{\mathbf{h}}^\dagger)\vec{P}_i + (\vec{\mathbf{h}} \cdot \vec{P}_i)\vec{\mathbf{h}}^\dagger + (\vec{\mathbf{h}}^\dagger \cdot \vec{P}_i)\vec{\mathbf{h}} + i\mathbf{g}\vec{P}_i \times \vec{\mathbf{h}}^\dagger - i\mathbf{g}^\dagger\vec{P}_i \times \vec{\mathbf{h}}\},$$

where Tr_j symbolizes the trace over the target spin space.

Since the Tr_j reduces the expressions in curly brackets to simple functions of k, $\cos\theta$, and $\hat{e}_l$ with $l = 1, 2,$ or 3, we may write

(4.109a) $$\frac{1}{2j+1}\,\mathrm{Tr}_j(\mathbf{g}\vec{\mathbf{h}}^\dagger) = \sum_l a_l(k, \cos\theta)\hat{e}_l,$$

(4.109b) $$\frac{i}{2j+1}\,\mathrm{Tr}_j(\vec{\mathbf{h}} \times \vec{\mathbf{h}}^\dagger) = \sum_l b_l(k, \cos\theta)\hat{e}_l,$$

(4.109c) $$\frac{1}{2j+1}\,\mathrm{Tr}_j(\mathbf{g}\mathbf{g}^\dagger - \vec{\mathbf{h}} \cdot \vec{\mathbf{h}}^\dagger) = c(k, \cos\theta),$$

(4.109d) $$\frac{1}{2j+1}\,\mathrm{Tr}_j[(\vec{\mathbf{h}} \cdot \vec{P}_i)\vec{\mathbf{h}}^\dagger + (\vec{\mathbf{h}}^\dagger \cdot \vec{P}_i)\vec{\mathbf{h}}] = \sum_{l,m} d_{lm}(\hat{e}_l \cdot \vec{P}_i)\hat{e}_m,$$

where the a_l may be complex and the b_l, c and d_{lm} are real functions of k and $\cos\theta$. Also, d_{lm} is symmetric in l and m. Eq. (4.107a) then takes the form

(4.110) $$\sigma = \sigma_0 + \sum_l (2\,\mathrm{Re}\,a_l - b_l)(\hat{e}_l \cdot \vec{P}_i),$$

[22] All products of Pauli spin matrices are linearized by using $(\vec{A} \cdot \vec{\sigma})\vec{\sigma} = \vec{A} + i\vec{\sigma} \times \vec{A}$, and the traces are reduced to $\mathrm{Tr}\,\mathbf{1} = 2(2j+1)$ and $\mathrm{Tr}\,\vec{\sigma} = 0$. Since the spin spaces of incident beam and target are kinematically independent, the traces over the two spaces may be taken independently.

and eq. (4.108) becomes

(4.111)
$$\sigma\vec{P}_f = \sum_l (2\ \mathrm{Re}\ a_l + b_l)\hat{e}_l + c\vec{P}_i + \sum_{l,m} d_{lm}(\hat{e}_l \cdot \vec{P}_i)\hat{e}_m + \sum_l 2\ \mathrm{Im}\ a_l(\hat{e}_l \times \vec{P}_i).$$

If parity is conserved, as might be expected when the scattering is due to strong or electromagnetic interactions, **t** is invariant under inversion so that by eq. (4.104) it is seen that **g** must be a scalar (rather than a pseudoscalar) and $\vec{\mathbf{h}}$ an axial vector because $\vec{\sigma}$ is an axial vector. Therefore $\mathbf{g}\vec{\mathbf{h}}^\dagger$, $\vec{\mathbf{h}}\mathbf{g}^\dagger$, and $\mathbf{h} \times \vec{\mathbf{h}}^\dagger$ are axial vectors. The only axial vector among the $\hat{e}_l$ is $\hat{e}_1$, and it therefore would follow from P invariance that

(4.112a)
$$a_2 = a_3 = b_2 = b_3 \equiv 0.$$

Similarly, since $\vec{P}_i$ is also an axial vector,

(4.112b)
$$d_{12} = d_{21} = d_{31} = d_{13} \equiv 0$$

if parity is conserved.

The cross section eq. (4.110) for the scattering of a polarized beam by an unpolarized target in the case of parity conservation is therefore

(4.113)
$$\sigma = \sigma_0 + (2\ \mathrm{Re}\ a_1 - b_1)(\hat{e}_1 \cdot \vec{P}_i).$$

It will be noted that *only the component of* $\vec{P}_i$ *normal to the scattering plane has an effect on the scattering cross section*, a result that is a direct consequence of parity conservation, since $\vec{P}_i$ and $\hat{e}_1$ are axial vectors while $\hat{e}_2$ and $\hat{e}_3$, which define the other components of $\vec{P}_i$ appearing in eq. (4.110), are polar vectors.

In order to discuss the polarizing power of an unpolarized target, let us consider the resulting polarization, eq. (4.111), in the scattering of an *unpolarized beam* ($\vec{P}_i = 0$). Again under the assumption of P invariance, we find

(4.114)
$$\sigma_0 \vec{P}_f = (2\ \mathrm{Re}\ a_1 + b_1)\hat{e}_1.$$

Thus it is found that, if $\vec{P}_i = 0$, $\vec{P}_f$ *is in the direction of the normal to the scattering plane*, a result that is another direct consequence of parity conservation: in the absence of any axial vector other than $\hat{e}_1$ with which to parameterize the system, $\vec{P}_f$, which is an axial vector, must be proportional to $\hat{e}_1$.[23] We therefore define the *polarizing power* $\mathscr{P}$ of an unpolarized target

[23] Polarization of the incident beam introduces another parameter and therefore, as shown by eqs. (4.110) and (4.111), the direction of $\vec{P}_f$ depends on the details of the dynamics when $\vec{P}_i \neq 0$, unless $\vec{P}_i$ is in the direction $\hat{e}_1$.

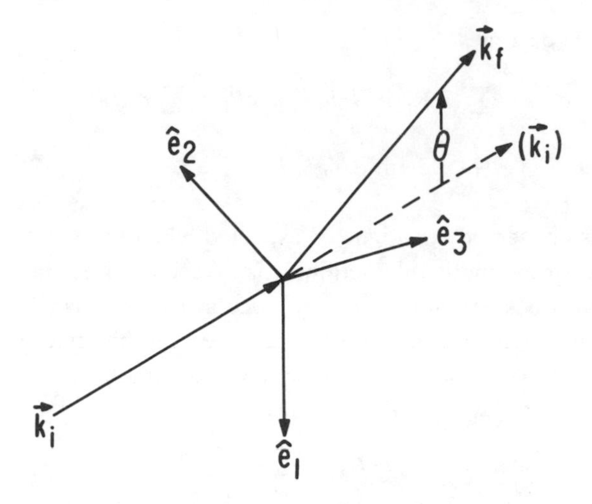

FIG. 4.1 The elastic scattering to the "left" of an incident beam of relative (center-of-mass) momentum $\vec{k}_i$ into a state of relative momentum $\vec{k}_f$ at angle θ with $\vec{k}_i$, showing the basic unit vectors $\hat{e}_l$. Scattering to the "right" may be obtained by rotating the figure 180° about the axis $\vec{k}_i$.

as the component of $\vec{P}_i$ in the direction of $\hat{e}_1$, whence

(4.115) $$\mathscr{P} = (2 \text{ Re } a_1 + b_1)/\sigma_0 .$$

At the same time, eq. (4.113) may be said to determine the *analyzing power* of the same target because it provides a means of measuring the polarization ($\vec{P}_i$) of an incident beam. Operationally, analyzing power is measured by comparing scattering to the left of the incident beam (positive θ) with scattering (by the same angle) to the right (negative θ). The two cases then differ in that $\hat{e}_1$ changes sign in going from one side to the other. If the number of particles scattered to the left and right, respectively (see fig. 4.1), are N_L and N_R, we define the scattering asymmetry to be

(4.116a) $$A = \frac{N_L - N_R}{N_L + N_R}.$$

A is therefore a directly measurable quantity that may also be expressed in terms of the corresponding cross sections:

(4.116b) $$A = \frac{\sigma_L - \sigma_R}{\sigma_L + \sigma_R}.$$

A theoretical expression for A is then found from eq. (4.113) to be

(4.117a)
$$A = \mathscr{A}(\hat{e}_1 \cdot \vec{P}_i),$$

where

(4.117b)
$$\mathscr{A} = (2 \operatorname{Re} a_1 - b_1)/\sigma_0$$

is defined as the "analyzing power."

4.8 Consequences of T Invariance for Polarization and Scattering of Spin $\frac{1}{2}$ Particles from Unpolarized Targets

The consequences of T invariance may be expressed in terms of the properties of $\mathbf{t}$ that follow from the condition eq. (4.83). Since $\mathbf{g}$ and $\vec{\mathbf{h}}$ can be written as a linear combination of irreducible tensor products of the $\mathbf{J}_l$, we may make use of eq. (3.32) and the behavior of $\vec{\mathbf{J}}$ under the transformation T to determine the time reversal properties of its matrix elements between the eigenvalues M of $\mathbf{J}_z$:

(4.118)
$$\langle M_i | \vec{\mathbf{J}} | M_f \rangle^* = \langle T(M_i) | T\vec{\mathbf{J}} | M_f \rangle = \langle T(M_i) | T\vec{\mathbf{J}}T^{-1} | T(M_f) \rangle,$$

whence

(4.119)
$$\langle -M_i | \vec{\mathbf{J}} | -M_f \rangle i^{2(M_f - M_i)} = -\langle M_f | \vec{\mathbf{J}} | M_i \rangle,$$

if the phase convention for the states $|\rangle$ corresponds to eq. (3.73). This is the general basis for eq. (4.94). A repetition of the application eq. (4.93) of the condition for T invariance, eq. (4.83), to the more general case leads then to the same conclusion, that $\mathbf{t}$ must be invariant under the substitution eq. (4.95) supplemented by

(4.120)
$$\vec{\mathbf{J}} \to -\vec{\mathbf{J}}.$$

From the form eq. (4.104) of $\mathbf{t}$, it is then evident that the invariance of $\mathbf{t}$ requires that

(4.121)
$$\mathbf{g} \to \mathbf{g} \quad \text{and} \quad \vec{\mathbf{h}} \to -\vec{\mathbf{h}}$$

under these substitutions. Therefore, when the substitutions are applied to the right- and left-hand sides of eqs. (4.109), we find in particular that[24]

(4.122)
$$a_2 = b_1 = b_3 = 0.$$

[24] This argument is a generalization of the one due to Wolfenstein (1956, p. 64) for the special case of P-invariant systems.

Thus for a T- and P-invariant system eqs. (4.115) and (4.117b) yield

(4.123) $$\mathscr{A} = \mathscr{P},$$

so the analyzing power equals the polarizing power.[25]

Since the assumption of T invariance plays an essential role in this argument, it provides a method for testing the validity of the assumption, often referred to as "the $\mathscr{P} - \mathscr{A}$ method." However, some caution is required in applying the method as a test for T violation. There are two points to be made: first that, for *spin 0* targets, $\mathscr{A} = \mathscr{P}$ follows from parity conservation whether or not the dynamics are T invariant, and second, that unless we have independent evidence for P invariance, it is necessary to verify that eq. (4.123) is independent of the question of P invariance in all applications other than the $j = 0$ case.

The first point is a consequence of the fact that, because of angular momentum conservation, the spin cannot flip on scattering by a spin 0 target. Therefore both $\mathscr{A}$ and $\mathscr{P}$ are determined by the difference in spin up and spin down cross sections. Explicitly, when $j = 0$ and parity is conserved, the vector $\vec{h}$ in eq. (4.89) is a simple function of k and $\cos\theta$ given by

(4.124) $$\vec{h} = g'(k, \cos\theta)\sin\theta\ \hat{e}_1$$

in the notation of eq. (4.92). Therefore eq. (4.109b) leads to $b_1 \equiv 0$, and we find

(4.125) $$\mathscr{P} = \mathscr{A} = 2\ \mathrm{Re}\ gg'\ \sin\theta.$$

This result is particularly useful for tests of T invariance, since spin 0 targets may therefore be used as standard polarizers and analyzers independent of the T-invariance question *as long as P invariance has been established independently* for the reaction process.

To deal with the second question, we must reexamine the argument leading to eqs. (4.112) that are based on the assumption of P invariance. Under the contrary assumption that parity is not conserved, **g** may be a linear combination of scalars and pseudoscalars, and $\vec{\mathbf{h}}$ may be a linear combination of polar and axial vectors. Therefore the conditions eq. (4.112) no longer apply, and from eqs. (4.110), (4.111), and (4.122) we find that, for a

[25] Wolfenstein (1949) was the first to state that $\mathscr{A} = \mathscr{P}$, and he indicates in his obituary of Ashkin (Wolfenstein 1982) that Ashkin called to his attention the need for a proof, whereupon they gave a proof together based on T invariance (Wolfenstein and Ashkin 1952). However, they assumed P invariance throughout their treatment, so that only the $\hat{e}_1$ component of polarization is produced or analyzed. See also Dalitz (1952) and Blin-Stoyle (1952). Baz' (1957), Shirokov (1957), Bell and Mandl (1958), Satchler (1958), and Biedenharn (1959) showed that, even when parity is not conserved, the result can be generalized to other directions of polarization.

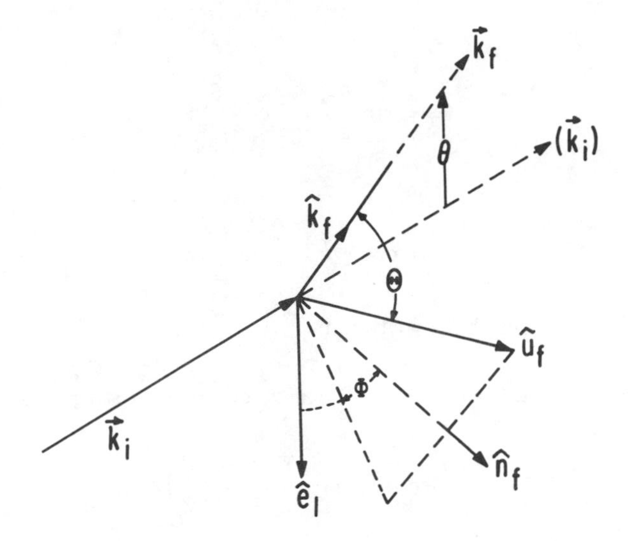

FIG. 4.2 The basic vectors $\hat{n}_f$, $\hat{e}_1$, $\hat{k}_f$ used to characterize the direction of polarization $\hat{u}_f$ after scattering of an unpolarized incident beam on an unpolarized target (center of mass).

T-invariant system,

(4.126) $$\sigma = \sigma_0 + ([2 \text{ Re } a_1\hat{e}_1 - b_2\hat{e}_2 + 2 \text{ Re } a_3\hat{e}_3] \cdot \vec{P}_i)$$

and, for the case of an unpolarized incident beam ($\vec{P}_i = 0$),

(4.127) $$\sigma_0 \vec{P}_f = 2 \text{ Re } a_1\hat{e}_1 + b_2\hat{e}_2 + 2 \text{ Re } a_3\hat{e}_3 .$$

It is evident that there is no longer any general restriction on the direction of $\vec{P}_f$ even when the incident beam is unpolarized. Furthermore, for a polarized incident beam, all three components of $\vec{P}_i$ contribute to the cross section. These complications that can arise in the absence of parity conservation lead to the need for new operational definitions of polarizing and analyzing power.

In order to define the polarizing power, it is useful to introduce in place of $\hat{e}_2$ and $\hat{e}_3$ the unit vectors $\hat{k}_f$ and $\hat{n}_f = \hat{e}_1 \times \hat{k}_f$, where $\hat{k}_f$ is in the direction $\vec{k}_f$ and $\hat{n}_f$ is normal to $\hat{k}_f$ and $\hat{e}_1$ as in figure 4.2. Then the unit vector in the direction of $\vec{k}_i$ is

(4.128) $$\hat{k}_i = \hat{k}_f \cos\theta + \hat{n}_f \sin\theta$$

and, from eqs. (4.90),

(4.129a) $$\hat{e}_1 = \hat{k}_f \times \hat{n}_f ,$$

while

$$\hat{e}_2 = \sin\frac{\theta}{2}\,\hat{k}_f - \cos\frac{\theta}{2}\,\hat{n}_f \tag{4.129b}$$

and

$$\hat{e}_3 = \cos\frac{\theta}{2}\,\hat{k}_f + \sin\frac{\theta}{2}\,\hat{n}_f. \tag{4.129c}$$

Now eq. (4.127) may be rewritten as

$$\sigma_0 \vec{P}_f = 2\text{ Re } a_1\hat{e}_1 + \left(2\text{ Re } a_3 \cos\frac{\theta}{2} + b_2 \sin\frac{\theta}{2}\right)\hat{k}_f + \left(2\text{ Re } a_3 \sin\frac{\theta}{2} - b_2\cos\frac{\theta}{2}\right)\hat{n}_f, \tag{4.130}$$

and if

$$\hat{u}_f = \hat{k}_f \cos\Theta + \hat{n}_f \sin\Theta\cos\Phi + \hat{e}_1 \sin\Theta\sin\Phi, \tag{4.131}$$

the polarizing power in the direction $\hat{u}_f$ will be $(\vec{P}_f \cdot \hat{u}_f)$, given by

$$\mathscr{P}(\Theta, \Phi) = \left[2\text{ Re } a_1 \sin\Theta\sin\Phi + \left(2\text{ Re } a_3\cos\frac{\theta}{2} + b_2\sin\frac{\theta}{2}\right)\cos\Theta + \left(2\text{ Re } a_3\sin\frac{\theta}{2} - b_2\cos\frac{\theta}{2}\right)\sin\Theta\cos\Phi\right]\Big/\sigma_0. \tag{4.132}$$

Note that this definition agrees with eq. (4.115) if we assume parity conservation and take $\Theta = \Phi = \pi/2$.

The polarization in the initial state rather than the final state is the object of interest in defining the analyzing power, so we now choose as unit vectors $\hat{e}_1$, $\hat{k}_i$, and $\hat{n}_i = \hat{k}_i \times \hat{e}_1$, where $\hat{n}_i$ is normal to $\hat{k}_i$ and $\hat{e}_1$ (fig. 4.3). Then eqs. (4.128) and (4.129) are replaced by

$$\hat{k}_f = \hat{k}_i\cos\theta + \hat{n}_i\sin\theta \tag{4.133}$$

and

$$\hat{e}_1 = \hat{n}_i \times \hat{k}_i, \tag{4.134a}$$

$$\hat{e}_2 = -\sin\frac{\theta}{2}\,\hat{k}_i + \cos\frac{\theta}{2}\,\hat{n}_i, \tag{4.134b}$$

$$\hat{e}_3 = \cos\frac{\theta}{2}\,\hat{k}_i + \sin\frac{\theta}{2}\,\hat{n}_i, \tag{4.134c}$$

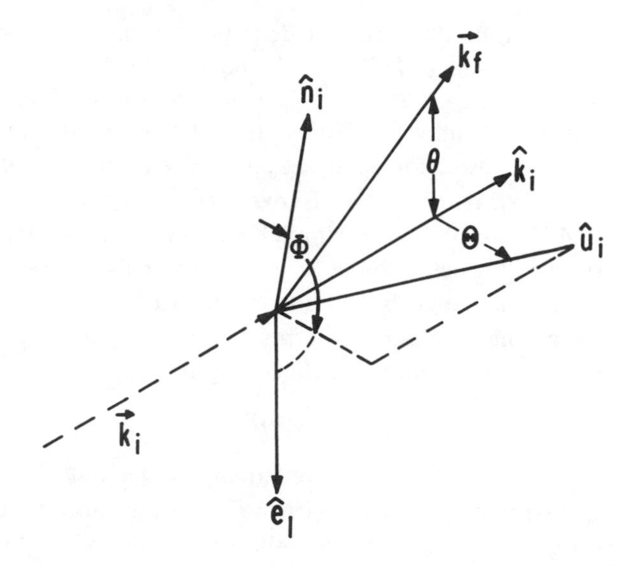

FIG. 4.3 The basic vectors $\hat{n}_i$, $\hat{e}_1$, $\hat{k}_i$ used to characterize the direction of polarization $\hat{u}_i$ of a polarized beam incident on an unpolarized target (center of mass) and scattered to the "left." Scattering to the "right" is obtained by rotating 180° about $\hat{k}_i$.

and the cross section, eq. (4.126), becomes

(4.135) $$\sigma = \sigma_0 + \left(\left[2 \text{ Re } a_1 \hat{e}_1 + \left(b_2 \sin\frac{\theta}{2} + 2 \text{ Re } a_3 \cos\frac{\theta}{2}\right)\hat{k}_i - \left(b_2 \cos\frac{\theta}{2} - 2 \text{ Re } a_3 \sin\frac{\theta}{2}\right)\hat{n}_i\right]\cdot \vec{P}_i\right).$$

If $\vec{P}_i$ is taken to be in the direction $\hat{u}_i$ defined by

(4.136) $$\hat{u}_i = \hat{k}_i \cos\Theta + \hat{n}_i \sin\Theta \cos\Phi + \hat{e}_1 \sin\Theta \sin\Phi,$$

where Θ and Φ are the *same* angles as those in eq. (4.131), we find

(4.137) $$\sigma = \sigma_0 + P_i\left[2 \text{ Re } a_1 \sin\Theta \sin\Phi + \left(2 \text{ Re } a_3 \cos\frac{\theta}{2} + b_2 \sin\frac{\theta}{2}\right)\cos\Theta + \left(2 \text{ Re } a_3 \sin\frac{\theta}{2} - b_2 \cos\frac{\theta}{2}\right)\sin\Theta \cos\Phi\right],$$

(4.138) $$\sigma = \sigma_0[1 + P_i \mathscr{P}(\Theta, \Phi)],$$

on the basis of eq. (4.132).

A formal but operationally incorrect definition of analyzing power suggested by eqs. (4.113) and (4.117b) would be to take it as equal to the coefficient of $\sigma_0 P_i$ in eq. (4.138). Then we would find in general that for a T-invariant system the analyzing power defined (incorrectly) in this way equals the polarizing power for polarization at given angles Θ, Φ whether or not the system is P invariant. The T invariance enters here in the use of eqs. (4.126) and (4.127) to obtain the simple relationship eq. (4.138).

The difficulty with this definition of $\mathscr{A}$ is that the analyzing power should be defined operationally—that is, in terms of the way it is to be measured. To retain an operational definition in terms of the measured asymmetry, eq. (4.116), the analyzing power must be defined in general by

(4.139)
$$A = \mathscr{A}(\Theta, \Phi)P_i$$

for polarization of magnitude P_i measured at angles Θ and Φ.

To relate this asymmetry to the polarizing power, we note that the scattering to the "right" is obtained from the scattering to the "left" by a rotation of 180° about the $\hat{k}_i$ axis (see fig. 4.1), which corresponds to

(4.140)
$$\Phi \rightarrow (\Phi + \pi).$$

Therefore, from eq. (4.116b) and eq. (4.138) we find

(4.141)
$$A = \frac{\mathscr{P}(\Theta, \Phi) - \mathscr{P}(\Theta, \Phi + \pi)}{2 + P_i[\mathscr{P}(\Theta, \Phi) + \mathscr{P}(\Theta, \Phi + \pi)]} P_i$$

or, from eqs. (4.132) and (4.139),

(4.142)
$$\mathscr{A}(\Theta, \Phi) = \frac{\mathscr{P}(\Theta, \Phi) - \left(2 \text{ Re } a_3 \cos \frac{\theta}{2} + b_2 \sin \frac{\theta}{2}\right) \cos \Theta}{1 + P_i\left(2 \text{ Re } a_3 \cos \frac{\theta}{2} + b_2 \sin \frac{\theta}{2}\right) \cos \Theta}.$$

Thus, in general, the analyzing power obtained by experiment is *not* equal to the polarizing power as normally defined.

We do find (Bell and Mandl 1958) that

(4.143)
$$\mathscr{A}\left(\Theta = \frac{\pi}{2}, \Phi\right) = \mathscr{P}\left(\Theta = \frac{\pi}{2}, \Phi\right);$$

that is, *even in the absence of parity conservation the analyzing power equals the polarizing power for transversely polarized beams by virtue of T invariance*. Unfortunately, the characteristic of parity nonconservation is that scattered beams usually have longitudinal components of polarization.

Therefore, *in using $\mathscr{P} - \mathscr{A}$ as a test of T invariance an implicit assumption is being made that the system is known to be P invariant.*

This reservation is important in considerations involving weak interactions, for which it is known that parity is not conserved. It is of particular importance when we consider polarization experiments with leptons at high energies, where the effects of weak interactions become important. Since there is independent evidence that the upper limit on parity nonconserving interactions in nuclear processes is very small ($\sim 10^{-6}$) relative to the strong interaction,[26] it is safe to assume that the $\mathscr{P} - \mathscr{A}$ test may be used for comparable nuclear reactions to test T invariance down to a level of the order of 10^{-5}.

The treatment of the $\mathscr{P} - \mathscr{A}$ test given here has referred only to elastic scattering. The result may be generalized to cases of inelastic scattering if the polarizing power for a reaction is compared with the analyzing power for the inverse reaction.[27] Some applications of this form of the test in nuclear reactions are reported in what follows.

4.9 Application to Polarization Experiments

Let us turn now to experimental applications of the $\mathscr{P} - \mathscr{A}$ test. The first uses of this test appear to be the experiments of Hillman, Johansson, and Tibell (1958) and Abashian and Hafner (1958) on nuclear elastic scattering. Hillman, Johansson, and Tibell obtained $\mathscr{P}/\mathscr{A}$ for protons on targets of H, Li, Be, and Al, while Abashian and Hafner measured the ratio just for *p-p* scattering.

The combined *p-p* experiments lead to

$$(\mathscr{P}/\mathscr{A}) \approx 0.95 \pm 0.05. \tag{4.144}$$

Use is made here of the ratio rather than the difference $\mathscr{P} - \mathscr{A}$ because it makes the significance of the stated experimental errors more apparent. The data taken for a larger number of values of energy and angle obtained by Hwang and others (1960) for *p-p* scattering give results consistent with eq. (4.144) and have comparable estimates of the error.

The measurements of $\mathscr{P}/\mathscr{A}$ for protons on targets other than protons by Hillman, Johansson and Tibell yield comparable results. Rosen and Brolley (1959) applied this test to elastic scattering of protons from Ni, Fe, and Co, obtaining $\mathscr{P}/\mathscr{A} \approx 1$ with estimated errors of the order of 15 to 20 percent. A determination of $\mathscr{P}/\mathscr{A}$ for protons on ^{13}C was made by Gross and others

[26] See, for example, Lobashev et al. (1967). For an experiment making use of the more direct but much less precise polarization methods, see Rosen and Brolley (1959) and Thorndike (1965).

[27] See note 21.

(1968) using a direct comparison with measurements on ^{12}C that, by taking advantage of the exact equality $\mathscr{P} = \mathscr{A}$ for the spin 0 nucleus, gives the ratio directly. This avoidance of absolute measurements of $\mathscr{P}$ and $\mathscr{A}$ reduces the errors considerably. Their result is

$$\mathscr{P}(p + {}^{13}\mathrm{C})/\mathscr{A}(p + {}^{13}\mathrm{C}) = 0.992 \pm 0.025. \quad \textbf{(4.145)}$$

The final example of elastic scattering is for the neutron-proton system. Bhatia and others (1982) make use of the assumption of charge symmetry of the neutron-proton interaction in analyzing their measurement, with the result

$$(\mathscr{P}/\mathscr{A})_{n,\,p} = 0.95 \pm 0.09 \quad \textbf{(4.146)}$$

The first specific T-invariance test based on *inelastic* nuclear scattering was carried out by Slobodrian and others (1981), who measured the polarizing power for the produced protons in $^7\mathrm{Li}(^3\mathrm{He}, p)^9\mathrm{Be}$ and $^9\mathrm{Be}(^3\mathrm{He}, p)^{11}\mathrm{B}$ and the analyzing power for incident polarized protons in the inverse reactions. They find that $\mathscr{P} - \mathscr{A}$ is a strong function of Θ, differing greatly from zero for both reactions. However, Hardekopf and others (1982) have repeated the measurement of the polarizing power in $^9\mathrm{Be}(^3\mathrm{He}, p)^{11}\mathrm{B}$ and find agreement within experimental error at all angles with the analyzing power obtained for the inverse reaction by Slobodrian and others. The polarizing power of both reactions has been measured more recently by Trelle and others (1984), and they also find results that are completely in accord with the measurements on analyzing power. We can conclude that there exists no substantiated evidence for T violation in these reactions.

There are other relationships between initial and final states of polarization in nuclear scattering or reactions that can, at least in principle, be exploited to test T invariance. Handler and others (1967) have used one of these to carry out a precision test on p-p scattering that may provide the best available limit on the level of T violation in nuclear forces. If the forces are T invariant, Sprung (1961) has shown (see also Phillips 1958; Thorndike 1965; Binstock 1981) that the following relationship is valid:

$$\tan\theta = [(\vec{P}_{f1}\cdot\hat{n}_{f1}) + (\vec{P}_{f2}\cdot\hat{k}_{f2})]/[(\vec{P}_{f1}\cdot\hat{k}_{f1}) - (\vec{P}_{f2}\cdot\hat{n}_{f2})], \quad \textbf{(4.147)}$$

where the subscripts 1 and 2 refer to experiments carried out with initial polarizations $(\vec{P}_{i1}\cdot\hat{k}_i) = 1$ and $(\vec{P}_{i2}\cdot\hat{n}_i) = 1$, respectively. The notation is the same as that used earlier (see eqs. 4.128 and 4.133).

By carrying out two related triple scattering experiments to determine $\vec{P}_i$ and $\vec{P}_f$ from p-p scattering at 430 MeV, Handler and others (1967) were able to test this relationship. Their result for the T-violating, parity-conserving amplitude $\mathbf{t}'$ is

(4.148) $$|\mathbf{t}'| \cos \alpha = 0.0020 \pm 0.010 \text{ mb}^{1/2},$$

where α is a phase related to the *p-p* phase shift. When the measured values of these phase shifts are introduced, they obtain an upper limit on the T-violating phase η of

(4.149) $$\eta \lesssim 5.5°.$$

Aprile and others (1981) have made enough polarization measurements in *p-p* scattering at 579 MeV to reconstruct the elements of the scattering matrix, including a T-violating term $\mathbf{t}'$. They place an upper limit of

(4.150) $$|\mathbf{t}'| < 0.1$$

on this term. However, because measurements at several values of the energy are needed to obtain the information on phase shifts that is required to extract η from this information on $\mathbf{t}'$, the limit eq. (4.148) remains the most severe constraint on T violation in *p-p* interactions.[28]

[28] This was pointed out to me by L. G. Pondrom (personal communication).

5 | Decay Processes

The validity of the assumption of universal time reversal invariance was first questioned as a result of the discovery of parity violation in weak interactions. The weak interactions manifest themselves in decay processes such as the beta decay of nuclei and the decay of strange particles, and the evidence for parity violation arose from experiments on such processes.

The theory of decay phenomena is implicitly included in the treatment of collisions and reactions that is summarized in chapter 4. However, because the term "decay" implies that the reaction process is a slow one, and therefore due to an interaction of small magnitude, its treatment can be simplified by invoking perturbation methods. Usually the decay rate can be calculated by elementary perturbation methods using the Fermi "golden rule" that gives the rate in terms of the modulus squared of the matrix element of the interaction causing the decay. We shall find that it is important to go one step beyond that simple concept and make use of the method introduced by Weisskopf and Wigner (1930) and generalizations thereof.

In this chapter we shall consider the general phenomenological quantum mechanical theory of decay processes based on those perturbation methods and the consequences of T invariance for decay phenomena. Several examples of attempts to obtain a measure of T violation in decay processes will be discussed, but the treatment of the only physical system for which substantial evidence for T violation has been found thus far, the K^0, $\bar{K}^0$ meson system, merits enough attention that it occupies the whole of chapter 9.

5.1 Time Reversal and the Decay Process

Weisskopf and Wigner (1930) showed that a quantum mechanical system that is unstable with respect to dissociation into two or more particles will decay according to the standard radioactive decay law $e^{-\Gamma t}$ if there is a small interaction having a nonvanishing matrix element between the initial system and the dissociated state. This is a direct and general consequence of the dynamic equation of quantum mechanics, eq. (3.5). It applies to photon emission and to beta decay as well as to other decays owing to weak interactions.

When viewed as a description of the "motion" of a system, exponential decay appears to defy the notion of time reversal invariance—that is, the notion that replacing t by $-t$ in the exponential decay law leads to a possible alternative behavior. However, as in all such cases, this apparent gross asymmetry in time has to do with the choice of initial conditions rather than with the properties of the interaction under time reversal. We have assumed that the initial system is the unstable system. But the unstable system must be formed by some process before the chosen initial time. One way of approaching the reversibility question is to choose an earlier initial time so that the details of formation enter into the description of the process.

For example, we might consider the formation of a neutron by a (p, n) reaction on a stable target nucleus N. Then the initial conditions describe the motion of the proton toward collision with the target nucleus N, and the process is described by the formation of another nucleus N' with the emission of the neutron, followed by emission from the neutron of an electron and an antineutrino:

$$p + N \rightarrow N' + n, \quad n \rightarrow p + e^- + \bar{\nu}. \tag{5.1}$$

Although the first step is a strong interaction process, the second is due to weak interactions and may be described as an exponential decay. On the other hand, the overall process may be viewed as a reaction initiated by a two-body collision and may be described by a collision matrix as in section 4.3. Then the consequence of T invariance of *all* the interactions is again simply the principle of detailed balance, eq. (4.83), relating the reaction

$$p + N \rightarrow N' + p + e^- + \bar{\nu}, \tag{5.2a}$$

with the reverse reaction

$$N' + p + e^- + \bar{\nu} \rightarrow p + N. \tag{5.2b}$$

From this viewpoint, the impossibility of reversing the exponential decay is replaced by the impossibility (actually, the improbability) of being able to set up the initial conditions required to implement the reverse reaction. Those initial conditions would involve the precise formation of the four wave packets associated with the four particles produced in reaction (5.2a), but with the group velocity of each reversed.

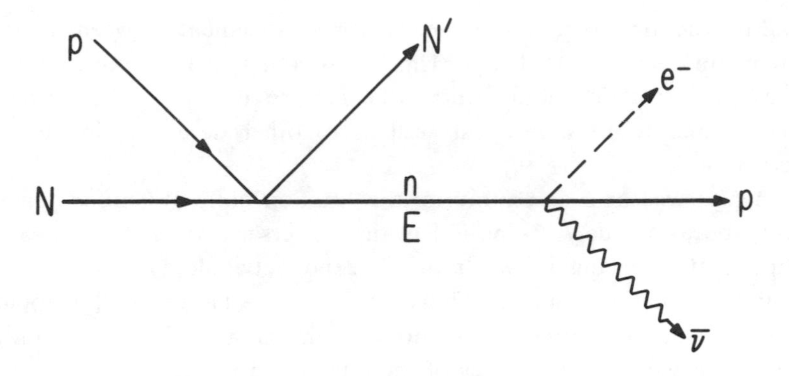

FIG. 5.1 Diagram of process eq. (5.2a). E is the energy of the neutron in its intermediate state between production and decay.

The connection between this description of the process in terms of a reaction matrix and the time-dependent description of an exponential decay can be made in a general way by using the fact that the neutron is stable in the approximation that the weak interactions are neglected so that the process may be described by the Feynman diagram, figure 5.1. The neutron line is represented in the matrix element as a propagator of energy E (including the rest mass m_0 of the neutron) and having a pole in the complex energy plane at a mass m^* that takes into account the corrections to m_0 owing to weak interactions. This pole would be on the real axis (i.e., the mass m^* would be real) if the mass m_0 were less than $m_p + m_e$, where m_p is the mass of the proton. Since m_0 is larger than that, the neutron is unstable, and the propagator has a pole with a negative imaginary part (on the second Riemann sheet, Peierls 1955) corresponding to a mass

$$m^* = m_n - i\Gamma/2, \tag{5.3}$$

where $m_n = m_0 + \Delta m$ is the mass at resonance energy, Γ is the width of the resonance, and Δm is a weak interaction correction to the mass that is of order Γ.

The diagram, figure 5.1, therefore yields a resonance reaction matrix element in a momentum ($\vec{k}$) representation having a pole at energy $E = m^* + \vec{k}^2/2m_0$. Information on the space-time behavior of the unstable particle is obtained by constructing colliding wave packets to describe the initial collision and convoluting their Fourier transforms with the reaction matrix to obtain wave packets describing the decay products. The result of this analysis is (Jacob and Sachs 1961) that the probability for the decay into the final state wave packets after a time interval t following the collision of the initial packets is proportional to $e^{-\Gamma t}$, where $\Gamma/2$ is the imaginary part of the mass m^*. That the value of Γ obtained in this way is the same as that obtained from Fermi's "golden rule" (or Weisskopf-Wigner perturbation theory) has been shown in a general way by Lipschutz (1966).

Since experiments testing detailed balance are not possible in this case, there remains the question of how to test for T violation in a weak decay process. The answer is that interference methods analogous to those described for nuclear electromagnetic transitions in section 4.2 may be used to determine a T-violating phase, η. This phase will show up most prominently in the transition rate as an interference term having the property that it changes sign under motion reversal, just as in the electromagnetic case (see table 4.2).

Let us consider as an example the beta decay of the neutron,

$$n \rightarrow p + e^- + \bar{\nu}, \tag{5.4}$$

and examine the way the decay rate depends on the spin $\vec{\sigma}_n$ of the neutron. The neutron is taken to be at rest; that is, we consider the process in the center-of-mass coordinate system. Then the momentum vectors of the three decay products are coplanar. For illustrative purposes we will assume that the decay rate depends on the neutron spin component normal to the decay plane in the form[1]

$$\Gamma \sim 1 + B(\vec{\sigma}_n \cdot \vec{k}_p \times \vec{k}_e), \tag{5.5}$$

where $\vec{k}_p$ and $\vec{k}_e$ are the proton and electron momentum, respectively.

It is clear that the second term changes sign under motion reversal. This behavior can be related to the time reversal question in a graphic way by the method introduced for the discussion of classical mechanics in section 2.3. For this purpose we return to the movie of the process illustrated in figure 5.1 and assume it to be taken in the neutron center-of-mass system. Also assume that the spin of the neutron can be seen in the movie and that it is as shown in figure 5.2*a*, which also shows the velocities of the associated decay products. Since the neutron spin is normal to the decay plane, the second term in eq. (5.5) is maximized.

Reversing the film sequence leads to the motion represented by the corresponding figure 5.2*b*. This figure may be rotated 180° around the neutron spin axis to restore the configuration of the decay products so it is the same as that in figure 5.2*a* without affecting the neutron spin, which is opposite to that in figure 5.2*a*. The procedure of section 2.3 is now to repeat the experiment with the initial conditions as shown in the reversed movie, figure 5.2*b*,

[1] The expression for the rate based on general four-fermion weak coupling theory is given by Jackson, Treiman, and Wyld (1957a). See also Holstein (1972).

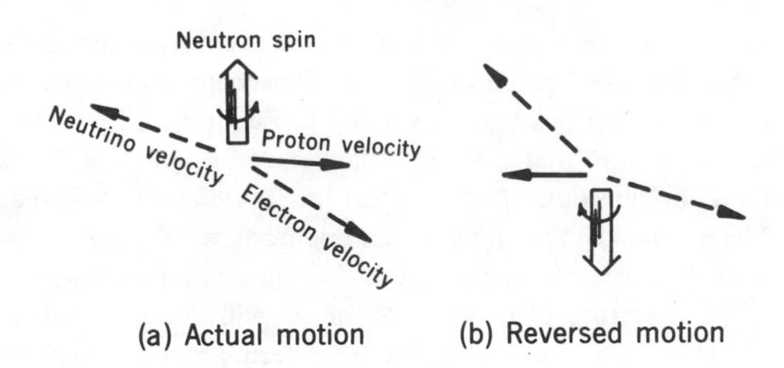

FIG. 5.2 (*a*) Motion of neutron spin before decay and velocities of decay products as recorded on film of figure 5.1. (*b*) Motion of neutron spin and velocities (momenta) of decay products as observed in reversed movie projection of film of figure 5.1.

that is, with the neutron spin reversed. Invariance of the dynamics under motion reversal would mean that the decay rate would be the same in both cases. But we have just seen that the relative sign of the two terms in eq. (5.5) is reversed in going from figure 5.2*a* to figure 5.2*b*. Therefore B is a measure of the violation of motion reversal, and it may be used to measure the violation of time reversal.

By now the reader should suspect that B is not a *direct* measure of time reversal, on the basis of our earlier warnings about the effects of final state interactions. That general subject, especially for the case of strong interactions, will finally be addressed in the next two sections. In the particular case of beta decay the final state interaction effects are due only to the Coulomb interaction between the proton and electron; they have been calculated in detail by Jackson, Treiman, and Wyld (1957b) and by Ebel and Feldman (1957).[2] They are quite small, so a measurement indicating that B has an appreciable value would show the existence of T violation of the weak interaction leading to beta decay.

Measurements of the magnitude of B for the beta decay of polarized neutrons have been made by several groups,[3] the first by Burgy and others (1958) and the most recent by Steinberg and others (1974). The latter give the value

$$B(n) = (-1.1 \pm 1.7) \times 10^{-3}. \tag{5.6}$$

All the results are consistent with $B = 0$, in other words, with T invariance of the weak interaction *responsible for this particular process*. A similar

[2] See also Brodine (1970).

[3] See also Clark and Robson (1960 and 1961) and Erozolimsky et al. (1968, 1970).

measurement for the beta decay of polarized ^{19}Ne has been carried out by Baltrusaitis and Calaprice (1977),[4] with the result

$$B(^{19}\text{Ne}) = (-0.5 \pm 1.0) \times 10^{-3}, \tag{5.7}$$

which is again consistent with $B = 0$, although in this case the estimated error is of the order of the Coulomb final state interaction effect (Callan and Treiman 1967).

5.2 Final State Interaction Effects

We now apply the general theory of collisions, as summarized in section 4.3, to the special case of decay processes. In this case the initial state describes a single (unstable) particle or "fragment," which will usually be taken to be at rest, and the final state consists of two or more particles and/or fragments. Since the interaction responsible for the transition is small enough for decay to occupy a measurable period of time, it will be treated as a perturbation, and its contribution to the total Hamiltonian of the system will be denoted by W for "weak." It should be kept in mind that since radiative electromagnetic interactions are small enough to be treated by perturbation methods they may be included in W, although in general we shall give more attention to what are known as the Weak Interactions (i.e., those responsible for beta decay, etc.) in what follows.

If, for example, the initial state $|i\rangle$ consists of the fragment A and the final state $|f\rangle$ consists of the pair of fragments B and C, we seek the transition rate for the process

$$A \to B + C \tag{5.8}$$

due to the perturbation W. According to Fermi's golden rule the decay rate, or transition probability, is given by

$$\Gamma_{fi} = \frac{2\pi}{\hbar} d_f |W_{fi}|^2, \tag{5.9}$$

where d_f is the density of final energy states and W_{fi} is the matrix element of W between zero-order (in the interaction W) states.

This seems straightforward enough, but if there is a strong interaction between fragments B and C we find ourselves in a quandary over how to calculate the matrix element. Let us assume that the total Hamiltonian is

$$H_T = H + W, \tag{5.10}$$

where H includes the strong interactions, in particular those acting between

[4] See also Calaprice et al. (1969, 1974).

B and C. Then according to eq. (4.70) we have *two* complete sets of zero-order final states, $|\psi_f^+\rangle$ and $|\psi_f^-\rangle$, and the question is, Which set is to be used to calculate the matrix element? Since from a physical viewpoint the final state is being radiated, the natural assumption is that the function satisfying the radiated (outgoing) wave condition, $|\psi_f^+\rangle$, should be used. However, as noted by Mott and Massey (1933, p. 258; 1949, p. 356; 1965, p. 797) and especially emphasized by Watson (1952), that is the wrong conclusion.

To arrive at the correct conclusion, we follow Watson and make use of the formalisms of section 4.5, where the strong interaction for channel a is denoted by V_a, and so on. Since "channel" i in the present case consists of a single fragment, there is no interaction V_i, and eq. (4.68a) is replaced simply by

(5.11a) $$H = H_i,$$

while eq. (4.68b) has the form

(5.11b) $$H = H_f + V_f,$$

V_f being the final state (strong) interaction. Note that H_i and H_f may also involve strong interactions acting within each fragment.

In the notation of section 4.5, the exact solution for the initial state strong interaction problem is $|\chi_i\rangle$, a "free particle" wave function for which there is no distinction between incoming and outgoing waves, and for the final state it is either $|\psi_f^+\rangle$ or $|\psi_f^-\rangle$, given by eq. (4.70) (with the subscript a replaced by f). These functions are now to be used as the zero-order functions in obtaining an amplitude for the reaction eq. (5.8). For this purpose we may start from the *exact* solutions for the total Hamiltonian H_T, which may be written as

(5.12a) $$H_T = H_i + W$$

or as

(5.12b) $$H_T = H_f + V_f + W.$$

If the exact solutions of

(5.13) $$H_T|\phi_a^\pm\rangle = E_a|\phi_a^\pm\rangle$$

are introduced, the exact amplitude for the transition from the initial to the final state may now be obtained from eqs. (4.73) and (5.12b):

(5.14) $$T_{fi}^+ = \langle\chi_f|V_f + W|\phi_i^+\rangle_{E_f=E_i}.$$

However, this form is inconvenient for the application of a perturbation approximation to W because the strong interaction V_f appears explicitly. Therefore we make use of the general relationship eq. (4.74) to write

$$T^+_{fi} = T^{-*}_{if} \tag{5.15}$$

with

$$T^-_{if} = \langle \chi_i | W | \phi^-_f \rangle_{E_f = E_i}, \tag{5.16}$$

according to eqs. (4.73) and (5.12a). Thus we find that the exact amplitude is

$$T^+_{fi} = \langle \phi^-_f | W | \chi_i \rangle_{E_f = E_i}, \tag{5.17}$$

where use has been made of the fact that W is a Hermitian operator.

In the spirit of perturbation theory, the exact state $|\phi^-_f\rangle$ may be expanded in powers of matrix elements of W starting from the strong interaction states $|\psi^-_f\rangle$ as zero-order functions. Therefore, in first order, the matrix element describing decay into the outgoing final state is

$$W_{fi} = \langle \psi^-_f | W | \chi_i \rangle_{E_f = E_i}, \tag{5.18}$$

and we see that it is the *incoming* state $|\psi^-_f\rangle$ rather than the *outgoing* state $|\psi^+_f\rangle$ that is to be used to calculate the matrix element that appears in the expression eq. (5.9) for the transition probability.

The simple form of the result eq. (5.18) depends primarily on the fact that in the initial state the strong interaction Hamiltonian does not split into an internal part and a channel interaction but instead has the form eq. (5.11a). In the case of a decay process, this is because the "incoming channel" consists of a single isolated system. However, there are cases in which eq. (5.11a) would hold even though the incoming channel consists of two colliding objects. That will happen when the only interaction between the colliding objects is weak enough to be treated as part of the perturbation W. Examples would include photodisintegration processes and disintegration induced by neutrinos. In these cases the reaction amplitude is also given by eq. (5.18), and much of what we will learn later about applications of this equation can be used with appropriate minor modifications.

Eq. (5.18) is the basis for determining the consequences of T invariance of the weak interaction W when there are final state interactions that are also T invariant. We invoke the by now familiar procedure based on eq. (3.32) to determine the complex conjugate of the matrix element:

$$\begin{aligned} W^*_{fi} &= \langle T(\psi^-_f) | TWT^{-1}T | \chi_i \rangle_{E_f = E_i} \\ &= e^{i(\phi_i - \phi_f)} \langle \psi^+_{f'} | W | \chi_{i'} \rangle_{E_f = E_i}, \end{aligned} \tag{5.19}$$

where f' and i' are the motion-reversed quantum numbers and the phase factor corresponds to that appearing in eq. (4.82).

To relate the right-hand side of eq. (5.19) to W_{fi}, we make use of the fact that the states $|\psi_c^-\rangle$ for all channels c form a complete set so that

$$\langle\psi_{f'}^+| = \sum_c \langle\psi_{f'}^+|\psi_c^-\rangle\langle\psi_c^-|. \tag{5.20}$$

By the definition, eq. (4.71), of the S matrix for the strong interactions (including final state interactions), this becomes

$$\langle\psi_{f'}^+| = \sum_c S_{cf'}^*\langle\psi_c^-|. \tag{5.21}$$

But $S_{cf'}$ is a unitary matrix that can in principle always be brought into diagonal form. Of course the solution of this eigenvalue problem would be very difficult, since it amounts to an exact solution of the Schroedinger equation for the strong interactions. The result would be the determination of the eigenvalues $e^{2i\delta_c}$ of S for each of the "eigenchannels" c. Here δ_c is the "eigenphase" of S.

These eigenchannels may be very complicated. For example, in a nuclear reaction two particles may collide to produce three or more particles. Then the eigenstates of the S matrix are mixtures of states describing different numbers of particles. However, in some cases, especially when the energy is not too high, such mixing does not occur and it is possible to define some of the eigenstates in such a way that the eigenphases can be measured. Two-particle elastic scattering is an example, and in that case the eigenphase is just the well-known "phase shift." If there are several two-particle channels involved in the elastic scattering, conservation laws (conservation of angular momentum or of isotopic spin) usually make it possible to determine the eigenchannels.

To simplify eq. (5.21), we assume that the states $|i\rangle$, $|f\rangle$, $|c\rangle$, etc., are the eigenstates of S. Then

$$S_{cf} = e^{2i\delta_f}\delta(c-f) \tag{5.22}$$

and

$$\langle\psi_{f'}^+| = e^{-2i\delta_f}\langle\psi_{f'}^-|, \tag{5.23}$$

where we have made use of $\delta_{f'} = \delta_f$ because the strong interactions are assumed to be T invariant. Therefore eq. (5.19) becomes

$$W_{fi}^* = e^{i(\phi_i - \phi_f)}e^{-2i\delta_f}W_{f'i'}. \tag{5.24}$$

It is convenient to introduce the transition amplitudes A_{fi} defined by

$$A_{fi} = \left(\frac{2\pi}{\hbar} d_f\right)^{1/2} W_{fi} \tag{5.25}$$

so that the transition probability, eq. (5.9), becomes simply

$$\Gamma_{fi} = |A_{fi}|^2, \tag{5.26}$$

and eq. (5.24) becomes

$$A_{fi}^* = e^{i(\phi_i - \phi_f)} e^{-2i\delta_f} A_{f'i'}. \tag{5.27}$$

This result suggests the introduction of a "reduced" decay amplitude a_{fi} defined by

$$a_{fi} = e^{-i\delta_f} A_{fi} \tag{5.28}$$

since, then, eq. (5.27) takes the simpler form

$$a_{fi}^* = e^{i(\phi_i - \phi_f)} a_{f'i'}. \tag{5.29}$$

For certain special systems, rotational invariance leads to the relationship

$$a_{fi} = e^{i(\phi_i - \phi_f)} a_{f'i'}, \tag{5.30}$$

in which case time reversal invariance implies that a_{fi} is real. The most obvious, but important, example is that of a spin 0 particle decaying into two spin 0 particles. Rotational invariance then guarantees that A_{fi} is independent of the sign of the relative momentum in the final state, which is the only kinematic variable.

We now consider the case of a T-violating weak interaction in a system subject to T-invariant strong interactions. Then the perturbation causing the decay may be put in the form

$$W = W^e + W^o, \tag{5.31}$$

where W^e and W^o are Hermitian operators and

$$TW^eT^{-1} = W^e, \quad TW^oT^{-1} = -W^o, \tag{5.32}$$

so that

$$TWT^{-1} = W^e - W^o. \tag{5.33}$$

If amplitudes A_{fi}^e and A_{fi}^o associated with W^e and W^o are defined in accordance with eq. (5.25), the decay amplitude is

$$A_{fi} = A_{fi}^e + A_{fi}^o. \tag{5.34}$$

The conditions eq. (5.32) and the assumption that the strong interactions are T invariant then allow us to apply the argument leading to eq. (5.27) to the even and odd amplitudes separately to obtain the results

(5.35a) $$A_{fi}^{e*} = e^{i(\phi_i - \phi_f)} e^{-2i\delta_f} A_{f'i'}^{e}$$

and

(5.35b) $$A_{fi}^{o*} = -e^{i(\phi_i - \phi_f)} e^{-2i\delta_f} A_{f'i'}^{o}.$$

In terms of the reduced amplitudes defined in accordance with eq. (5.28),

(5.36) $$a_{fi} = a_{fi}^{e} + a_{fi}^{o},$$

these conditions become

(5.37a) $$a_{fi}^{e*} = e^{i(\phi_i - \phi_f)} a_{f'i'}^{e}$$

and

(5.37b) $$a_{fi}^{o*} = -e^{i(\phi_i - \phi_f)} a_{f'i'}^{o}.$$

Some insight into the implications of this result can be obtained by again considering the simplest case for which eq. (5.30) is satisfied, that is, when the motion-reversed states can be transformed into the original states by a rotation. Then eqs. (5.37) become

(5.38) $$a_{fi}^{e*} = a_{fi}^{e}, \quad a_{fi}^{o*} = -a_{fi}^{o},$$

so that a_{fi}^{e} is real and a_{fi}^{o} is purely imaginary. Hence

(5.39) $$a_{fi} = \pm|a_{fi}^{e}| \pm i|a_{fi}^{o}|,$$

and the magnitude of the phase η of a_{fi} is given by

(5.40) $$\tan \eta = |a_{fi}^{o}|/|a_{fi}^{e}|.$$

Since the ratio $|a_{fi}^{o}|/|a_{fi}^{e}|$ clearly is a measure of the relative contribution of the odd term in the interaction, W, η is a phase measuring the degree of T violation in this particular case.

Experimental implications of eqs. (5.37) in the general case will manifest themselves in measurements of partial decay rates Γ_{fi}. When eq. (5.26) is expressed in terms of the reduced amplitudes, we find that

(5.41) $$\Gamma_{fi} = |a_{fi}^{e}|^2 + |a_{fi}^{o}|^2 + 2\,\mathrm{Re}(a_{fi}^{e*} a_{fi}^{o}).$$

Eqs. (5.37) have implications for *motion reversal* because $|i'\rangle$, $|f'\rangle$ are the states obtained from $|i\rangle$, $|f\rangle$ by motion reversal. Thus the decay rate for the motion-reversed system is found to be

(5.42) $$\Gamma_{f'i'} = |a_{fi}^{e}|^2 + |a_{fi}^{o}|^2 - 2\,\mathrm{Re}(a_{fi}^{e*} a_{fi}^{o}),$$

and a difference between the two decay rates is produced by the interference term. A measure of the degree of violation of *motion reversal* is given by

(5.43)
$$\Delta = \frac{|\Gamma_{f'i'} - \Gamma_{fi}|}{|\Gamma_{f'i'} + \Gamma_{fi}|} = \frac{2|\mathrm{Re}(a^{e*}_{fi}\, a^{o}_{fi})|}{|a^{e}_{fi}|^2 + |a^{o}_{fi}|^2}.$$

Thus

(5.44)
$$\Delta \sim \mathrm{Re}(a^{o}_{fi}/a^{e}_{fi}),$$

and one can hope to measure the degree of T violation by measuring terms in the decay rate that are odd under motion reversal.

However, in the special case represented by eqs. (5.30) and (5.40), $\Delta = 0$ independent of the magnitude of η because a^{o}_{fi}/a^{e}_{fi} is purely imaginary. This is simply the consequence of the assumption that rotational symmetry guarantees invariance under motion reversal in the special case. The point may be illustrated by noting that in order for Δ to be different from zero the dependence of the decay rate on the momentum and spin vectors must include at least three such vectors. The reason is that a product of an odd number of vectors that are odd under T is required to produce a term that changes sign under motion reversal, and rotational invariance of the rate limits such terms to scalars, such as the triple scalar product. The special case of a spin 0 particle decaying into two spin 0 particles may be cited again as the most important example. There is only one relevant kinematic variable, the relative momentum of the decay products, and the decay rate depends only on its magnitude, which is invariant under motion reversal. Therefore there is no dependence on the direction of motion.

There arises in this important case the question of how to measure T violation, since motion reversal is irrelevant. The resolution of this question, which turns out to be a crucial one requiring detailed investigation, is left until chapter 9, where the seminal case (belonging to the particular category in question) of the neutral K^0 mesons is treated.

To be able to carry out the analysis of K^0 decay, we must complete our treatment of the effect of final state interactions in the general case. The final state interactions do not make an appearance in eq. (5.43) because the introduction of the reduced amplitudes a_{fi} eliminates the eigenphase δ_f. However, we must keep in mind that this step was made possible by the assumption that the final state $|f\rangle$ is a single eigenstate of the reaction matrix for the strong interactions. We usually do not have the option of choosing the decay channel in this convenient way, and we shall find, by considering specific examples, how the final state interactions introduce terms mimicking the result eq. (5.43) when the decay channel is not an eigenstate of the S matrix but a linear combination of eigenstates.

5.3 *T*-Violation Mimicry

It is the purpose of this section to show how the appearance of the final state phase shift δ_f, despite the assumption of T invariance underlying the derivation of eq. (5.27), can lead to the appearance of terms in the transition rate that are odd under motion reversal and therefore may give the impression that there *is* T violation. This mimicry of T violation can be related to the appearance in the offending terms of a factor $\sin(\delta_1 - \delta_2)$, where δ_1 and δ_2 are the eigenphases of two states that are mixed to form $|f\rangle$. The antiunitarity of T causes all phase factors to be changed to their complex conjugates under T, which therefore not only subjects all kinematic variables to motion reversal but also changes the sign of $(\delta_1 - \delta_2)$. The result is that although they *do* change sign under motion reversal, the offending terms *do not* change sign under T, and there is no contradiction.

If the interaction W does violate T invariance, a phase or phases such as η, given by eq. (5.40), will appear in the amplitude in such a way as to change the interference term so that it is proportional to $\sin(\delta_1 - \delta_2 + \eta)$. If the eigenphases δ_1 and δ_2 can be measured, as they sometimes can, or if they can be calculated reliably, as in the case of Coulomb scattering, then a measurement of the interference term makes it possible to determine η. We have already become familiar with this procedure in our discussion of experimental results in chapter 4. It remains to be demonstrated that these terms have the alleged form.

To illustrate the point, let us assume that the final state is a linear combination of two eigenstates of the S matrix

$$|f\rangle = c_0^*|0\rangle + c_1^*|1\rangle, \tag{5.45}$$

where the asterisk denotes the conjugate complex, the notation $|0\rangle$ and $|1\rangle$ being chosen for reasons that will become apparent later.

From eqs. (5.18) and (5.25) we then find that

$$A_{fi} = c_0 A_{0i} + c_1 A_{1i}. \tag{5.46}$$

If all interactions are T invariant, we may invoke the arguments leading to eq. (5.27) to write

$$A_{li}^* = e^{i(\phi_i - \phi_f)} e^{-2i\delta_l} A_{l'i'} \tag{5.47}$$

where $l = 0$ or 1 and δ_l is the eigenphase of the S matrix associated with the state $|l\rangle$. It is convenient to introduce reduced amplitudes for the eigenchannels:

$$a_{li} = e^{-i\delta_l} A_{li}, \tag{5.48}$$

so that

$$a_{li}^* = e^{i(\phi_i - \phi_f)} a_{l'i'} \tag{5.49}$$

and the total amplitude is given by

$$A_{fi} = c_0 e^{i\delta_0} a_{0i} + c_1 e^{i\delta_1} a_{1i}. \tag{5.50}$$

The decay rate is then given by

$$\Gamma_{fi} = c_0^2 |a_{0i}|^2 + c_1^2 |a_{1i}|^2 + 2c_0 c_1 \operatorname{Re}[a_{0i}^* a_{1i} e^{i(\delta_1 - \delta_0)}], \tag{5.51}$$

since the eigenstates $|l\rangle$ of the S matrix are total angular momentum eigenstates and the phase convention is fixed in accordance with eq. (3.73) so that the coefficients c_0 and c_1 may be taken to be real. Under motion reversal the rate is transformed by use of eq. (5.49) to

$$\Gamma_{f'i'} = c_0^2 |a_{0i}|^2 + c_1^2 |a_{1i}|^2 + 2c_0 c_1 \operatorname{Re}[a_{0i} a_{1i}^* e^{i(\delta_1 - \delta_0)}]. \tag{5.52}$$

Then the change under motion reversal corresponding to eq. (5.43) is

$$\Delta = \frac{2c_0 c_1 \operatorname{Im}(a_{0i}^* a_{1i}) \sin(\delta_1 - \delta_0)}{c_0^2 |a_0|^2 + c_1 |a_1|^2 + 2c_0 c_1 \operatorname{Re}(a_{0i}^* a_{1i}) \cos(\delta_1 - \delta_0)}. \tag{5.53}$$

Thus we confirm that *even for T-invariant systems* the final state interactions lead to a violation of invariance under *motion reversal*, except in the unlikely event that $\delta_1 = \delta_0$.

We may now turn to the case of T violation by the perturbation W, eq. (5.31). Then, as for eqs. (5.36) and (5.37), we may write

$$a_{li} = a_{li}^e + a_{li}^o, \tag{5.54}$$

with

$$a_{li}^{e*} = e^{i(\phi_i - \phi_f)} a_{l'i'}^e \tag{5.55a}$$

and

$$a_{li}^{o*} = -e^{i(\phi_i - \phi_f)} a_{l'i'}^o. \tag{5.55b}$$

The resulting expression for Δ will differ from eq. (5.53) by terms proportional to $\operatorname{Re}(a_{0i}^{e*} a_{1i}^o + a_{1i}^{e*} a_{0i}^o)\cos(\delta_1 - \delta_0)$ and $\operatorname{Re}(a_{li}^{e*} a_{li}^o)$, each of which is a measure of T violation. Little purpose would be served by writing out the general expression, because the various phase factors involved can be determined only in specific cases. Therefore we now turn to consideration of a simple, but important, example.

The example to be considered is that of the decay of a spin $\frac{1}{2}$ particle into another spin $\frac{1}{2}$ particle and a spin 0 particle. The initial state $|m_i\rangle$ is specified by the spin quantum number $m_i = \pm\frac{1}{2}$. The final state is specified by the spin quantum number m_f of the spin $\frac{1}{2}$ particle and the relative momentum $\vec{k}$ of the two particles, and it may be a linear combination of states having total angular momentum $\frac{1}{2}$ and specified by some additional quantum number l.

As in the case of scattering we may introduce transition matrix amplitudes $\mathbf{A}(\vec{k})$ with matrix elements given by (compare eq. 4.88)

(5.56) $$A_{fi} = \langle m_f | \mathbf{A}(\vec{k}) | m_i \rangle.$$

Rotational invariance of the amplitude requires that the matrices $\mathbf{A}(\vec{k})$ have the form

(5.57) $$\mathbf{A}(\vec{k}) = F(k)\mathbf{1} + G(k)(\vec{\sigma} \cdot \hat{k}).$$

The second term would be excluded by parity conservation. However, since it will be of particular interest to consider decay owing to Weak Interactions, which do not conserve parity, we shall use the *P*-violating expression eq. (5.57).

Eq. (5.57) is in the desired form eq. (5.46), with F and G playing the roles of the $l = 0$ and $l = 1$ terms. Since the total angular momentum of the final state is $j = \frac{1}{2}$, the eigenstates of the strong interaction S matrix may have either orbital angular momentum $l = 0$ (S state) or $l = 1$ (P state). But these states have opposite parity, so the even parity matrix element F can refer only to the $l = 0$ state and the odd parity matrix element G only to the $l = 1$ state. Therefore we may write F and G in terms of reduced amplitudes $f(k)$ and $g(k)$:

(5.58a) $$F(k) = fe^{i\delta_0}$$

and

(5.58b) $$G(k) = ge^{i\delta_1},$$

where δ_0 is the S wave phase shift and δ_1 is the P wave phase shift for scattering of the spin $\frac{1}{2}$ particle by the spin 0 particle of the final state. Parity conservation of the strong interactions guarantees that the $l = 0$ and $l = 1$ states are each eigenstates of the S matrix.

Having the amplitude in the matrix form, eq. (5.57), puts us in a position to analyze decay experiments for intially polarized particles in terms of the density matrix, which takes the standard form, eq. (4.97), for both the initial and final states. If the initial state of polarization of the decaying particle is $\vec{\mathrm{P}}_i$, the corresponding density of initial states is

(5.59) $$\rho_i = \tfrac{1}{2}[\mathbf{1} + (\vec{P}_i \cdot \vec{\sigma})],$$

and the density matrix of final states is

(5.60) $$\rho_f = \mathbf{A}\rho_i \mathbf{A}^\dagger/\Gamma,$$

where Γ is the partial decay rate into a final state of momentum $\vec{k}$ averaged over initial spin states:

$$\Gamma = \mathrm{Tr}(\mathbf{A}\rho_i \mathbf{A}^\dagger). \tag{5.61}$$

By inserting eq. (5.57) for **A**, we express the measurable quantities in terms of $F(k)$ and $G(k)$:

$$\begin{aligned}\Gamma\rho_f = \tfrac{1}{2}\{|F|^2[1+(\vec{P}_i\cdot\vec{\sigma})]+|G|^2[1+2(\vec{P}_i\cdot\hat{k})(\hat{k}\cdot\vec{\sigma})-(\vec{P}_i\cdot\vec{\sigma})]\\ + 2\,\mathrm{Re}(FG^*)[(\vec{P}_i\cdot\hat{k}) + (\hat{k}\cdot\vec{\sigma})] - 2\,\mathrm{Im}(FG^*)(\vec{P}_i \times \hat{k}\cdot\vec{\sigma})\},\end{aligned} \tag{5.62}$$

$$\Gamma = |F|^2 + |G|^2 + 2\,\mathrm{Re}(FG^*)(\vec{P}_i\cdot\hat{k}), \tag{5.63}$$

$$\begin{aligned}\Gamma\vec{P}_f = (|F|^2 - |G|^2)\vec{P}_i + 2[(\vec{P}_i\cdot\hat{k})|G|^2\\ + \mathrm{Re}(FG^*)]\hat{k} - 2\,\mathrm{Im}(FG^*)(\vec{P}_i \times \hat{k}).\end{aligned} \tag{5.64}$$

The last term in eq. (5.62), being a pseudoscalar, is a measure of parity violation, as are the next-to-last and last terms in eq. (5.64), since $\vec{P}_i$ is an axial vector and $\hat{k}$ is a polar vector. At the same time the last term in eq. (5.64) is also a measure of motion reversal violation, since polarization vectors and momentum vectors are odd under motion reversal. Thus it is to the terms proportional to $\mathrm{Im}(FG^*)$ that we look for tests of T invariance.

The origin of these terms can be understood in terms of the orbital angular momentum of the particles in the final state. The nonconservation of parity allows a mixture of the $l = 0$ and $l = 1$ states of opposite parity; the interference between them leads to the parity-violating terms in the transition rate. The consequences of T invariance are expressed by eq. (5.49) or, in this special case,

$$\langle m_f|\mathbf{a}_l(\vec{k})|m_i\rangle^* = i^{2(m_i-m_f)}\langle -m_f|\mathbf{a}_l(-\vec{k})|-m_i\rangle \tag{5.65}$$

where $\mathbf{a}_0 = f\mathbf{1}$ and $\mathbf{a}_1 = g(\vec{\sigma}\cdot\vec{k})$ so that

$$f_l^*(k)\delta_{m_i m_f} = f(k)\delta_{m_i m_f} \tag{5.66a}$$

and

$$g^*(k)\langle m_i|\vec{\sigma}\cdot\vec{k}|m_f\rangle = -i^{2(m_i-m_f)}g(k)\langle -m_f|\vec{\sigma}\cdot\vec{k}|-m_i\rangle, \tag{5.66b}$$

since $\vec{\sigma}$ is Hermitian. From the property of the Pauli matrices eq. (4.94), it follows that both $f(k)$ and $g(k)$ are *real* functions. Then the measurable quantities in eqs. (5.62) and (5.64) are found to be

$$|F|^2 = f^2, \tag{5.67a}$$

$$|G|^2 = g^2, \tag{5.67b}$$

$$\mathrm{Re}\,FG^* = fg\cos(\delta_1 - \delta_0), \tag{5.67c}$$

and

$$\text{(5.67d)} \qquad \text{Im } FG^* = fg \sin(\delta_1 - \delta_0).$$

As we found earlier, the term violating motion reversal is Im FG^*. Therefore we again find, as in eq. (5.53), that there is a term mimicking T violation in this T-invariant system, and it is proportional to $\sin(\delta_1 - \delta_0)$.

To establish a way the effects of a T-violating term in W can be separated from the mimic, we consider the consequence of adding a T-odd term W^o to the T-invariant weak interaction, which we again denote by W^e. Comparing eqs. (5.55b) and (5.65) makes it clear that adding the odd term will introduce imaginary parts in the reduced amplitudes. Thus T violation manifests itself by the introduction of complex phase factors in f and g. Only the relative phase of the two functions is relevant to the calculation of rates, and this T-violating phase, which I again denote by η, is constant because the only variable k on which it could depend is determined by energy conservation in the decay process. Then the modification of eqs. (5.67) is simply that obtained by making the substitution

$$\text{(5.68)} \qquad \delta_1 - \delta_0 \rightarrow \delta_1 - \delta_0 + \eta.$$

A measurement of $\vec{P}_f$ as a function of $\vec{P}_i$, and the direction $\hat{k}$ of the relative momentum of the decay products may be used to test T invariance in this decay process because it makes possible a determination of $\tan(\delta_1 - \delta_0 + \eta)$. If δ_0 and δ_1 are known by means of independent strong interaction (scattering) experiments, the measured value of $\tan(\delta_1 - \delta_0 + \eta)$ may be used either to determine η or to place an upper limit on its magnitude.[5]

[5] Arash, Moravcsik, and Goldstein (1985) show that "it is impossible to construct, in any reaction in atomic, nuclear, or particle physics, a null experiment that would unambiguously test the validity of time-reversal invariance independently of dynamical assumptions." The connection with the considerations above is evident: to establish $\eta = 0$, it is necessary to carry out two experiments, one determining $\delta_1 - \delta_0 + \eta$ and the other providing the value of $\delta_1 - \delta_0$, which is dynamic information. Even in the case of a reaction such as $K^\pm \rightarrow \pi^0 + \mu^\pm + \nu$, for which we believe we know that the final state interactions are negligible, their result is correct in principle because we are making a dynamic assumption. We shall find in chapter 9 that the experiments on T violation making use of the K^0, $\bar{K}^0$ system also are consistent with their result, although they are based on a "null experiment" (yielding a non-null result) on CP invariance. It will be shown that it is necessary either to make the dynamic assumption of CPT invariance or to add information from other reactions to establish the connection with T invariance. However, insofar as the overall question of T invariance is concerned, the theorem is not as general as it might appear. We have found that there exists a quite general class of null experiments that are able in each case to unambiguously test T invariance, namely, the measurements of the electric dipole moments of isolated systems (sec. 4.2). This is not a violation of the theorem of Arash, Moravcsik, and Goldstein because the experiments do not involve a reaction. Because of their lack of ambiguity, the electric dipole moment measurements provide the most fundamental tests of T invariance. A closely related and equally fundamental test would be a measurement of the splitting of the Kramers degeneracy in the absence of external magnetic fields.

Eqs. (5.63) and (5.64) may be rewritten in the convenient form (Lee and Yang 1957):

(5.69a) $$\Gamma(\hat{k}) = \Gamma_0[1 + \alpha(\vec{P}_i \cdot \hat{k})]$$

and

(5.69b) $$[1 + \alpha(\vec{P}_i \cdot \hat{k})]\vec{P}_f = [(\vec{P}_i \cdot \hat{k}) + \alpha]\hat{k} + \beta(\vec{P}_i \times \hat{k}) + \gamma[\hat{k} \times (\vec{P}_i \times \hat{k})],$$

with

(5.70a) $$\Gamma_0 = |F|^2 + |G|^2,$$

(5.70b) $$\Gamma_0 \alpha = 2\mathrm{Re}(FG^*) = 2fg \cos(\delta_1 - \delta_0 + \eta),$$

(5.70c) $$\Gamma_0 \beta = -2\,\mathrm{Im}(FG^*) = 2fg \sin(\delta_1 - \delta_0 + \eta),$$

(5.70d) $$\Gamma_0 \gamma = |F|^2 - |G|^2.$$

Thus,

(5.71) $$\beta/\alpha = \tan(\delta_1 - \delta_0 + \eta).$$

Measuring the partial decay rate $\Gamma(\hat{k})$ of the polarized particle as a function of $\hat{k}$ makes it possible to determine the "helicity" parameter α, which measures parity violation, and measuring the final state polarization normal to the plane formed by the vectors $\vec{P}_i$ and $\hat{k}$ yields the value of β, whereby $\tan(\delta_1 - \delta_0 + \eta)$ can be found by means of eq. (5.71).

This very interesting experiment has been carried out, first by Cronin and Overseth (1963) and more recently by Overseth and Roth (1967) and by Cleland and others (1972) for the decay process

(5.72) $$\Lambda^0 \to p + \pi^-.$$

Here Λ^0 is the neutral strange baryon having spin $\frac{1}{2}$ and π^- is the negative pion having spin 0. This case is of particular interest because the only evidence we have for violation of T invariance arises in the decay of strange particles into nonstrange particles—that is, in the decay of neutral K mesons—and that evidence is indirect. A determination of η in the decay eq. (5.72) would provide direct evidence for T violation in a decay process involving the same kind of transition and therefore, presumably, the same mechanism.

The result of the experiment by Overseth and Roth was

(5.73a) $$\beta/\alpha = -0.16 \pm 0.10,$$

leading to

(5.73b) $$\delta_1 - \delta_0 + \eta = -9^\circ \pm 5.5^\circ.$$

The $j = \frac{1}{2}$ (π, p) phase shifts are measured directly. There are two sets of phase shifts corresponding to isotopic spin $I = \frac{1}{2}$ and $I = \frac{3}{2}$ of the pion-nucleon system. It is assumed that the decay eq. (5.72) is subject to the $\Delta I = \frac{1}{2}$ rule, so that only the $I = \frac{1}{2}$ phase shifts need to be considered, since the Λ^0 has $I = 0$. Then the pion-nucleon scattering experiments give

(5.73c) $$\delta_1 - \delta_0 = -6.5^\circ \pm 1.5^\circ,$$

which is consistent with the result eq. (5.73b) with $\eta = 0$. Therefore the experiment provides no evidence for T violation and, in fact, because of the relatively large estimated error, provides only a rather high upper limit of 10° or so on the magnitude of η. The most recent measurements of β, obtained by Cleland and others (1972), is in accord with eq. (5.73a).

To push such experiments to the level at which T violation occurs in the K^0, $\bar{K}^0$ decays ($\eta \approx 0.1^\circ$) would require an enormous increase in precision in both the scattering experiments and the decay experiments. These results illustrate how the final state interaction effects place all such experiments in double jeopardy: the high precision required for T-violation experiments is not easy to attain, but analysis of the experiments suffering from final state interaction effects also requires measuring eigenphases of the S matrix with comparable precision, a very difficult goal because the strong interaction experiments usually involve interactions between many different channels.

The K^0, $\bar{K}^0$ experiments do not suffer from this difficulty because they are not experiments on motion reversal. Instead they provide a direct test of CP invariance, and they establish CP violation for this particular system. The resulting evidence for T violation is indirect, requiring an analysis that demands a detailed understanding of the phenomenology of the K^0, $\bar{K}^0$ system. Because of its detailed and technical nature, the discussion of this subject and its implications for further work on other physical systems will be presented in chapters 9 and 10, after we have considered the implications of P, T, and C in the context of the more general structure of the theory.

6 | Improper Transformations in Relativistic Classical Field Theories

Our analysis of the physical implications of time reversal and the *CP* transformation has, up to this point, been based on phenomenological models. These models have a firm basis in terms of effective interactions and general quantum mechanical concepts, such as the Wigner transformation properties of quantum mechanical states presented in chapter 3. However, some attention must now be given to the theory at a more fundamental level, that is, at the level of relativistic quantum field theories, in order to establish a basis for including in phenomenological theory concepts from particle theory such as charge conjugation, the *CPT* theorem, and quark models. Furthermore, it is only through understanding the phenomena at this level that we can hope to arrive at a fundamental theory.

The foundations of quantum field theory are laid by application of the correspondence principle (in its most general sense) to classical fields. Here the term "classical" is used to include not only the traditional classical fields such as the electromagnetic field, but also fields like the Schroedinger wave function, the one-particle Dirac spinor function, and the Klein-Gordon scalar field. In order to establish the definitions and uses of the improper transformations *P* and *T* in quantum field theory, it is necessary to establish their meaning in the context of the classical field theories.

The relativistic fields are classified according to their behavior under (proper) Lorentz transformations, that is, as tensors or spinors of a given rank. The transformations of both tensors and spinors may be expressed in terms of the behavior of tensor products of spinors in two dimensions (van der Waerden 1932), but it is convenient to distinguish tensor fields from spinor fields because the former are associated with particles of integral (or zero) spin and the latter with particles of half-integral spin. Of course the ultimate questions of physics concern the fields to be associated with the fundamental particles—for example, the spinor fields (quark fields) from which tensor fields (meson fields) that describe composite particles having integral spin may be formed, or the gauge (electromagnetic, intermediate vector boson) fields that appear to be associated with elementary (non-composite) particles.

6.1 Classical Tensor Fields: Transformation under *P*

For the description of Lorentz covariant fields it will be advantageous to use the notation

(6.1a) $$(\vec{\phi}, i\phi_0) \equiv (\phi_1, \phi_2, \phi_3, \phi_4)$$

for a general four-vector and

(6.1b) $$x = (\vec{x}, ict) \equiv (x_1, x_2, x_3, x_4)$$

for the four-vector of position and time, the imaginary fourth component being used (rather than a metric tensor) because of the central role complex conjugation plays in the transformation T. For the same reason the Pauli (1958) representation for the gamma matrices describing linear transformations of the four-component Dirac spinor field will be used.[1] Note that in this notation each of the matrices γ_μ and γ_5 is Hermitian (and unitary). Tensor components will be labeled $T_{\mu\nu\lambda\ldots}$, where μ, ν, λ, ... each run through the values 1 to 4 corresponding to the four-vector components, eq. (6.1). Components ψ_α of Dirac spinors, when needed, will carry indices α, β, ... from the beginning of the Greek alphabet, and α, β, ... each run through the values 1 to 4, which, however, are not to be confused with the vector indices.

Throughout this and the following chapter we shall measure time in units of length—that is, replace $x_0 = ct$ by t, where c is the speed of light. Similarly, we shall measure mass, energy, and momentum in units of inverse length, calling $k_0 = E/\hbar c$ the "energy" and $\vec{k} = \vec{p}/c$ the "momentum."[2]

Among the possible tensor fields of all ranks $T_{\mu\nu\lambda\ldots}(x)$, there are only a

[1] The connection between the gamma matrices in Pauli metric and the Bjorken and Drell (1964) metric that is used more commonly now may be found in Adler and Dashen (1968, pp. 7 ff.). Although my choice of notation differs from that of Bjorken and Drell (1964, 1965) for the reasons given in the text, their book on relativistic quantum field theory is an excellent source for obtaining the background in field theory as it is treated here. I do not use the manifestly covariant forms of commutation and anticommutation relations in defining the kinematics of the fields because that would jumble the dynamics with the kinematics. Instead, I use equal-time commutation and anticommutation relations, which will be found to simplify the treatment of the transformations, especially T. Thus I am taking advantage of the fact that covariance is well established and am choosing a particular but arbitrary reference frame in which to work, thereby avoiding some irrelevant complications.

[2] In the vernacular of field theory we say that "the units are chosen so that $\hbar = c = 1$." Note that in these units the dimensionless constant $e^2/\hbar c$ becomes just e^2 so that e, the charge on the electron, is a dimensionless quantity.

few cases that need be considered here in order to establish the physical principles that are of interest. These are the irreducible tensors (i.e., those corresponding to particles of given spin):

(I) $\phi(x)$, a tensor of rank 0, scalar field,
(II) $\phi_\mu(x)$, a tensor of rank 1, "transverse" vector field,
(III) $\phi_{\mu\nu}(x) = -\phi_{\nu\mu}(x)$, an antisymmetric tensor of rank 2,
(IV) $\chi_\mu(x) = iT_{\nu\lambda\eta}(x)$, $\mu\nu\lambda\eta$ cyclic and $T_{\Pi(\nu\lambda\eta)} = (-1)^\Pi T_{\nu\lambda\eta}$, where Π is a permutation of $\nu\lambda\eta$, an antisymmetric tensor of rank 3, a pseudovector field, and
(V) $\chi(x) = iT_{\mu\nu\lambda\eta}$, $T_{\Pi(\mu\nu\lambda\gamma)} = (-1)^\Pi T_{\mu\nu\lambda\eta}$, an antisymmetric tensor of rank 4, a pseudoscalar field.

The factors of i are introduced in cases V and IV in order that the pseudoscalar field and the space components of the pseudovector field be real if the tensors are constructed from products of four-vectors having real space components.

The behavior of each type of field under proper Lorentz transformations is determined by its tensor character and is indicated by the descriptive terms "scalar," "vector," and so forth. Its behavior under the improper transformations P and T is also determined in part by its tensor character, since the appearance of the index μ on the tensor implies a linear transformation corresponding to that of x_μ. However, as we have seen in chapter 3, the information concerning the transformations P and T obtained in this way must be supplemented to take into account the underlying quantum mechanics even in the case of what we are calling "classical fields," since they include the Klein-Gordon and Dirac fields.

Again, since we are considering a situation in which the equations of motion (field equations) may not be invariant under P and T, it is necessary to separate the kinematic from the dynamic aspects of the theory in order to define *kinematically admissible* transformations. We shall treat the classical tensor fields as the relativistic generalization of the Schroedinger wave function for a single particle so that the commutation relations, eq. (3.3), for the spatial comments $\vec{x}$ and $\vec{p}$ of the position and momentum define the kinematics of the fields. To further separate the kinematics from the dynamics, the transformation conditions on the fourth component p_0 of p, which is the energy, are established for the case of the *free particle*—that is, the case of no forces or interaction. This step corresponds to that taken in the case of Newtonian mechanics in section 1.4, which led from eq. (1.4) to the "kinematic" constraint eq. (1.6).

The tensor fields corresponding to the free particle cases are referred to as "free fields."

The improper tranformation P is that associated with the inversion

(6.2) $$P: \quad x_\mu \rightarrow \bar{x}_\mu = -\varepsilon_\mu x_\mu,$$

where the symbol ε_μ is defined by[3]

(6.3) $$\varepsilon_1 = \varepsilon_2 = \varepsilon_3 = 1, \varepsilon_4 = -1.$$

The corresponding transformation of the energy-momentum vector of the free particle is the same:

(6.4) $$P: \quad p_\mu \rightarrow \bar{p}_\mu = -\varepsilon_\mu p_\mu.$$

To formulate the kinematically admissible transformations for the fields, we must consider the behavior of the free fields $T_{\mu\nu\gamma\ldots,0}(x)$, which are assumed to satisfy the Klein-Gordon equation

(6.5) $$(\Box - m^2)T_{\mu\nu\lambda\ldots,0} = 0,$$

with the D'Alembertian $\Box$ defined by

(6.6) $$\Box = \sum_\mu \frac{\partial^2}{\partial x_\mu^2} = \nabla^2 - \frac{\partial^2}{\partial t^2}.$$

Eq. (6.5) follows from: (1) the free particle condition

(6.7a) $$p^2 = m^2,$$

where

(6.7b) $$p^2 = -\Sigma_\mu p_\mu^2,$$

(2) the commutation relations eq. (3.3), and (3) the substitution for the *free particle* energy

(6.7c) $$p_4 \rightarrow i\frac{\partial}{\partial x_4}.$$

The relationships eqs. (6.5) through (6.7c) are invariant under the transformations eqs. (6.2) and (6.4) when the tensor $T_{\mu\nu\lambda\ldots,0}$ transforms as the product $x_\mu x_\nu x_\lambda \ldots$; that is,

(6.8a) $$P: \quad \phi_0(x) \rightarrow \phi_0^P(\bar{x}) = \phi_0(x),$$

(6.8b) $$P: \quad \phi_{\mu,0}(x) \rightarrow \phi_{\mu,0}^P(\bar{x}) = -\varepsilon_\mu \phi_{\mu,0}(x),$$

[3] The repeated index appearing in $\varepsilon_\mu x_\mu$ does *not* imply a sum over μ as in the usual summation notation.

(6.8c) $$P: \quad \phi_{\mu\nu,\,0}(x) \rightarrow \phi^{P}_{\mu\nu,\,0}(\bar{x}) = \varepsilon_{\mu}\,\varepsilon_{\nu}\,\phi_{\mu\nu,\,0}(x),$$

(6.8d) $$P: \quad \chi_{\mu,\,0}(x) \rightarrow \chi^{P}_{\mu,\,0}(\bar{x}) = \varepsilon_{\mu}\,\chi_{\mu,\,0}(x),$$

(6.8e) $$P: \quad \chi_{0}(x) \rightarrow \chi^{P}_{0}(\bar{x}) = -\chi_{0}(x).$$

Eq. (6.8d) follows for fields of type IV because $-\varepsilon_{\nu}\,\varepsilon_{\lambda}\,\varepsilon_{\eta} = \varepsilon_{\mu}$ when $\mu\nu\lambda\eta$ are cyclic, and eq. (6.8e) follows for type V because $\varepsilon_{\mu}\,\varepsilon_{\nu}\,\varepsilon_{\lambda}\,\varepsilon_{\eta} = -1$. The difference in sign between eqs. (6.8a) and (6.8e) and between eqs. (6.8b) and (6.8d) is the cause for the prefix "pseudo" in describing the transformation properties of types IV and V. Note that the space components of the vector field type II form a polar vector and the space components of the pseudovector field type IV form an axial vector. Furthermore, the three independent space-time components $(\phi_{14}, \phi_{24}, \phi_{34})$ of the type III field form a polar vector (electric field) and the three independent space-space components $(\phi_{12}, \phi_{23}, \phi_{31})$ form an axial vector (magnetic field).

To determine the behavior of the interacting fields it is necessary to consider the dynamic behavior of the fields, that is, the effect of their interactions. The effect of the interaction is to modify eq. (6.5) to the inhomogeneous form

(6.9a) $$(\Box - m^2)T_{\mu\nu\lambda\cdots} = -j_{\mu\nu\lambda\cdots}(x),$$

where $j_{\mu\nu\lambda\cdots}(x)$ is a source term ("current density") describing the interactions with (charged) particles. This source function is a tensor of the same rank as $T_{\mu\nu\lambda\cdots}$, thereby guaranteeing the covariance of the equation under proper Lorentz transformation. If the interactions are P invariant, $j_{\mu\nu\lambda\cdots}(x)$ will transform in the same way as the free fields, eq. (6.8), as will the interacting fields $T_{\mu\nu\lambda\cdots}(x)$. When the interactions violate P invariance, $j_{\mu\nu\lambda\cdots}(x)$, and therefore, in general, the $T_{\mu\nu\lambda\cdots}(x)$ do *not* satisfy the same conditions as the free fields, eqs. (6.8).

The way the behavior of the interacting fields under P is determined by the behavior of the free fields will be illustrated here for the scalar and pseudoscalar fields. (To simplify the notation, $\phi(x)$, will be used for both cases henceforth except where it is necessary to make a clear distinction, as in table 7.1.) Eq. (6.9a) is then

(6.9b) $$(\Box - m^2)\phi(x) = -j(x),$$

where $j(x)$ is invariant under *proper* Lorentz transformations.

When the functions $\phi(x) = \phi(\vec{x}, t)$ and $\dot{\phi}(\vec{x}, t)$ are given for all $\vec{x}$ at some particular time τ, a solution to eq. (6.9b) may be written as the integral equation[4]

(6.10a)
$$\phi(x) = \phi_\tau(x) + \int d^4y\, \Delta_\tau(x, y) j(y),$$

where $\phi_\tau(x)$ is a solution of the homogeneous equation, that is, (6.5) for the free field, satisfying the same boundary conditions:

(6.10b)
$$\phi_\tau(\vec{x}, \tau) = \phi(\vec{x}, \tau),$$

(6.10c)
$$\dot{\phi}_\tau(\vec{x}, \tau) = \dot{\phi}(\vec{x}, \tau);$$

that is, $\phi_\tau(x)$ is the particular free field that matches $\phi(x)$ at $t = \tau$. The function $\Delta_\tau(x, y)$ is the Green's function for eq. (6.9b), which is the solution of

(6.11a)
$$(\Box - m^2)\Delta_\tau(x, y) = -\delta_4(x - y),$$

satisfying the boundary conditions

(6.11b)
$$\Delta_\tau(\vec{x}, \tau; y) = 0$$

(6.11c)
$$\dot{\Delta}_\tau(\vec{x}, \tau; y) = 0,$$

where the dot indicates the derivative with respect to the (real) time component of the four-vector x. The function $\delta_4(x - y)$ is the real four-dimensional delta function $\delta(x_0 - y_0)\delta(x_1 - y_1)\delta(x_2 - y_2)\delta(x_3 - y_3)$.

From these properties it is evident that Δ_τ is invariant when x and y undergo the transformation eq. (6.2). Therefore the way $\phi(x)$ is transformed under P is determined by the way $\phi_\tau(x)$ and $j(y)$ in eq. (6.10a) behave under P. The former is given by eq. (6.8a) for a scalar field and eq. (6.8e) for a pseudoscalar field. The way $j(y)$ transforms under P depends on the nature of the interactions. If they are invariant under P, $j(y)$ will also transform in accordance with either eq. (6.8a) or eq. (6.8e); that is, it will be invariant (even) under P for the scalar case and odd (change of sign) in the pseudoscalar case.

If the interactions are *not* P invariant, $j(y)$ will be a mixture of even and odd terms so that $\phi(x)$, given by eq. (6.10a), will also be a linear combination of fields having even and odd parity even when, at $t = \tau$, it has a given parity (e.g., when at $t = \tau$ it is a function that can be written as a linear combination of spherical harmonics of *either* even l *or* odd l). Thus, in general, $\phi(x)$ describes a field of mixed parity (a time-dependent linear combination of fields of opposite parity).

[4] See Yang and Feldman (1950). Note that I use the *real* four-dimensional volume element $d^4x = dx_0 dx_1 dx_2 dx_3$ throughout.

Usually $j(x)$ will be a function of $\phi(x)$ and other fields with which $\phi(x)$ interacts. These other fields will be denoted generically by $\psi(x)$, which may or may not be the Dirac field described later. The determination of $\phi^P(x)$, the transformed $\phi(x)$, requires a knowledge of $\psi^P(x)$, and vice versa, and the kinematically admissible transformation of the free field $\psi_0(x)$ associated with $\psi(x)$ is needed to determine $\psi^P(x)$. The way the transformations of the free fields determine $\phi^P(x)$ and $\psi^P(x)$ can be made evident when the coupling between the fields is so weak that it is a good approximation to use the lowest order of a perturbation expansion—that is, to replace $j(x)$ in eq. (6.10a) with $j_\tau(x)$, where $j_\tau(x)$ is obtained from $j(x)$ by replacing $\phi(x)$ with $\phi_\tau(x)$ (and $\psi(x)$ by a corresponding free field $\psi_\tau(x)$). Then, since $j(x)$ is a tensor of the same rank as $\phi(x)$ (see eq. 6.9a), the transformation of $j_\tau(x)$ under P follows from eq. (6.8a) or eq. (6.8e) and the corresponding equation for $\psi_\tau(x)$.

In principle this approximation can be iterated for stronger coupling, and the transformation of the set of iterated equations can also be expressed in terms of the transformations of the free fields. Thus we find that the transformation $\phi(x) \to \phi^P(x)$ (and $\psi(x) \to \psi^P(x)$) is governed by the transformations eq. (6.8a) and eq. (6.8e) of $\phi_\tau(x)$ and their equivalent for $\psi_\tau(x)$.

If $j_\tau(x)$ transforms in the same way as $\phi_\tau(x)$, that is, $j_\tau^P(\bar{x}) = \pm j_\tau(x)$ for the scalar or pseudoscalar fields, respectively, then by iteration of the integral equation for $\phi(x)$ (and $\psi(x)$) it will be found that $\phi(x)$ and $j(x)$ are also transformed in that same way. This property of $j_\tau(x)$ is therefore the required condition for P invariance of the interactions to which the field $\phi(x)$ is subjected.

The extension of the foregoing analysis to tensor fields of higher rank is straightforward.

6.2 Classical Tensor Fields: Transformation under T

This lengthy discussion of the behavior of the fields under P is meant to serve as preparation for the treatment of time reversal, which makes use of many of the same concepts. The time reversal transformation is, in the notation of eq. (6.2),

(6.12a) $$T: \quad x_\mu \to x'_\mu = \varepsilon_\mu x_\mu .$$

The corresponding transformation of the energy-momentum vector of the free particle is

(6.12b) $$T: \quad p_\mu \to p'_\mu = -\varepsilon_\mu p_\mu .$$

The two equations are consistent with the commutation relation eq. (3.3) if and only if the operator K that takes the conjugate complex of the fields is included in the operator T acting on the classical fields, as is shown by eq. (3.17).

This introduction of K is not so evident on the basis of the relativistic equations of motion of the free fields, eq. (6.5), as it was for the Schroedinger equation (3.21), because $\Box - m^2$ is a real operator. The reason for the difference from the Schroedinger equation is that eq. (6.5) is of second order in the time and is therefore unchanged by time reversal. However, we shall see that the dynamics of the *quantized* field equations are determined by a universal equation that is of first order in t and has the form of eq. (3.21), although with a greatly enlarged meaning. Therefore, the need for introducing K into the transformation T will be apparent from the dynamic equations of the quantized free field as well as from the commutation relations.

From these considerations it should be clear that the time reversal transformation of the classical *free* scalar field is to be taken to be the same as that for the Schroedinger wave function of a spinless particle, eq. (3.39):

(6.13a) $$T: \quad \phi_0(x) \rightarrow \phi'_0(x') = K\phi_0(x)K^{-1},$$

where the factor K^{-1} has been introduced here to make it clear that in later use of $\phi(x)$ as an operator (chap. 7) K acts only on ϕ. As an example of the meaning of this equation we note that for a scalar plane wave solution $\phi_k(x)$ of momentum $\vec{k}$, $\phi_{k'}(x')$ is a solution of momentum $-\vec{k}$.

To determine the corresponding behavior of tensors of higher rank, it is necessary to be more specific about the particular tensor fields to be considered. It must be recognized that in our notation, for every appearance of the index $\mu = 4$ on the tensor field the field carries a factor i, since the proper Lorentz transformations are real number transformations of the physical variables.

Let us consider the case of the transverse vector field, which is the one to be associated with a particle of spin 1. An example of such a vector field is the gauge field, like the vector potential of electrodynamics, $A_\mu(x)$. Note that the source function for the vector potential $\vec{A}(x)$ of electrodynamics is the current density, which is odd under motion reversal, hence $\vec{A}$ is also odd. We use this as a guide for the behavior of the transverse field and, keeping in mind the factor i in $\phi_{4,0}$ (eq. 6.1a), therefore write

(6.13b) $$T: \quad \phi_{\mu,0}(x) \rightarrow \phi'_{\mu,0}(x') = -K\phi_{\mu,0}(x)K^{-1}.$$

We shall find later that this is the only behavior that is consistent with that of vector fields constructed from products of spinors.

An example of the antisymmetric tensor fields that is of interest is the field derived from the gauge field ϕ_μ,

$$\phi_{\mu\nu} = \frac{\partial \phi_\nu}{\partial x_\mu} - \frac{\partial \phi_\mu}{\partial x_\nu}, \tag{6.14}$$

which, for electrodynamics, includes the magnetic and electric fields.[5] From eqs. (6.13b) and (6.14) it follows that

$$T: \quad \phi_{\mu\nu,\,0}(x) \rightarrow \phi'_{\mu\nu,\,0}(x') = -K\phi_{\mu\nu,\,0}(x)K^{-1}. \tag{6.13c}$$

In considering the pseudofields it is simplest to start from the pseudoscalar of type V, $\chi(x) = iT_{1234}$. Since the factor i has been extracted, $\chi(x)$ can be a real field. Then χ_0 transforms like the product $iw_1x_2y_3z_4 = -w_1x_2y_3z_0$, where w, x, y, z are four-vectors of distinct space-time points. Therefore the free fields have odd signatures under time reversal:

$$T: \quad \chi_0(x) \rightarrow \chi'_0(x') = -K\chi_0(x)K^{-1}. \tag{6.13d}$$

Similarly, it is found that the free pseudovector field $\chi_{\mu,\,0}(x) = iT_{\nu\lambda\eta,\,0}$ transforms like $ix_\nu y_\lambda z_\eta$ and therefore with the same signature as $\chi_0(x)$:

$$T: \quad \chi_{\mu,\,0}(x) \rightarrow \chi'_{\mu,\,0}(x') = -K\chi_{\mu,\,0}(x)K^{-1}. \tag{6.13e}$$

The behavior of interacting tensor fields under T will be illustrated, as for the behavior under P, by considering the example of the scalar and pseudoscalar fields, denoted jointly by $\phi(x)$. Again the integral equation of motion, eq. (6.10a), provides the starting point for determining the transformed field $\phi'(x')$ where

$$T: \quad \phi(x) \rightarrow \phi'(x'). \tag{6.15}$$

It is implicit in what has gone before that $\phi'(x')$ is linearly related to $K\phi(x)K^{-1}$. This is another case in which the statement will be more easily justified in the quantized version of the theory.

Since $j(x)$ is a function of the field $\phi(x)$ and other fields interacting with it, we can also write

$$T: \quad j(x) \rightarrow j'(x'), \tag{6.16}$$

where $j'(x')$ is linearly related to $Kj(x)K^{-1}$. Therefore, from eq. (6.10a) we

[5] The vector potentials of classical electromagnetic theory are real functions so that the introduction of K might appear to be unnecessary. However, it is sometimes convenient to use complex fields (e.g., for quantization), and then K is necessary. Furthermore, the classical non-Abelian gauge fields are generally complex. For example, the non-Abelian nature of the Yang-Mills field is associated with the commutation relations for isotopic spin, whose time reversal properties are discussed in section 3.5.

have

$$\phi'(x') = \phi'_{-\tau}(x') + \int d^4y\, \Delta^*_\tau(x, y)j'(y'). \tag{6.17}$$

The reason for writing $(\phi_\tau)' = \phi'_{-\tau}$ is made evident by applying T to the boundary condition eqs. (6.10b) and (6.10c). More directly, the result follows from the relationship

$$\Delta^*_\tau(x, y) = \Delta_{-\tau}(x', y'), \tag{6.18}$$

which can be obtained from the explicit form[6] of $\Delta_\tau(x, y)$ and converts eq. (6.17) to

$$\phi'(x') = \phi'_{-\tau}(x') + \int d^4y'\, \Delta_{-\tau}(x', y')j'(y'). \tag{6.19}$$

In section 6.1 the free fields $\phi_\tau(x)$ for an *arbitrary* fixed value of τ were treated as the kinematic variable in discussing the transformation P. In order to define a suitable variable for defining a kinematically admissible transformation T, this arbitrariness is eliminated because the dependence of $\phi_\tau(x)$ (and $\psi_\tau(x)$) on the index τ is governed by the dynamics and, in general, T changes the sign of τ, as shown by eq. (6.19). The way the dynamics enters in determining the dependence of $\phi_\tau(x)$ on τ is evident from the conditions eqs. (6.10b) and (6.10c), since the time dependence of the interacting fields $\phi(x)$ is determined by the interactions. Thus we arrive at the conclusion that the only viable choice of the free field for defining the kinematically admissible transformation T is $\phi_0(x)$, the free field $\phi_\tau(x)$ defined for the particular choice $\tau = 0$, which is the "symmetry point" for T.

The kinematically admissible T is then defined in accordance with eq. (6.13a) or eq. (6.13d) as

$$\phi'_0(x') = \pm K\phi_0(x)K^{-1}, \tag{6.20}$$

which is consistent with eq. (6.19).

In general the relationship between $j'(y')$ and $j(y)$ will be determined by the behavior under T of the fields on which $j(y)$ depends as well as on any complex coefficients that appear in it explicitly (because of the operator K). As in the case of P, discussed earlier, it is clear that eq. (6.19) can be iterated in terms of the free fields by starting from the approximation in which $j(x)$ is replaced by $j_0(x)$ (i.e., $j_\tau(x)$ at $\tau = 0$). Then the transformation from $\phi(x)$ to

[6] $\Delta_\tau(x, y) = \theta(t_y - \tau)\Delta_R(x - y)[+\theta(\tau - t_y)]\Delta_A(x - y)$, where $\theta(t) = 0$ for $t < 0$ and $\theta(t) = 1$ for $t > 0$, and $\Delta_R(x)$ and $\Delta_A(x)$ are the retarded and advanced Green's functions with $\Delta^*_R(x) = \Delta_A(x')$. See Yang and Feldman (1950). For later reference (see eq. 6.33a) note that the Green's function for the Dirac equation is $S_\tau(x, y) = (\sum_\mu \gamma_\mu(\partial/\partial x_\mu) - m)\Delta_\tau(x, y)$.

$\phi'(x')$ is determined by eq. (6.20) and the corresponding transformation of $\psi_0(x)$. The condition that the interactions be T invariant is that $j_0(x)$ transform in the same way as a free field of the same signature:

(6.21a) $$j'_0(x') = \pm K j_0(x) K^{-1},$$

so that, after iteration,

(6.21b) $$\phi'(x') = \pm K\phi(x)K^{-1},$$

for scalar or pseudoscalar fields, respectively. The extension of these results to tensor fields of higher rank is again straightforward.

6.3 The Classical Dirac Spinor Field: Transformation under *P* and *T*

The classical (unquantized) Dirac field is a four-component entity $\psi_\alpha(x)$, $\alpha = 1$ to 4, which is usually written as a single-column matrix $\psi(x)$. The corresponding single-row matrix is written as the transpose $\tilde{\psi}(x)$ of $\psi(x)$. The free field equation with a particle of mass m and corresponding to eq. (6.7a) is the Dirac equation

(6.22) $$\sum_\mu \gamma_\mu \frac{\partial \psi_0}{\partial x_\mu} + m\psi_0 = 0,$$

where the 4×4 gamma matrices satisfy the anticommutation relations

(6.23) $$\gamma_\mu \gamma_\nu + \gamma_\nu \gamma_\mu = 2\delta_{\mu\nu}.$$

As already noted, the Pauli (1958) form of the gamma matrices

(6.24) $$\vec{\gamma} = \begin{pmatrix} 0 & -i\vec{\sigma} \\ i\vec{\sigma} & 0 \end{pmatrix}, \qquad \gamma_4 = \begin{pmatrix} 1 & 0 \\ 0 & -1 \end{pmatrix},$$

expressed as matrices whose elements are the 2×2 matrices, 0, 1, and $\vec{\sigma}$, will be used here, the advantage being that each of the matrices is Hermitian.

Under the proper Lorentz tranformations

(6.25) $$x_\mu \rightarrow x_\mu^L = \sum_\nu a_{\mu\nu} x_\nu,$$

the spinor $\psi(x)$ undergoes a linear, unimodular transformation

(6.26) $$L: \quad \psi(x) \rightarrow \psi^L(x^L) = S(a)\psi(x),$$

where $S(a)$ is a 4×4 matrix in the spinor space. The relationship between S and the matrix $a = (a_{\mu\nu})$ is governed by the condition[7]

(6.27a) $$S^{-1}\gamma_\mu S = \sum_\nu a_{\mu\nu}\gamma_\nu,$$

[7] See, e.g., Bjorken and Drell (1964, p. 20).

which follows from the requirement that the Dirac equation be covariant, and by the condition

(6.27b) $$\det S = 1,$$

which fixes the arbitrary constant left free by eq. (6.27a). The spinor transformation S is not in general unitary. Instead, it satisfies the condition

(6.27c) $$\gamma_4 S^{-1} = S^{\dagger}\gamma_4,$$

which follows from eq. (6.27a) and the relationships between $a^{*}_{\mu\nu}$ and $a_{\mu\nu}$.

The requirement for the kinematic admissibility of the improper transformation P is that, in addition to its consistency with the commutation relations eq. (3.3), it leave the *free field* Dirac equation invariant. That is the analogue of the requirement eq. (1.9) on Newton's equation of motion. Therefore the transformation S_P of $\psi_0(x)$ associated with P must satisfy eq. (6.27a) with the improper Lorentz transformation P defined by

(6.28) $$a_{\mu\nu} = -\varepsilon_{\mu}\delta_{\mu\nu}.$$

Thus

(6.29) $$S_P^{-1}\gamma_{\mu}S_P = -\varepsilon_{\mu}\gamma_{\mu}$$

or

(6.30a) $$\vec{\gamma}S_P + S_P\vec{\gamma} = 0,$$

(6.30b) $$\gamma_4 S_P - S_P\gamma_4 = 0.$$

From eqs. (6.30) it follows that

(6.31a) $$S_P = \lambda\gamma_4,$$

and eq. (6.27c) becomes

(6.31b) $$|\lambda|^2 = 1,$$

so that λ is a phase factor.

Since λ must also be consistent with the unimodular condition eq. (6.27b),

(6.31c) $$\lambda^4 = 1,$$

whence the choice of phase factor is limited to

(6.31d) $$\lambda = \pm 1, \ \pm i.$$

The set of these possibilities for which λ is imaginary is a matter of a choice of phase,[8] because, in contrast to the case of a proper transformation, for

[8] The implications of such a choice of phase are discussed in Wick, Wightman, and Wigner (1952). Their work was at least in part a response to the observation by Yang and Tiomno (1950) that eq. (6.31d) suggests the possible existence of two distinct and uncoupled classes of fermions, those with intrinsic parity ± 1 and those with intrinsic "parity" $\pm i$.

which the phase is fixed by continuation from the identity transformation, there is some ambiguity in the phase associated with an improper transformation. The usual choice is to take λ to be a real number so that there are just two possible classes of Dirac fields, those with even and with odd "intrinsic" parity. The definition of "parity" is made in such a way that it corresponds to the parity of the corresponding Schroedinger wave function in the nonrelativistic limit to the Dirac function for a particle, multiplied by the intrinsic parity. For the corresponding antiparticle in the same physical state, defined by charge conjugation of the field, the resulting field will be found in section 7.4 to have opposite intrinsic parity.

The transformation induced on the free spinor field by P may now be expressed in the form

(6.31e) $$P:\ \psi_0(x) \rightarrow \psi_0^P(\bar{x}) = \pm\gamma_4\psi_0(x).$$

To determine the behavior of the interacting fields under the kinematically admissible transformation eq. (6.31e), it is again necessary to consider the dynamic equation for $\psi(x)$, just as in the case of the scalar fields. The equation of motion for the Dirac field, in contrast to eq. (6.9b) for the scalar fields, is taken to be

(6.32) $$\sum_\mu \gamma_\mu \frac{\partial\psi(x)}{\partial x_\mu} + m\psi(x) = f(x),$$

where $f(x)$ is a spinor source function with spinor components $f_\alpha(x)$. As in the case of $j(x)$, $f(x)$ will be a function of $\psi(x)$ and other fields (e.g., $\phi(x)$) interacting with $\psi(x)$.

Eq. (6.32) may also be written as an integral equation

(6.33a) $$\psi(x) = \psi_\tau(x) - \int d^4y\ S_\tau(x, y) f(y),$$

where $\psi_\tau(x)$ is a solution of the homogeneous Dirac equation, that is, eq. (6.22), satisfying the boundary condition at $t = \tau$

(6.33b) $$\psi_\tau(\vec{x}, t) = \psi(\vec{x}, \tau).$$

$S_\tau(x, y)$ is the 4×4 matrix Green's function satisfying

(6.34a) $$\sum_\mu \gamma_\mu \frac{\partial S_\tau}{\partial x_\mu} + mS_\tau = -\delta_4(x - y)$$

and the boundary condition

(6.34b) $$S_\tau(\vec{x}, \tau;\ y) = 0.$$

Note that in contrast to the scalar (or tensor) case, the Dirac equation is of first order and therefore the solution is determined by specifying the function $\psi(\vec{x}, \tau)$. Boundary conditions on the time derivatives do not occur as in the case of the scalar fields, eqs. (6.10c) and (6.11c).

The transformation of eq. (6.33a),

(6.35) $$P\colon\quad \psi(x) \rightarrow \psi^P(\bar{x}),$$

is determined by eq. (6.31e) and by

(6.36) $$P\colon\quad f(x) \rightarrow f^P(\bar{x}),$$

where f^P is determined by the way the fields appearing in $f(x)$ transform. Again this transformation may be found in principle by iteration starting from $f_\tau(x)$, which is obtained by expressing $f(x)$ in terms of the free fields. Then $f_\tau^P(\bar{x})$ will be determined by the kinematically admissible transformation eq. (6.31e) and corresponding transformations for other fields appearing in $f_\tau(x)$. If the interactions are invariant under P,

(6.37a) $$f_\tau^P(\bar{x}) = \pm\gamma_4 f_\tau(x),$$

and it follows from iteration of eq. (6.33a) that *for a P-invariant system*

(6.37b) $$\psi^P(\bar{x}) = \pm\gamma_4\, \psi(x),$$

where the sign is determined by the intrinsic parity of the free field.

We turn now to consideration of the transformation T, eqs. (6.12). The kinematic conditions imposed in this case on the classical Dirac field are, in addition to the consistency with the commutation relations eqs. (3.3), the invariance of the free field Dirac equation (6.22). Therefore

(6.38) $$T\colon\quad \psi_0(x) \rightarrow \psi_0'(x') = S_T K \psi_0(x) K^{-1},$$

where S_T is a unimodular 4×4 matrix determined by the requirement that $\psi_0'(x')$ satisfy the Dirac equation

(6.39) $$\sum_\mu \gamma_\mu \frac{\partial \psi_0'}{\partial x_\mu'} + m\psi_0' = 0.$$

The requirement that eq. (6.39) be equivalent to eq. (6.22) is

(6.40) $$S_T^{-1} \gamma_\mu S_T = \gamma_\mu^*,$$

where the asterisk denotes the conjugate complex of the matrix and is equivalent to the transpose $\tilde{\gamma}_\mu$, since the gamma matrices are Hermitian in our representation, eq. (6.24). In this representation

(6.41) $$\gamma_1^* = -\gamma_1,\quad \gamma_2^* = \gamma_2,\quad \gamma_3^* = -\gamma_3,\quad \gamma_4^* = \gamma_4,$$

so that the conditions eq. (6.40) become

(6.42) $$S_T\gamma_1 + \gamma_1 S_T = S_T\gamma_2 - \gamma_2 S_T = S_T\gamma_3 + \gamma_3 S_T = S_T\gamma_4 - \gamma_4 S_T = 0,$$

whose solution is

(6.43) $$S_T = \lambda'\gamma_1\gamma_3$$

by virtue of the anticommutation relations, eq. (6.23).

Again the phase factor λ' must be determined by some convention. The Wigner convention introduced in chapter 3 may be retained by writing

(6.44a) $$S_T = (\gamma_3\gamma_1 - \gamma_1\gamma_3)/2i = \sigma_2,$$

where we follow the standard practice of using the same notation for the 4×4 matrices

(6.44b) $$\vec{\sigma} = [\vec{\gamma} \times \vec{\gamma}]/2i$$

as is used for the 2×2 Pauli sigma matrices that appear as elements of the 4×4 matrix

(6.44c) $$\vec{\sigma} = \begin{pmatrix} \vec{\sigma} & 0 \\ 0 & \vec{\sigma} \end{pmatrix}.$$

Thus the kinematically admissible transformation of the free fields is just the extension

(6.45) $$\psi_0'(x') = \sigma_2 K\psi_0(x)K^{-1}$$

of eq. (3.46) for the nonrelativistic particle of spin $\frac{1}{2}$.

The determination of the behavior of the interacting spinor field under T proceeds as before on the basis of the equation of motion, eq. (6.33a). If

(6.46a) $$T\colon\ \psi(x) \rightarrow \psi'(x')$$

and

(6.46b) $$T\colon\ f(x) \rightarrow f'(x'),$$

then eq. (6.33a) transforms to

(6.46c) $$\psi'(x') = \psi'_{-\tau}(x') - \int d^4y'\, S_{-\tau}(x', y')f'(y').$$

To obtain eq. (6.46c), use has been made of the property[9]

(6.46d) $$\sigma_2 S_\tau^*(x, y)\sigma_2 = S_{-\tau}(x', y'),$$

[9] See note 6.

which is again to be expected on the basis of the transformation of the boundary condition, eq. (6.33b). The field variable used to define a kinematically admissible T must again be defined at $\tau = 0$. The operational definition of T as given by eq. (6.45) is therefore

(6.47) $$\psi'_0(x') = \sigma_2 K\psi_0(x)K^{-1}.$$

The earlier arguments showing that $\psi'(x')$ is determined by the transformations of the free fields apply again.

If the interactions are T invariant,

(6.48a) $$f'_0(x') = \sigma_2 Kf_0(x)K^{-1},$$

so that the transformation of the interacting field *in the case of T invariance* is

(6.48b) $$\psi'(x') = \sigma_2 K\psi(x)K^{-1}.$$

7 | Improper Transformations in Relativistic Quantum Field Theories

The physical basis for defining the transformations under P and T of the quantized fields is taken to be the behavior established for the classical fields in chapter 6. The quantized fields admit to another improper transformation in addition to P and T, namely, charge conjugation C. Although C has a clear-cut meaning only for the quantized fields, the physical basis of its definition resides in the form of the classical field equations in the presence of external electromagnetic fields.

To describe the kinematically admissible improper transformations P, T, and C in relativistic quantum field theories, it is necessary to start from a knowledge of the formal structure of the theory, just as it was in making the transition from Newtonian to quantum mechanics in chapter 3. The three elements of the formal structure of any dynamic theory that are set forth in the beginning of chapter 3 must therefore be identified specifically for quantum field theories.

7.1 The Formal Structure of Quantum Field Theory

The identification of the three elements in section 3.1 describing the formal structure of quantum mechanics may be carried over to quantum field theories with some changes of wording. These changes of wording, though small, involve very significant changes in mathematical structure and interpretation. Their significance will be highlighted in the following statement of the formal structure of quantum field theory, which will otherwise be a repetition of the statements made in section 3.1.

A. Manifold

1. Every state of the system is represented by a ray $|\Psi\rangle$ in a unitary *functional* space, and the inner product between two such states is denoted by $\langle\Psi_1|\Psi_2\rangle$. This space will be referred to as the "great big space" (GBS) because it is capable of describing the state (in the usual sense of quantum mechanics) of an arbitrary number of particles, or a superposition of such states.

2. Each measurable physical quantity (observable) or function thereof is represented by a Hermitian linear operator in this functional space. These operators are "blown up" versions of the corresponding one-particle operators of quantum mechanics; that is, they may be envisioned as matrices acting on the basis vectors of the GBS. For example, the matrix representing an observable of the single-particle theory becomes, for states associated with a specific number N of particles, the sum of the single-particle observables operating in the great big space formed by taking the outer product of the N one-particle spaces (Weyl 1931, pp. 246 ff.; March 1951, pp. 154 ff.). Since the number of "degrees of freedom" in field theory is infinite, the connection between the set of observables and the structure of the GBS is not as direct as is the case for the quantum mechanics of a fixed number of particles. The unitary functional space is determined by the properties of the fields, which are operators on the GBS. Thus, for example, the scalar field $\phi(x)$ is an operator defined at each space-time point x, and $\vec{x}$ serves as a parameter (on the same footing as t) describing the point at which the operator is defined. Observables are given as functions of these field operators. The charge density is such an observable and may serve to define the position of a particle, but $\vec{x}$ itself is not an observable. The blown-up observables **Q** and transformations **U** operating in the functional space will be denoted by boldface symbols.
3. The numerical values that can result from measurements of an observable are its eigenvalues.
4. The expectation value of an observable **Q** for a system in the state $\mathbf{\Psi}$ is $\langle \mathbf{\Psi} | \mathbf{Q} | \mathbf{\Psi} \rangle$ if the ray is normalized so that $\langle \mathbf{\Psi} | \mathbf{\Psi} \rangle = 1$.

B. Kinematics

1. The kinematic variables are the field operators (e.g., $\phi(x)$) and canonically conjugate field operators ($\pi(x)$) to be defined for each type of field.
2. The kinematic conditions are equal-time commutation or anti-commutation relations among the field operators establishing the algebra that underlies the unitary functional space.
3. Any nondynamic supplementary conditions (such as the Lorentz conditions for the vector potential) are kinematic constraints.

C. Dynamics

1. The dynamics govern the way the field operators or the states vary with time. They are determined by the Hamiltonian operator **H**.
2. The variation of the states with time is given in the Schroedinger

representation by

$$i\frac{\partial|\Psi\rangle}{\partial t} = \mathbf{H}\,|\,\Psi(t)\rangle. \tag{7.1}$$

3. Alternatively, the variation of the field operators and observables with time is given in Heisenberg representation by equations of the form

$$-i\dot{\phi}(x) = [\mathbf{H}, \phi(x)], \quad -i\dot{\pi}(x) = [\mathbf{H}, \pi(x)] \tag{7.2a}$$

for the fields and

$$-i\dot{\mathbf{Q}} = [\mathbf{H}, \mathbf{Q}] \tag{7.2b}$$

for the observables. Note that eq. (7.2b) follows from eq. (7.2a) when $\mathbf{Q}$ is given as a function of the field operators with no explicit time dependence.

In eq. (7.2a), the scalar fields $\phi(x)$ and $\pi(x)$ may be replaced by any of the tensor or spinor fields and their conjugate fields. When $\mathbf{H}$ corresponds to the classical Hamiltonian for the fields, the equations for $\dot{\phi}$ and $\dot{\pi}$ may be combined to give field equations (equations of motion) of the classical form, eq. (6.9b). The time derivatives appearing in this and subsequent equations are to be interpreted as commutators with the Hamiltonian, as in eq. (7.2a). In the absence of interactions $j(x) \equiv 0$, and eq. (6.9b) becomes the free field equation, eq. (6.5). In general the source term $j(x)$ is also an operator, since it depends on the field operator $\phi(x)$ or on operators associated with other fields with which ϕ interacts.

For the Dirac field, the equation of motion following from eq. (7.2a) has the same form as the classical eq. (6.32), where again the time derivative is to be interpreted as the commutator with the Hamiltonian and the source function $f(x)$ is necessarily a Dirac four-component spinor operator with components $f_\alpha(x)$.

The Lorentz invariance of the equations of motion is manifest for eqs. (6.9b) and (6.32), but the invariance of the equal-time commutation or anticommutation relations is not so apparent, since the assumption of equal time implies a specification of a Lorentz reference frame. Consideration of this requirement is outside the scope of this book, but the demonstration that it is satisfied can be found in standard textbooks on relativistic quantum field theories (e.g., Bjorken and Drell 1965).

The fundamental requirement that a transformation of the GBS (or of the kinematic operators in that space) be kinematically admissible is that the equal-time commutation or anticommutation relations among the field operators be invariant under the transformation. A supplementary requirement is that the equations of motion of the *free* field operator be invariant. Both of these requirements will be satisfied by a unitary transformation **U** of the functional space that is consistent with the equations of motion of the free field. However, the specific form of **U** will be governed by the requirement that the field operators, or at least the matrix elements of the field operators, transform in the same way as the classical fields do under the prescribed physical transformation.

Since the way the classical fields transform under, for example, the transformation P is different for different types of field, it is necessary to specify the type of classical field and the way it transforms in order to determine the corresponding unitary transformation **U**. In the case of time reversal T, we expect that the transformation of the great big space will be antiunitary and the determination of the factor **U** will again require a knowledge of the type of classical field. Therefore specific types of fields, the scalar (or pseudoscalar) field and the Dirac spinor field, will be treated in the following consideration of improper transformations. The behavior of other tensor fields should be made evident by these examples.

7.2 Quantized Scalar and Pseudoscalar Fields: Transformation under P and T

The kinematic conditions on the scalar or pseudoscalar field are given by the equal-time commutation relations

$$[\pi(\vec{x}, t), \phi(\vec{y}, t)] = -i\delta(\vec{x} - \vec{y}), \tag{7.3a}$$

$$[\phi(\vec{x}, t), \phi(\vec{y}, t)] = [\phi(\vec{x}, t), \phi^\dagger(\vec{y}, t)] = 0, \tag{7.3b}$$

$$[\pi(\vec{x}, t), \pi(\vec{y}, t)] = [\pi(\vec{x}, t), \pi^\dagger(\vec{y}, t)] = 0, \tag{7.3c}$$

$$[\pi^\dagger(\vec{x}, t), \phi(\vec{y}, t)] = 0, \tag{7.3d}$$

where $\pi(x)$, the field conjugate to $\phi(x)$, is defined by[1]

$$\pi(x) = \dot{\phi}^\dagger(x). \tag{7.4}$$

The free field operators $\phi_0(x)$ and $\pi_0(x)$ satisfy the same relationships.

If the unitary transformation of the GBS associated with P is denoted by **P**, then it must satisfy the requirement that it transform the free fields in accordance with the behavior of the classical fields, eqs. (6.8a) and (6.8e). That is, for the free fields $\phi_0(x)$ (i.e., corresponding to $\tau = 0$), which will

[1] See, e.g., Bjorken and Drell (1965, pp. 38 ff.). I assume here that the interaction Lagrangian does not contain derivatives of $\phi(x)$. Keep in mind that I am using $\phi(x)$ to denote *either* the scalar *or* the pseudoscalar field here.

henceforth be taken as the kinematic variables,

(7.5) $$\phi_0^P(\bar{x}, t) = \mathbf{P}_\pm \phi_0(\bar{x}, t)\mathbf{P}_\pm^{-1} = \pm\phi_0(-\bar{x}, t),$$

for the scalar and pseudoscalar cases, respectively. The interchange of x and $\bar{x}$ in eqs. (6.8) to obtain eq. (7.5) is a matter of convenience, and similar allowable interchanges will be made later.

$\mathbf{P}_\pm$ may be constructed explicitly in terms of the free field operators by making use of the commutation relations. For this purpose we first develop the general method, since it will be applied to the other improper transformations also. To that end, we consider a unitary transformation on the free field operators having the form

(7.6) $$\mathbf{U} = e^{i\theta\Gamma},$$

where Γ is a Hermitian operator and θ is a real number. The resulting transformation of an operator function $\mathbf{Q}$ of the field operators is given by

(7.7) $$\mathbf{Q}_\theta = \mathbf{UQU}^{-1} = e^{i\theta\Gamma}\mathbf{Q}e^{-i\theta\Gamma},$$

which can be evaluated by means of a formal expansion in powers of θ:

(7.8) $$\mathbf{Q}_\theta = \sum_n \frac{(i\theta)^n}{n!}\left[\frac{\partial^n \mathbf{Q}_\theta}{\partial(i\theta)^n}\right]_{\theta=0}.$$

The derivatives appearing here may be expressed as n^{th}-order commutators by virtue of eq. (7.7), since

(7.9a) $$\frac{\partial \mathbf{Q}_\theta}{\partial(i\theta)} = \Gamma\mathbf{Q}_\theta - \mathbf{Q}_\theta\Gamma,$$

and, by repetition:

(7.9b) $$\left[\frac{\partial^n \mathbf{Q}_\theta}{\partial(i\theta)^n}\right]_{\theta=0} = [\Gamma, [\Gamma, \ldots [\Gamma, \mathbf{Q}] \ldots]]_n,$$

where the commutator with $\mathbf{Q}$ is taken n times.

For the improper transformations we are concerned with just two operators $\mathbf{Q}$ that are transformed into each other, for example the field operators $\phi_0(x)$ and $\phi_0^P(x)$. Let us denote them by $\mathbf{Q}^+$ and $\mathbf{Q}^-$ in order to obtain a generic form for the transformation. If there exists a Hermitian operator Γ having the property

(7.10) $$[\Gamma, \mathbf{Q}^\pm] = \eta^{\pm 1}\mathbf{Q}^\mp,$$

where η is a complex number, then eqs. (7.8) and (7.9b) lead to the result

(7.11) $$\mathbf{Q}_\theta^\pm = \cos\theta\,\mathbf{Q}^\pm + i\eta^{\pm 1}\sin\theta\,\mathbf{Q}^\mp.$$

This result may now be used to construct the unitary transformation $\mathbf{P}_{\pm}$ satisfying eq. (7.5). Consider the Hermitian operator

(7.12a) $$\mathbf{\Gamma} = i\int d^3y[\pi_0(\vec{y}, t^0)\phi_0(-\vec{y}, t^0) - \phi_0^\dagger(-\vec{y}, t^0)\pi_0^\dagger(\vec{y}, t^0)],$$

where t^0 is a fixed but arbitrary value of t. From the commutation relations eq. (7.3a) it is found that

(7.12b) $$[\mathbf{\Gamma}, \phi_0(\vec{x}, t^0)] = \phi_0(-\vec{x}, t^0),$$

and if we take $\mathbf{Q}^{\pm} = \phi_0(\pm\vec{x}, t^0)$, eq. (7.11) with $\theta = \pi/2$ is equivalent to

(7.12c) $$\mathbf{P}'\phi_0(\vec{x}, t^0)\mathbf{P}'^{-1} = i\phi_0(-\vec{x}, t^0),$$

where

(7.12d) $$\mathbf{P}' = e^{i\pi\Gamma/2}.$$

Now consider the Hermitian operator

(7.13a) $$\mathbf{\Gamma}' = i\int d^3y[\pi_0(\vec{y}, t^0)\phi_0(\vec{y}, t^0) - \phi_0^\dagger(\vec{y}, t^0)\pi_0^\dagger(\vec{y}, t^0)]$$

with the property

(7.13b) $$[\mathbf{\Gamma}', \phi_0(\vec{x}, t^0)] = \phi_0(\vec{x}, t^0).$$

Then, eq. (7.11) yields (note that $\mathbf{Q}^+ = \mathbf{Q}^- = \phi_0(\vec{x}, t^0)$ in this case)

(7.13c) $$\mathbf{P}''_{\pm}\phi_0(\vec{x}, t^0)\mathbf{P}''^{-1}_{\pm} = \mp i\phi_0(\vec{x}, t^0),$$

where

(7.13d) $$\mathbf{P}''_{\pm} = e^{\mp i\pi\Gamma'/2}.$$

Therefore the transformations eq. (7.5) at $t = t^0$ are given by

(7.14) $$\mathbf{P}_{\pm} = \mathbf{P}''_{\pm}\mathbf{P}',$$

with $\mathbf{P}'$ defined by eqs. (7.12a) and (7.12d).

Since the commutators of $\mathbf{\Gamma}$ and $\mathbf{\Gamma}'$ with $\pi_0(\vec{x}, t^0)$ equal $-\pi_0(-\vec{x}, t^0)$ and $-\pi_0(\vec{x}, t^0)$, respectively (compare eqs. 7.12b and 7.13b), it is also found that

(7.15a) $$\mathbf{P}_{\pm}\pi_0(\vec{x}, t^0)\mathbf{P}_{\pm}^{-1} = \pm\pi_0(-\vec{x}, t^0)$$

or, from the definition eq. (7.4) of π,

(7.15b) $$\mathbf{P}_{\pm}\phi_0(\vec{x}, t^0)\mathbf{P}_{\pm}^{-1} = \pm\phi_0(-\vec{x}, t^0).$$

To determine the transformation of $\phi_0(x)$ for arbitrary $t \neq t^0$, we note that the time dependence of $\mathbf{P}_{\pm}\phi_0(x)\mathbf{P}_{\pm}^{-1}$ is given by the same second-order differential equation as applies to $\phi_0(x)$, namely, eq. (6.5), where t^0 is held fixed in the definition of $\mathbf{P}_\pm$. Furthermore, $\mathbf{P}_\pm\phi_0(x)\mathbf{P}_\pm^{-1}$ satisfies the initial conditions eqs. (7.5) and (7.15b) at $t = t^0$. Therefore eq. (7.5) is satisfied for all values of t.

Eq. (7.14) defines unitary transformations of the GBS associated with $\phi_0(x)$. Specifically, if the basis states of the GBS are the free particle states generated by the particular free fields $\phi_0(x)$ and $\pi_0(x)$, application of the transformation to the states is straightforward. Therefore the transformation owing to the space inversion P of the matrix elements of any operator may be obtained by applying the unitary transformation $\mathbf{P}$ made up of $\mathbf{P}_\pm$ combined with the corresponding transformation of other fields that interact with $\phi(x)$. In particular, for the interacting field operators,

(7.16a) $$\phi^P(x) = \mathbf{P}\phi(x)\mathbf{P}^{-1}$$

(7.16b) $$\pi^P(x) = \mathbf{P}\pi(x)\mathbf{P}^{-1}.$$

To determine the effect of the interactions on this transformation, we may again use eqs. (6.10) for the field operators, where $j(x)$ is an operator function of the field operators, because the form of the equation of motion is identical with that for the classical fields. Therefore

(7.17a) $$\phi^P(x) = \phi_0^P(x) + \int d^4y\Delta_0(x, y)j^P(y),$$

where

(7.17b) $$\phi_0^P(x) = \mathbf{P}\phi_0(x)\mathbf{P}^{-1}$$

and

(7.17c) $$j^P(x) = \mathbf{P}j(x)\mathbf{P}^{-1}.$$

Given the form of $j(x)$ in terms of the relevant fields, $j_0(x)$ is obtained by substitution of the appropriate free fields, and the transformation of $j_0(x)$ under P is given by

(7.18) $$j_0^P(x) = \mathbf{P}j_0(x)\mathbf{P}^{-1},$$

where $\mathbf{P}$ is the appropriate unitary transformation for all the relevant fields. Thus, as in the case of classical fields, a knowledge of the form of $j_0(x)$ in terms of the free fields is sufficient to determine the transformation of $\phi(x)$, given by eq. (7.17a). If the interactions are P invariant,

(7.19) $$\mathbf{P}j_0(\vec{x}, t)\mathbf{P}^{-1} = \pm j_0(-\vec{x}, t),$$

and

(7.20)
$$\phi^P(\vec{x}, t) = \pm\phi(-\vec{x}, t)$$

for scalar and pseudoscalar fields, respectively.

This treatment of the transformation **P** of the field operators serves as a paradigm of other improper transformations. Thus the key to determining the effect of T is to seek a transformation **T** of the GBS that yields a transformation of the kinematic variables (the free field operators) that is the same as that expected of the classical free fields.

The kinematic requirement that the equal-time commutation relations eqs. (7.3) be invariant under the transformation **T** leads to the result that **T** is antiunitary because we require that eq. (7.4) defining the canonically conjugate field $\pi(x)$ apply also to the transformed fields. That is, if[2]

(7.21)
$$\phi'(x') = \mathbf{T}\phi(x)\mathbf{T}^{-1}$$

and

(7.22)
$$\pi'(x') = \frac{\partial}{\partial t'}\,\phi'^{\dagger}(x'),$$

substitution of $t' = -t$ yields

(7.23)
$$\begin{aligned}\pi'(x') &= -\frac{\partial}{\partial t}\,\mathbf{T}\phi^{\dagger}(x)\mathbf{T}^{-1}\\ &= -\mathbf{T}\pi(x)\mathbf{T}^{-1}.\end{aligned}$$

Therefore

(7.24)
$$\mathbf{T}[\pi(\vec{x}, t), \phi(\vec{y}, t)]\mathbf{T}^{-1} = -[\pi'(\vec{x}, -t), \phi'(\vec{y}, -t)]$$

or, according to eq. (7.3a),

(7.25)
$$-\mathbf{T}i\delta(\vec{x} - \vec{y})\mathbf{T}^{-1} = i\delta(\vec{x} - \vec{y})$$

if ϕ' and π' are to satisfy the same commutation relations as ϕ and π. It follows that **T** is antiunitary,

(7.26)
$$\mathbf{T} = \mathbf{U}K = K\mathbf{U}^*,$$

where $\mathbf{U}^*$, the conjugate complex of **U**, is unitary since **U** is unitary. This result is in accord with the kinematic requirement of invariance under T of

[2] Note the difference in the way the argument of $\phi(x)$ is treated in eq. (7.21) as compared with eq. (7.5). The reason is that replacing $\vec{x}$ by $-\vec{x}$ is a linear transformation on ϕ while replacing t by $-t$ requires the nonlinear operator K, and it must be built into **T** as in eq. (7.26). Compare eq. (7.81).

the equation

(7.27a) $$i\frac{\partial|\mathbf{\Psi}_0(t)\rangle}{\partial t} = \mathbf{H}_0|\mathbf{\Psi}_0(t)\rangle,$$

where $\mathbf{H}_0$ is the *free* field Hamiltonian for which

(7.27b) $$\mathbf{T}\mathbf{H}_0\mathbf{T}^{-1} = \mathbf{H}_0.$$

The argument is identical with that based on eq. (3.21).

The determination of $\mathbf{U}$ again is effected on the basis of the kinematic requirements on the free fields ϕ_0, eqs. (6.13a) and (6.13d). Thus for the scalar field operator

(7.28a) $$\phi_0'(x) = \mathbf{T}\phi_0(x')\mathbf{T}^{-1} = K\phi_0(x')K^{-1},$$

(7.28b) $$\pi_0'(x) = -\mathbf{T}\pi_0(x')\mathbf{T}^{-1} = -K\pi_0(x')K^{-1},$$

and for the pseudoscalar field operator

(7.28c) $$\phi_0'(x) = \mathbf{T}\phi_0(x')\mathbf{T}^{-1} = -K\phi_0(x')K^{-1},$$

(7.28d) $$\pi_0'(x) = -\mathbf{T}\pi_0(x')\mathbf{T}^{-1} = K\pi_0(x')K^{-1}.$$

The equation for the scalar case is satisfied by[3]

(7.29a) $$\mathbf{U}_+ = 1.$$

The determination of $\mathbf{U}$ for the pseudoscalar case makes use of the method developed to construct $\mathbf{P}_\pm$. By direct application of eq. (7.11) with $\mathbf{Q}_+ = \mathbf{Q}_- = \phi_0(\vec{x}, t^0)$ and $\theta = \pi$, we find

(7.29b) $$\mathbf{U}_-^* = \exp\left\{-\pi\int d^3y[\pi_0(\vec{y}, t^0)\phi_0(\vec{y}, t^0) - \phi_0^\dagger(\vec{y}, t^0)\pi_0^\dagger(\vec{y}, t^0)]\right\}.$$

Thus

(7.30a) $$\mathbf{U}_\pm^*\phi_0(\vec{x}, t^0)\mathbf{U}_\pm^{*-1} = \pm\phi_0(\vec{x}, t^0),$$

and

(7.30b) $$\mathbf{U}_\pm^*\pi_0(\vec{x}, t^0)\mathbf{U}_\pm^{*-1} = \pm\pi_0(\vec{x}, t^0)$$

for the scalar and pseudoscalar cases, respectively, and we may again use

[3] That the operator K does indeed change $x = (\vec{x}, t)$ into $x' = (\vec{x}, -t)$ in $\phi_0(x)$ can be shown directly by expanding $\phi_0(x)$ in plane waves with coefficients that are the creation and annihilation operators $a^\dagger(\vec{k})$ and $a(\vec{k})$. See eq. (7.53a) and Bjorken and Drell (1965, p. 39). Since all matrix elements of $a(\vec{k})$ are real, the only effects of K are to change the plane wave factor e^{-ik_0t} into e^{ik_0t} and to change the sign of the associated particle momentum, $\vec{k}$. Since $\vec{k}$ is the variable of integration, its change of sign has no effect on $\phi(x)$.

these equations as initial conditions at $t = t^0$ for the second-order differential equation satisfied by $\mathbf{U}^*\phi_0(x)\mathbf{U}^{*-1}$, eq. (6.5), to show that they hold for all t. It follows that $\mathbf{T} = \mathbf{U}K$, with $\mathbf{U}$ given by eq. (7.29a) or eq. (7.29b), provides the desired transformations, eqs. (7.28), of $\phi_0(x)$ for all values of t.

The effect of time reversal on the interacting field operator $\phi(x)$ is now given by eq. (7.21). Applying this transformation to eq. (6.10a) leads to a relationship having the same form as eq. (6.19) with

(7.31) $$j'(x') = \mathbf{T}j(x)\mathbf{T}^{-1}.$$

Here $\mathbf{T}$ must include, in addition to the factor $\mathbf{U}$ given above, the appropriate unitary factor for any other field operators $\psi(x)$ that appear in $j(x)$. Thus the results of the transformation take on the same appearance as they do for the classical fields. In particular, transformations having the form of eqs. (6.13a) and (6.13d) apply to the interacting field operators for T-invariant systems.

7.3 Quantized Dirac Spinor Field: Transformation under *P* and *T*

The kinematic conditions for spinor field operators are given by equal-time anticommutation relations. The notation used for writing these relations requires some care, because the operator ψ has four spinor components on which the gamma matrices operate, and each component is itself a field operator (an infinite matrix). We shall continue to write the four field operators $\psi_\alpha(x)$ as a column matrix of operators. The tilde will be used to denote the corresponding four-component row matrix of the *same* operators:

(7.32a) $$\tilde{\psi}(x) = [\psi_1(x), \psi_2(x), \psi_3(x), \psi_4(x)].$$

On the other hand, the dagger denotes the Hermitian conjugate over all, which is the row matrix of operators:

(7.32b) $$\psi^\dagger(x) = [\psi_1^\dagger(x), \psi_2^\dagger(x), \psi_3^\dagger(x), \psi_4^\dagger(x)].$$

The anticommutator of two operators A and B is denoted by

(7.33) $$\{A, B\} = AB + BA,$$

and the equal-time anticommutator relations for spinor components are[4]

(7.34a) $$\{\psi_\alpha(\vec{x}, t), \psi_\beta(\vec{y}, t)\} = 0$$

and

(7.34b) $$\{\psi_\alpha(\vec{x}, t), \psi_\beta^\dagger(\vec{y}, t)\} = \delta_{\alpha\beta}\,\delta(\vec{x} - \vec{y}).$$

Note that $i\psi_\alpha^\dagger(x)$ plays the role of the conjugate spinor field $\pi_\alpha(x)$. These

[4] See Bjorken and Drell (1965, p. 52).

kinematic conditions are invariant under any unitary *or antiunitary* transformation of the GBS associated with the operators $\psi(x)$.

Again the kinematically admissible transformations of the GBS must be defined in terms of free field operators $\psi_0(x)$ satisfying the same equation as the classical fields, eq. (6.22). The unitary transformation associated with P is again denoted by **P**, and the corresponding transformation of the field operator is constrained to be the same as that of the classical fields, eq. (6.31e). We take $\psi_0(x)$ to be the kinematic field variable and find

(7.35) $$\psi_0^P(\vec{x}, t) = \mathbf{P}\psi_0(\vec{x}, t)\mathbf{P}^{-1} = \gamma_4 \psi_0(-\vec{x}, t),$$

where we have arbitrarily chosen the positive sign in eq. (6.31e), thereby assigning positive "intrinsic" parity to the particles generated by ψ.

The construction of **P** may be carried out by means of eqs. (7.10) and (7.11). However, in this case the $\mathbf{Q}^{\pm}$ are spinor field operators, and the transformation must introduce the factor γ_4. Since we shall find the results to be useful later in constructing other transformations, let us seek a $\mathbf{\Gamma}$ satisfying eq. (7.10) where

(7.36a) $$\mathbf{Q}^+ = \psi_0(\vec{x}, t^0), \quad \mathbf{Q}^- = \gamma\psi_0(-\vec{x}, t^0),$$

and γ is any 4×4 constant Hermitian matrix satisfying

(7.36b) $$\gamma^2 = 1.$$

Consider the commutator of $\psi_{0\beta}(\vec{x}, t)$ with the inner product

(7.37a) $$\psi_0^\dagger(\vec{y}, t)\psi_0(\vec{z}, t) = \sum_\alpha \psi_{0\alpha}^\dagger(\vec{y}, t)\psi_{0\alpha}(\vec{z}, t).$$

From eqs. (7.34) we find

(7.37b) $$\begin{aligned}[\psi_0^\dagger(\vec{y}, t)\psi_0(\vec{z}, t), \psi_{0\beta}(\vec{x}, t)] &= -\sum_\alpha \{\psi_{0\alpha}^\dagger(\vec{y}, t), \psi_{0\beta}(\vec{x}, t)\}\psi_{0\alpha}(\vec{z}, t) \\ &= -\delta(\vec{x} - \vec{y})\psi_{0\beta}(\vec{z}, t).\end{aligned}$$

Therefore

(7.38) $$\mathbf{\Gamma} = -\int d^3y\psi_0^\dagger(\vec{y}, t^0)\gamma\psi_0(-\vec{y}, t^0)$$

is a Hermitian operator satisfying the required condition.

From eq. (7.11) it then follows that

(7.39a) $$\mathbf{P}\psi_0(\vec{x}, t^0)\mathbf{P}^{-1} = \gamma_4 \psi_0(-\vec{x}, t^0)$$

with

(7.39b) $$\mathbf{P} = e^{i\pi\Gamma'/2}e^{i\pi\Gamma/2},$$

where $\mathbf{\Gamma}$ is given by eq. (7.38), $\mathbf{\Gamma}'$ by

(7.39c)
$$\mathbf{\Gamma}' = \int d^3y \psi_0^\dagger(\vec{y}, t^0)\psi_0(\vec{y}, t^0),$$

and

(7.39d)
$$\gamma = \gamma_4 .$$

In this case the time dependence of $\psi_0(x)$ is given by the first-order eq. (6.22) and $\psi_0^P(x)$, given by eq. (7.35) (using eq. 7.39b with t^0 held fixed in $\mathbf{\Gamma}$ and $\mathbf{\Gamma}'$), satisfies the same differential equation with the initial condition at $t = t^0$ given by eq. (7.39a). Therefore eq. (7.39b) determines $\mathbf{P}$ in eq. (7.35) for all values of t.

Again, in order to determine the transformation of the interacting fields $\psi(x)$, we identify $\psi_0(x)$ to be the free field having $\tau = 0$ and appearing in eq. (6.33a), which may still be taken as the integral equation for the field operator $\psi(x)$. Then the behavior of $\psi(x)$ under P is given by

(7.40)
$$\psi^P(x) = \mathbf{P}\psi(x)\mathbf{P}^{-1},$$

where $\mathbf{P}$ includes the transformations of all the relevant fields, and it is governed by eq. (7.35) for $\psi_0(x)$ and by the way

(7.41)
$$f^P(x) = \mathbf{P}f(x)\mathbf{P}^{-1}$$

is related to $f(x)$. As in the case of the classical spinor field, the behavior of the source function $f_0(x)$, made up of free field operators, may be used to determine, through iteration, the behavior of $\psi(x)$. If $f_0(x)$ transforms in accordance with eq. (7.35), $\psi(x)$ will also transform in this way, and the corresponding system is P invariant.

We turn now to the construction of $\mathbf{T}$ for the spinor field operators. Although the anticommutation relations are invariant under *either* a unitary or an antiunitary transformation, we know that

(7.42a)
$$\mathbf{T} = \mathbf{U}K$$

because of the requirement eq. (7.27). It will again be useful to rewrite eq. (7.42a) as

(7.42b)
$$\mathbf{T} = K\mathbf{U}^*.$$

The determination of $\mathbf{U}$ follows the established pattern based on the behavior of the kinematic variables $\psi_0(x)$. These field operators are constrained to transform like the classical fields, that is, in accordance with eq. (6.47), which we rewrite in the form

(7.43)
$$\mathbf{T}\psi_0(x')\mathbf{T}^{-1} = \psi_0'(x) = K\sigma_2^*\psi_0(x')K^{-1}.$$

Since σ_2^* satisfies the condition eq. (7.36b), we may use (compare eq. 7.38)

(7.44a) $$\mathbf{\Gamma}^* = -\int d^3y \psi_0^\dagger(\vec{y}, t^0)\sigma_2^* \psi_0(\vec{y}, t^0)$$

to construct

(7.44b) $$\mathbf{U}^* = e^{-i\pi\mathbf{\Gamma}^*/2} e^{-i\pi\mathbf{\Gamma}'^*/2},$$

where $\mathbf{\Gamma}'$ is given by eq. (7.39c). Then

(7.45a) $$\mathbf{U}^*\psi_0(\vec{x}, t^0)\mathbf{U}^{*-1} = \sigma_2^* \psi_0(\vec{x}, t^0)$$

and

(7.45b) $$\mathbf{T}\psi_0(\vec{x}, t^0)\mathbf{T}^{-1} = \sigma_2 K\psi_0(\vec{x}, t^0)K^{-1},$$

with **T** given by eq. (7.42a), in agreement with the requirement of eq. (6.45).

Again the argument can be made on the basis of the first-order time dependence of the Dirac equation that, having satisfied this condition at $t = t^0$, it is valid for $\psi_0(\vec{x}, t)$ at all t. Therefore this transformation **T** may be applied to the interacting field operator $\psi(x)$ to obtain its time reversal properties. The result takes exactly the same form as the classical result eq. (6.46c), where now the transformation of the source function (operator) is given by

(7.46) $$f'(x') = \mathbf{T}f(x)\mathbf{T}^{-1}.$$

Thus the discussion of the behavior of the classical field at the end of section 6.3 applies just as well to the case of the quantized field.

7.4 Charge Conjugation

The improper transformation that replaces the field operator associated with a particle by the field operator associated with the antiparticle is called "charge conjugation" and denoted generically by C. The associated unitary transformation of the GBS, denoted by **C**, is an involutional operator as defined in chapter 3, since, when **C** is applied to a state and then repeated, the original state is restored. Therefore it must satisfy eq. (3.25),

(7.47) $$\mathbf{C}^2 = \eta_C' \mathbf{1},$$

where η_C' is a constant phase factor that will later be found to be equal to one.

To determine **C** for a given type of field, we consider the equation of motion of the field operator in the presence of an *external* electromagnetic field. This external field will be described by the four-vector potential $A_\mu(x)$, which is a classical field fixed by external conditions that are *not* subject to the transformation of interest. Since the classical Hamiltonian equations of motion for a charged particle in the presence of an electromagnetic field are obtained by making the substitution

(7.48a) $$p_\mu \rightarrow p_\mu - eA_\mu(x)$$

in the classical Hamiltonian (defined in the absence of the electromagnetic field), the equations of motion of the field operator are obtained by means of the substitution[5]

(7.48b) $$\frac{\partial}{\partial x_\mu} \rightarrow \frac{\partial}{\partial x_\mu} - ieA_\mu(x)$$

in eq. (6.9a) for tensor field operators and in eq. (6.32) for Dirac spinor field operators.

The kinematically admissible transformation **C** must be, as usual, independent of the internal dynamics of the fields. Thus the determination of **C** is to be based on consideration of the effect on the free field equations, namely eqs. (6.5) and (6.22), of the substitution eq. (7.48b).

In the case of the scalar (or pseudoscalar) field, this substitution yields the equations of motion

(7.49a) $$\Sigma_\mu \left[\frac{\partial}{\partial x_\mu} - ieA_\mu(x)\right]^2 \phi_A(x) = m^2 \phi_A(x),$$

where the field operator $\phi_A(x)$ becomes a free field operator when $A_\mu = 0$. The Hermitian conjugate of eq. (7.49a) is

(7.49b) $$\Sigma_\mu \left[\frac{\partial}{\partial x_\mu} + ieA_\mu(x)\right]^2 \phi_A^\dagger(x) = m^2 \phi_A^\dagger(x).$$

Therefore the field operator $\phi_A^\dagger(x)$ describes particles whose electric charge has the opposite sign from that of the particles described by $\phi_A(x)$. If the latter are called "particles," the former are their associated "antiparticles."

The external electromagnetic field was required here only to make this identification. By setting $A_\mu = 0$, we see that when the free field operator $\phi_0(x)$ is associated with the free particles, the operator $\phi_0^\dagger(x)$ is associated with the corresponding free antiparticles. These associations can be confirmed by expanding the free fields in terms of creation and annihilation operators for one-particle states and showing that $\phi_0(x)$ annihilates particles and creates antiparticles, while $\phi_0^\dagger(x)$ does the converse.[6] Thus we

[5] See Bjorken and Drell (1964, p. 10).
[6] See note 1.

define charge conjugation C as the transformation

(7.50a) $$C\colon\ \phi_0(x) \to \phi_0^C(x) = \eta_C\, \phi_0^\dagger(x)$$

and, in accordance with eq. (7.4),

(7.50b) $$C\colon\ \pi_0(x) \to \pi_0^C(x) = \eta_C^*\, \pi_0^\dagger(x),$$

where η_C is a constant phase factor unrelated to η_C'. The operator ϕ_0^C is called the "charge conjugate field," and π_0^C is the canonically conjugate field defined by eq. (7.4). From these definitions it follows that $\phi_0^C(x)$ and $\pi_0^C(x)$ satisfy the equal-time commutation relations eqs. (7.3).

This transformation of the field operator to its Hermitian conjugate should not be confused with the nonlinear transformation K associated with T. In fact, the relationship between the field operators ϕ_0^C and ϕ_0 is linear, as can be confirmed by writing $\phi_0(x)$ in terms of two independent Hermitian field operators ϕ_{0R} and ϕ_{0I}:

(7.51) $$\phi_0(x) = \phi_{0R}(x) + i\phi_{0I}(x),$$

where ϕ_{0R} and ϕ_{0I} are the quantized versions of the real and imaginary parts of the classical fields. In terms of these Hermitian fields eq. (7.50a) reads

(7.52a) $$\begin{aligned} \phi_{0R}^C &= \cos\chi\phi_{0R} + \sin\chi\phi_{0I}, \\ \phi_{0I}^C &= \sin\chi\phi_{0R} - \cos\chi\phi_{0I}, \end{aligned}$$

with χ defined by

(7.52b) $$\eta_C = e^{i\chi}$$

Eq. (7.52a) is clearly a linear relationship between field operators that can be generated by a unitary transformation **C** of the GBS.

The determination of **C** is best made by expressing the free field operator $\phi_0(x)$ and $\pi_0(x)$ in terms of the creation and annihilation operators $a^\dagger(k)$ and $a(k)$ of particles and $\bar{a}^\dagger(k)$ and $\bar{a}(k)$ of antiparticles of four-momentum k, for example,

(7.53a) $$\phi_0(x) = (2\pi)^{-3/2} \int d^3k(2k_0)^{-1/2}[a(\vec{k})e^{i\Sigma_\mu k_\mu x_\mu} + \bar{a}^\dagger(\vec{k})e^{-i\Sigma_\mu k_\mu x_\mu}].$$

The operators satisfy the commutation relations[7]

(7.53b) $$[a(\vec{k}), a^\dagger(\vec{k}')] = [\bar{a}(\vec{k}), \bar{a}^\dagger(\vec{k}')] = \delta(\vec{k} - \vec{k}'),$$

[7] See note 3. Formulating **C** in terms of $\phi_\tau(x)$ and $\pi_\tau(x)$ would involve the use of a Γ containing the complicated operator $[\nabla^2 - m^2]^{1/2}$ and application of the equations of motion. The creation and annihilation operators provide a much more direct approach to our objective, which is simply to show that the unitary transformation **C** can be constructed in terms of the free field operators.

and all other commutators vanish. The definition of **C** is then

(7.54) $$\mathbf{C}a(\vec{k})\mathbf{C}^{-1} = \eta_C\bar{a}(\vec{k}),\ \mathbf{C}\bar{a}(\vec{k})\mathbf{C}^{-1} = \eta_C^{-1}a(\vec{k}),$$

the second equation following from eq. (7.47). Because η_C is a phase factor, $\eta_C^{-1} = \eta_C^*$.

Eq. (7.11) may be exploited again to show that

(7.55a) $$\mathbf{C} = e^{i\pi\Gamma'_C/2}e^{i\pi\Gamma_C/2},$$

where

(7.55b) $$\Gamma_C = \int d^3k'[\eta_C a^\dagger(\vec{k}')\bar{a}(\vec{k}') + \eta_C^* \bar{a}^\dagger(\vec{k}')a(\vec{k}')]$$

and

(7.55c) $$\Gamma'_C = -\int d^3k'[a^\dagger(\vec{k}')a(\vec{k}') + \bar{a}^\dagger(\vec{k}')\bar{a}(\vec{k}')].$$

The phase factor η'_C appearing in eq. (7.47) can be determined by considering charge conjugation of a state of N free particles:

(7.56) $$|N \text{ particles}\rangle \sim a^\dagger(\vec{k}_1)a^\dagger(\vec{k}_2)\ldots a^\dagger(\vec{k}_N)|0\rangle,$$

where $|0\rangle$ is the vacuum state that is assumed to satisfy

(7.57a) $$\mathbf{C}|0\rangle = |0\rangle.$$

Then

(7.57b) $$\mathbf{C}|N \text{ particles}\rangle = \eta_C^{*N}|N \text{ antiparticles}\rangle$$

and

(7.57c) $$\mathbf{C}^2|N \text{ particles}\rangle = \eta_C^{*-N}\eta_C^{*N}|N \text{ particles}\rangle$$

for every state of the system. Therefore

(7.58a) $$\mathbf{C}^2 = \mathbf{1},$$

and the phase factor η'_C in eq. (7.47) is found to be

(7.58b) $$\eta'_C = 1$$

for the scalar and pseudoscalar fields.

The phase factor η_C is governed by the choice of relative phase between the sector of Hilbert space associated with particles and that associated with antiparticles. This phase is arbitrary because the commutation relations eq. (7.53b) do not determine the relative phases of the operators $a(\vec{k})$

and $\bar{a}(\vec{k})$. The usual convention is to take $\eta_C = \pm 1$, for example, with the choice $\eta_C = +1$ applying to pions and $\eta_C = -1$ to K^0, $\bar{K}^0$ systems. However, it will be convenient in this chapter and the next to take $\eta_C = 1$ for scalar and pseudoscalar fields, a choice that corresponds to taking a unique η_C (in eq. 7.67) for all spinor fields out of which the tensor fields are composed. This connection will become evident in chapter 8.

The behavior of the other tensor fields under C may be obtained in the same way, because the transformation does not act on the tensor indices. Care must be taken, however, to keep in mind that there is a factor i associated with every tensor index $\mu = 4$, and this factor is not to be conjugated under C, as is evident from the procedure leading from eq. (7.49a) to eq. (7.49b). Therefore the generalization of the transformation eq. (7.50a) to the general tensor takes the form

(7.59a) $$T^C_{\mu\nu\ldots\eta,\tau=0}(x) = \eta_C\,\varepsilon_\mu\,\varepsilon_\nu\,\ldots\,\varepsilon_\eta\,T^\dagger_{\mu\nu\ldots\eta,\tau=0}\,,$$

with ε_μ given by eq. (6.3).

The choice of phase factor η_C may be made independently for tensors of different rank. However, as we have already remarked for scalar fields, the phase of each will be found (in chap. 8) to be specified when the corresponding factor for spinor fields is chosen to be the same for each spinor field out of which the tensor field is composed. In the case of the vector field the resulting value of the phase can be anticipated by making a physical argument based on the behavior of the vector field $A_\mu(x)$ of electrodynamics. Note that we are dealing here with the quantized internal electromagnetic field of the total system that is being subjected to C and *not* with the external field used in eq. (7.49a) to define what is meant by C.

Since $A_\mu(x)$ is generated by electric charges that change sign under C, it also must change sign. On this basis we arrive at the value $\eta_C = -1$ for vector fields or

(7.59b) $$\phi^C_{\mu,0}(x) = -\varepsilon_\mu\,\phi^\dagger_{\mu,0}(x).$$

It follows that the electromagnetic field strength tensors $\phi_{\mu\nu}$ and therefore the second-rank antisymmetric tensors in general have $\eta_C = -1$ or

(7.59c) $$\phi^C_{\mu\nu,0}(x) = -\varepsilon_\mu\,\varepsilon_\nu\,\phi^\dagger_{\mu\nu,0}(x),$$

which is again consistent with the results obtained from the spinor representation of the second-rank tensors. For the pseudovector fields

(7.59d) $$\chi^C_{\mu,0}(x) = \varepsilon_\mu\,\chi^\dagger_{\mu,0}(x)$$

has the phase that will be found to be consistent with the spinor representation.

We turn now to the determination of **C** for the free spinor field $\psi_0(x)$, which is defined by

$$C:\quad \psi_0(x) \to \psi_0^C(x) = \mathbf{C}\psi_0(x)\mathbf{C}^{-1}. \tag{7.60}$$

The definition of $\psi_0^C(x)$ is obtained again by starting from the equation for the field describing the "free" charged particle in the presence of an external electromagnetic field. This equation is obtained from the equation for the free field, eq. (6.22), by means of the substitution eq. (7.48b):

$$\sum_\mu \gamma_\mu\left[\frac{\partial}{\partial x_\mu} - ieA_\mu(x)\right]\psi_A(x) + m\psi_A(x) = 0. \tag{7.61a}$$

When we remember that $\vec{A}(x)$ is real and $A_4(x)$ is imaginary, the Hermitian conjugate of eq. (7.61a) (see eq. 7.32b) is found to be

$$\sum_\mu \varepsilon_\mu\left[\frac{\partial}{\partial x_\mu} + ieA_\mu(x)\right]\psi_A^\dagger(x)\gamma_\mu + m\psi_A^\dagger(x) = 0. \tag{7.61b}$$

The factor ε_μ can be eliminated by multiplying from the right-hand side by γ_4 and using the anticommutation of the γ_μ to obtain:

$$\sum_\mu \left[\frac{\partial}{\partial x_\mu} + ieA_\mu(x)\right]\bar{\psi}_A(x)\gamma_\mu - m\bar{\psi}_A(x) = 0, \tag{7.61c}$$

where

$$\bar{\psi}_A(x) = \psi_A^\dagger(x)\gamma_4 \tag{7.62}$$

is a row matrix of four field operators (the four spinor components).

The transformed spinor $\psi_A^C(x)$ of eq. (7.60) is a column matrix that should satisfy (7.61a) except that the sign of the electric charge is reversed:

$$\sum_\mu \gamma_\mu\left[\frac{\partial}{\partial x_\mu} + ieA_\mu(x)\right]\psi_A^C(x) + m\psi_A^C(x) = 0, \tag{7.63}$$

and the elements of ψ_A^C can be written as linear combinations of the elements of $\bar{\psi}_A$ or, in matrix notation (see eq. 7.32a),

$$\psi_A^C(x) = \gamma_C\tilde{\bar{\psi}}_A(x) = \gamma_C\tilde{\gamma}_4\tilde{\psi}_A^\dagger, \tag{7.64}$$

where γ_C is a 4×4 matrix determined by the condition that eq. (7.63) follow from eq. (7.61c). This condition is obtained by transposing the four-component spinor equation (7.61c):

(7.65) $$\sum_\mu \tilde{\gamma}_\mu\left[\frac{\partial}{\partial x_\mu} + ieA_\mu(x)\right]\tilde{\bar{\psi}}_A(x) - m\tilde{\bar{\psi}}_A(x) = 0,$$

which becomes eq. (7.63) if

(7.66a) $$\gamma_C^{-1}\gamma_\mu\gamma_C = -\tilde{\gamma}_\mu.$$

Since $\tilde{\gamma}_\mu = \gamma_\mu^*$, eq. (6.41) may be used to write

(7.66b) $$\gamma_C\gamma_1 - \gamma_1\gamma_C = \gamma_C\gamma_2 + \gamma_2\gamma_C = \gamma_C\gamma_3 - \gamma_3\gamma_C = \gamma_C\gamma_4 + \gamma_4\gamma_C = 0$$

in place of eq. (7.66a), and the solution of this set of commutation and anticommutation relations is

(7.66c) $$\gamma_C = \eta_C\gamma_2\gamma_4,$$

where η_C is a phase factor.

The transformation **C** of the GBS satisfying the condition imposed by eq. (7.64) with $A_\mu = 0$,

(7.67) $$\mathbf{C}\psi_0(x)\mathbf{C}^{-1} = \eta_C\gamma_2\tilde{\bar{\psi}}_0^\dagger,$$

is, as in the case of tensor fields, most easily formulated in terms of the creation and annihilation operators for the spin 1/2 particles. For this purpose it is necessary to be specific about the way the phases of the creation and annihilation operators are defined. Our definitions are based on the following expansion of the free fields $\psi_0(x)$ in terms of the spinor plane wave solutions of eq. (6.22):

(7.68a) $$\psi_0(x) = (2\pi)^{-3/2}\sum_s \int d^3k k_0^{-1/2}[b(\vec{k}, s)u_s(\vec{k})e^{i\Sigma_\mu k_\mu x_\mu} + \bar{b}^\dagger(\vec{k}, s)v_s(\vec{k})e^{-i\Sigma_\mu k_\mu x_\mu}],$$

where

(7.68b) $$\left(\sum_\mu \gamma_\mu k_\mu - im\right)u_s(\vec{k}) = 0$$

and

(7.68c) $$k_0 = [m^2 + (\vec{k}\cdot\vec{k})]^{1/2}.$$

The classical spinor fields $v_s(\vec{k})$ are defined by[8]

(7.69) $$v_s(\vec{k}) = \gamma_2\tilde{u}_s^\dagger(\vec{k}),$$

[8] Compare Bjorken and Drell (1964, p. 69) and (1965, pp. 116 ff.), where a different definition is used. I prefer the definition eq. (7.69) because it avoids the burden of carrying an extra phase throughout the analysis. Note that the index $s = \pm 1$ is taken here to be *twice* the spin magnetic quantum number.

so that eq. (7.67) becomes

(7.70) $$\mathbf{C}b(\vec{k}, s)\mathbf{C}^{-1} = \eta_C \bar{b}(\vec{k}, s), \quad \mathbf{C}\bar{b}(\vec{k}, s)\mathbf{C}^{-1} = \eta_C^{-1} b(\vec{k}, s).$$

The $b^\dagger(\vec{k}, s)$ are the creation operators for spinor particles, and the $\bar{b}^\dagger(\vec{k}, s)$ are the creation operators for their antiparticles of the same momentum and spin. They satisfy the anticommutation relations that follow from eqs. (7.34):

(7.71a) $$\{b(\vec{k}, s), b^\dagger(\vec{k}', s')\} = \{\bar{b}(\vec{k}, s), \bar{b}^\dagger(\vec{k}', s')\} = \delta_{ss'}\,\delta(\vec{k} - \vec{k}'),$$

(7.71b) $$\{b(\vec{k}, s), b(\vec{k}', s')\} = \{\bar{b}(\vec{k}, s), \bar{b}(\vec{k}', s')\} = 0,$$

if the $u_s(\vec{k})$ are normalized in such a way that

(7.71c) $$\bar{u}_s(\vec{k})\gamma_\mu u_{s'}(\vec{k}) = -ik_\mu \delta_{ss'}$$

with

(7.71d) $$\bar{u}_s(\vec{k}) = u_s^\dagger(\vec{k})\gamma_4 .$$

From eq. (7.71a) it follows as for eq. (7.37b) that the commutator

(7.72) $$[b^\dagger(\vec{k}', s')\bar{b}(\vec{k}', s'), b(\vec{k}, s)] = -\delta_{ss'}\,\delta(\vec{k} - \vec{k}')\bar{b}(\vec{k}', s').$$

Therefore the Hermitian operator (note $\eta_C^{-1} = \eta_C^*$)

(7.73a) $$\Gamma = \sum_{s'} \int d^3k'[\eta_C b^\dagger(\vec{k}', s')\bar{b}(\vec{k}', s') + \eta_C^{-1}\bar{b}^\dagger(\vec{k}', s')b(\vec{k}', s')]$$

satisfies the conditions

(7.73b) $$[\Gamma, b(\vec{k}, s)] = -\eta_C \bar{b}(\vec{k}, s), \quad [\Gamma, \bar{b}(\vec{k}, s)] = -\eta_C^{-1} b(\vec{k}, s).$$

Also,

(7.73c) $$\Gamma' = -\sum_{s'} \int d^3k'[b^\dagger(\vec{k}', s')b(\vec{k}', s') + \bar{b}^\dagger(\vec{k}', s')\bar{b}(\vec{k}', s')]$$

satisfies

(7.73d) $$[\Gamma', b(\vec{k}, s)] = b(\vec{k}, s), \; [\Gamma', \bar{b}(\vec{k}, s)] = \bar{b}(k, s),$$

so that the unitary transformation eq. (7.67) is given by

(7.73e) $$\mathbf{C} = e^{i\pi\Gamma'/2}\, e^{i\pi\Gamma/2}.$$

The phase factor η_C for the spinor fields depends on the convention that is chosen for the relative phases of the particle and antiparticle states. The usual convention is to take $\eta_C = 1$ for spinors.[9] Irrespective of the choice of this phase, it can be shown by repetition of the argument used for scalar fields leading to eq. (7.58b) that

(7.74) $$\eta'_C = 1$$

also for the spinor fields. This combined with the earlier result establishes the basis for the comment following eq. (7.47) that $\eta'_C = 1$ in general.

Having established the behavior of the free field operators under C, we can follow the procedure used for the other improper transformations of the interacting fields. The transformations of these fields are given by

(7.75a) $$\phi^C(x) = \mathbf{C}\phi(x)\mathbf{C}^{-1},$$

(7.75b) $$\psi^C(x) = \mathbf{C}\psi(x)\mathbf{C}^{-1},$$

and so forth. When the unitary transformations are applied to the integral equations for the field operators, the behavior of the source terms (current density operators) will determine whether the system is invariant under C. This question can be formulated in the manner used to deal with P invariance and T invariance. We invoke iteration of the integral equations of motion for the field operators, such as eqs. (6.10a) and (6.33a), starting from insertion of the free fields, and find that they are invariant if and only if

(7.76a) $$\mathbf{C}j_0(x)\mathbf{C}^{-1} = j_0^\dagger(x),$$

(7.76b) $$\mathbf{C}f_0(x)\mathbf{C}^{-1} = \gamma_2\,\tilde{f}_0^\dagger(x),$$

so that the interacting fields transform in the same way as the free fields. Similar equations are satisfied for the tensors of higher rank.

The claim made in section 6.3 that particle and antiparticle spinor fields have the opposite intrinsic parity remains to be demonstrated. Since it has been assumed that $\psi_0(x)$ describes the field operator associated with a particle, the field operator associated with its antiparticle is

(7.77) $$\psi_0^C = \gamma_2\,\tilde{\psi}_0^\dagger.$$

If the intrinsic parity of the particle is taken to be positive,

(7.78a) $$\psi_0^P(\vec{x}, t) = \gamma_4\,\psi_0(-\vec{x}, t).$$

Thus

(7.78b) $$[\psi_0^C(\vec{x}, t)]^P = \gamma_2\gamma_4\,\tilde{\psi}_0^\dagger(-\vec{x}, t),$$

where use has been made of eq. (6.41) and $\gamma_\mu^* = \tilde{\gamma}_\mu$. By use of this same

[9] However, in order to match the phase conventions that will be adopted in chapters 9 and 10 for meson fields that are composites of quark (spinor) fields, the assignment $\eta_C = -1$ must be made for the s quark and the b quark. Again this is only a matter of convention.

equation and the anticommutation of γ_2 with γ_4 we find

(7.78c) $$\gamma_2 \gamma_4 \tilde{\psi}_0^\dagger(-\vec{x}, t) = -\gamma_4 \psi_0^C(-\vec{x}, t),$$

from which it follows that the intrinsic parity of the antiparticle is opposite that of the particle.

7.5 Summary

It has been established that the kinematically admissible transformation of the fields associated with P, T, and C can be constructed directly in terms of the free field operators. The application of these transformations to the free field operators leads by design to the same form as the transformations of the classical fields. Therefore the transformations can be characterized in the latter terms—that is, by an appropriate signature ($\pm$ sign) for the tensor fields and, in addition for the spinor fields, an appropriate gamma matrix or product of gamma matrices. Furthermore, the operator K is always associated with T, and the field is converted into its Hermitian conjugate by C.

A summary of the expressions for the effects of **C**, **P**, and **T** on the free fields is presented in table 7.1 for the cases that have been considered in this chapter. The entries in the table corresponding to eqs. (7.5) and (7.50) for the scalar and pseudoscalar fields, to eqs. (7.35) and (7.67) for the spinor fields, and to corresponding equations for the other tensor fields require no further explanation. However, it remains to be demonstrated that the application of **T** to these fields as expressed in eqs. (7.28a), (7.28c), and (7.43) can be put into the particularly simple form shown in table 7.1 by an appropriate choice of the representation in which the unitary factor in **T** is expressed. We have already found that, in the simple case of ordinary quantum mechanics, the change in the form of **U** from one representation to another is not given by a unitary transformation. Compare, for example, eqs. (3.39), (3.40), and (3.72). It is this kind of change that is required to convert the original equations of transformation into the more convenient form shown in the table.

Let us consider the application of eq. (7.28a) or (7.28c) to the scalar or pseudoscalar field when it is written in terms of the creation and annihilation operators, eq. (7.53a). We have

(7.79a) $$\begin{aligned}\phi_0'(x') &= \pm K\phi_0(x)K^{-1} \\ &= \pm(2\pi)^{-3/2}\int d^3k(2k_0)^{-1/2} \\ &\quad \cdot [a(\vec{k})e^{-i\Sigma_\mu k_\mu x_\mu} + \bar{a}^\dagger(\vec{k})e^{i\Sigma_\mu k_\mu x_\mu}],\end{aligned}$$

TABLE 7.1 Transformed Forms of Tensor and Spinor Fields for Improper Transformations

Field	$\phi_0(x)$	$\phi_{\mu,0}(x)$	$\phi_{\mu\nu,0}(x)$	$\chi_{\mu,0}(x)$	$\chi_0(x)$[a]	$\psi_0(x)$
P:	$\phi_0(-\vec{x}, t)$	$-\varepsilon_\mu \phi_{\mu,0}(-\vec{x}, t)$	$\varepsilon_\mu \varepsilon_\nu \phi_{\mu\nu,0}(-\vec{x}, t)$	$\varepsilon_\mu \chi_{\mu,0}(-\vec{x}, t)$	$-\chi_0(-\vec{x}, t)$	$\pm\gamma_4 \psi_0(-\vec{x}, t)$[b]
T:	$\phi_0(\vec{x}, -t)$	$-\phi_{\mu,0}(\vec{x}, -t)$	$-\phi_{\mu\nu,0}(\vec{x}, -t)$	$-\chi_{\mu,0}(\vec{x}, -t)$	$-\chi_0(\vec{x}, -t)$	$\sigma_2 \psi_0(\vec{x}, -t)$
C:[c]	$\phi_0^\dagger(x)$	$-\varepsilon_\mu \phi_{\mu,0}^\dagger(x)$	$-\varepsilon_\mu \varepsilon_\nu \phi_{\mu\nu,0}^\dagger(x)$	$\varepsilon_\mu \chi_{\mu,0}^\dagger(x)$	$\chi_0^\dagger(x)$	$\gamma_2 \tilde{\psi}_0^\dagger(x)$

[a] In making reference to the text it should be noted that while ϕ has been used there to designate either the scalar field or the pesudoscalar field, the specific notation, χ, of section 6.1, case V, is used here for the pseudoscalar field to differentiate it clearly from the scalar field ϕ.

[b] The $\pm$ signs are associated with particle and antiparticle fields, respectively.

[c] The choice of the η_C associated with each type of tensor field is based on the convention that all spinor fields are assigned the same η_C. See note 9.

because $a(\vec{k})$ and $\bar{a}(\vec{k})$ have real matrix elements in the usual occupation number representation. By changing the integration variable from $\vec{k}$ to $-\vec{k}$ we find

(7.79b)
$$\phi'_0(x') = \pm(2\pi)^{-3/2}\int d^3k(2k_0)^{-1/2} \cdot [a(-\vec{k})e^{i\Sigma_\mu k_\mu x'_\mu} + \bar{a}^\dagger(-\vec{k})e^{-i\Sigma_\mu k_\mu x'_\mu}],$$

since k_0 is unaffected by this change. Therefore, if the operators $a'(\vec{k})$, $\bar{a}'(\vec{k})$ are defined to be those associated with $\phi'_0(x')$ in the manner of eq. (7.53a), then, in the same occupation number representation,

(7.80a)
$$a'(\vec{k}) = \mathbf{T}a(\vec{k})\mathbf{T}^{-1} = \pm a(-\vec{k}),$$

(7.80b)
$$\bar{a}'(\vec{k}) = \mathbf{T}\bar{a}(\vec{k})\mathbf{T}^{-1} = \pm \bar{a}(-\vec{k}).$$

Thus, if the number of particles having momentum $\vec{k}$ in a certain state is $N(\vec{k})$, the corresponding number in the time-reversed state is $N'(\vec{k}) = N(-\vec{k})$, as would be expected, since the momentum changes sign under T. The required representation of $\mathbf{U}$ in $\mathbf{T} = \mathbf{U}K$ satisfying eqs. (7.80) can be obtained by the method used to obtain eqs. (7.55).[10]

In this representation of $\mathbf{T}$ it follows immediately from eqs. (7.80) that

(7.81)
$$\mathbf{T}\phi_0(x)\mathbf{T}^{-1} = \pm\phi_0(x')$$

for the scalar or pseudoscalar case, respectively. The result applies *only* to the special set of field operators ϕ_0 *unless* all interactions are T invariant, a case that will be discussed further in the next chapter.

Similar results for the spinor field are found by applying eq. (7.43) to eq. (7.68a):

(7.82)
$$\psi'_0(x') = (2\pi)^{-3/2}\sum_s \int d^3k k_0^{-1/2}[b(\vec{k}, s)\sigma_2 u_s^*(\vec{k})e^{-i\Sigma_\mu k_\mu x_\mu} + [\bar{b}^\dagger(\vec{k}, s)\sigma_2 v_s^*(\vec{k})e^{i\Sigma_\mu k_\mu x_\mu}].$$

It is a straightforward matter to show by means of eq. (7.68b) and the anticommutation relations for the gamma matrices that

(7.83a)
$$\sigma_2 u_s^*(\vec{k}) = i^s u_{-s}(-\vec{k})$$

[10] Compare the representations eqs. (7.29) of $\mathbf{U}$ corresponding to matrices associated with the "functional Hilbert space" based on functions of $\vec{x}$. That the matrix $\mathbf{U}$ takes on an entirely different form in the momentum space occupation number representation corresponds to the difference between a configuration space and a momentum space representation of $\mathbf{U}$ in ordinary quantum mechanics, as described in chapter 3. Compare eqs. (3.39) and (3.40).

and

(7.83b) $$\sigma_2 v_s^*(\vec{k}) = i^s v_{-s}(-\vec{k}),$$

where the Wigner phase convention, eq. (3.73), has been assumed for the solutions of eq. (7.68b). Note again[11] that $s = 2m_s$, where $m_s = \pm 1/2$ is the spin magnetic quantum number.

Repetition of the steps followed for the scalar fields leads then to the result

(7.84a) $$b'(\vec{k}, s) = \mathbf{T} b(\vec{k}, s)\mathbf{T}^{-1} = i^s b(-\vec{k}, -s)$$

(7.84b) $$\bar{b}'(\vec{k}, s) = \mathbf{T} \bar{b}(\vec{k}, s)\mathbf{T}^{-1} = i^s \bar{b}(-\vec{k}, -s)$$

in the occupation number representation. Furthermore, it follows from eqs. (7.84) and (7.83) and the antiunitarity of **T** that

(7.85) $$\mathbf{T}\psi_0(x)\mathbf{T}^{-1} = \sigma_2 \psi_0(x'),$$

which is again a special property of $\psi_0(x)$ in this representation.

Eqs. (7.81) and (7.85), and their counterparts for other tensor fields, are those given in table 7.1.

[11] See note 8.

8 | Some Consequences of Lorentz Invariance and of *T* Invariance

Invariance under the proper Lorentz transformations is implicit in our formulation of quantum field theories in terms of covariant tensor and spinor field equations. The *invariant* is the action, the integral over time of the Lagrangian, from which the classical field equations are derived.[1] The Lagrangian density $\mathscr{L}$ may be expressed as the sum of the two terms $\mathscr{L}_0$, the free field Lagrangian, and $\mathscr{L}_I$, the interaction. The invariance of $\mathscr{L}_0$ not only under proper Lorentz transformations but also under P, T, and C has been assumed here as a kinematic constraint.

The questions about P invariance, T invariance, and C invariance that we have been addressing are concerned with the behavior of $\mathscr{L}_I$ (or some term in $\mathscr{L}_I$) under these improper transformations. In this chapter we consider the implications of the various invariance properties for the form of some common simple interaction terms. These simple examples will be used to illustrate Pauli's method of proof of the CPT theorem.[2] Subsequent sections of this chapter will be concerned with some important consequences of CPT invariance and T invariance, including a general discussion of the constraints imposed on the S matrix and their application to such special cases as the electromagnetic and weak interaction form factors.

8.1 Transformations of Lagrangians under *P*, *T*, and *C*: The *CPT* Theorem

Lorentz invariant functions associated with the classical tensor fields are obtained by the contraction of products of tensors, that is, by taking inner

[1] See Bjorken and Drell (1965, chap. 11).

[2] Pauli (1955) proves the theorem by use of the irreducible representations of the homogeneous Lorentz group associated with tensor products of two-component spinor fields (van der Waerden 1932). The Wigner phase convention (see eq. 3.73 and note 8 of chap. 3) for the spinors then establishes a common phase convention under time reversal for the tensor fields and Dirac spinor fields. For a more general proof of the CPT theorem see Jost (1957) and Streater and Wightman (1964).

products of tensors. Invariant functions of the Dirac spinor fields may be obtained in a similar fashion in terms of tensors that are bilinear in the spinor fields. Manifestly invariant forms of both $\mathscr{L}_0$ and $\mathscr{L}_I$ are constructed in this way for the classical fields. Quantization of the fields then leads to operators $\mathscr{L}_0$ and $\mathscr{L}_I$ that are functions of the field operators. However, some care must be taken in replacing products of functions with products of operators, since the order of factors makes a difference in the latter case but not in the former.

The order of factors is determined by the requirement that the vacuum expectation value (expectation value in the vacuum state $|0\rangle$) of the product vanishes—that is, by "normal ordering" the factors. Normal ordering means that annihilation operators appear on the right-hand side of creation operators so that when the product of operators is applied to the vacuum state the result is the null state. (Note also that creation operators operating to the left on $\langle 0|$ produce the null state.)

The annihilation and creation operators associated with the fields $\phi(x)$ and $\psi(x)$ at a given instant $t = \tau$ are identified by separating the free fields $\phi_\tau(x)$ and $\psi_\tau(x)$ into their positive and negative frequency parts, $\phi_\tau^+(x)$, $\psi_\tau^+(x)$ and $\phi_\tau^-(x)$, $\psi_\tau^-(x)$. In the Fourier analysis of the free fields the factors e^{-ik_0t} (positive frequency) and e^{ik_0t} (negative frequency) occur with the annihilation and creation operators, respectively.[3] This is a Lorentz invariant separation, since the Lorentz transformations do not include the crossing of the light cone required to transform k_4 into $-k_4$ when the four-vector k is timelike, as it is in the Fourier expansion in terms of free fields associated with particles on the mass shell.

Each interacting field at a given but arbitrary time $t = \tau$ is then the sum of two terms[4]

(8.1a) $$\phi(\vec{x}, \tau) = \phi^+(\vec{x}, \tau) + \phi^-(\vec{x}, \tau),$$

(8.1b) $$\psi(\vec{x}, \tau) = \psi^+(\vec{x}, \tau) + \psi^-(\vec{x}, \tau),$$

where

(8.1c) $$\phi^\pm(\vec{x}, \tau) = \phi_\tau^\pm(\vec{x}, \tau)$$

and

(8.1d) $$\psi^\pm(\vec{x}, \tau) = \psi_\tau^\pm(\vec{x}, \tau).$$

[3] See Bjorken and Drell (1965, pp. 31, 91–93). Note that my treatment of the interacting fields by use of the positive and negative frequency parts of the free fields ϕ_0 and ψ_0 is simply a shorthand for their treatment by means of Fourier expansion (1965, pp. 91–95).

[4] See note 3.

The terms $\phi^+(x)$ and $\psi^+(x)$ defined in this way contain only annihilation operators, and the terms $\phi^-(x)$ and $\psi^-(x)$ contain only creation operators for particle states defined by the free fields $\phi_\tau(x)$ and $\psi_\tau(x)$ with $\tau = t$.

The normal-ordered product $:AB:$ of two field operators $A(x)$ and $B(y)$ is defined in terms of $A^\pm$ and $B^\pm$, the positive or negative frequency parts of A and B, as

(8.2) $$:AB: = \pm :BA: = A^+B^+ \pm B^-A^+ + A^-B^+ + A^-B^-,$$

where the plus sign applies to fields satisfying commutation relations (tensor fields) and the minus sign to those satisfying anticommutation relations (spinor fields). Normal ordering of products of more than two factors can be carried out in a similar fashion, always taking into account the minus sign that is introduced by the interchange of factors that are subject to anticommutation relations. Factors that commute, such as a product of free tensor fields with a product of free spinor fields at equal times, may be separated by writing each factor as a normal-ordered product.

The free field part of the Lagrangian is that Lagrangian for which the Euler-Lagrange equations would take the form eq. (6.5) or eq. (6.22). In the case of the charged scalar or pseudoscalar field it is

(8.3) $$\mathscr{L}_0[\phi(x)] = -\sum_\mu : \frac{\partial \phi^\dagger}{\partial x_\mu} \frac{\partial \phi}{\partial x_\mu} : - m^2 :\phi^\dagger \phi:,$$

and there are similar expressions (contracted on all tensor indices) for tensor fields of other ranks. For the Dirac spinor field it may be taken to be

(8.4) $$\mathscr{L}_0[\psi(x)] = -\frac{1}{2} \sum_\mu \left[:\bar{\psi}\gamma_\mu \frac{\partial \psi}{\partial x_\mu}: - :\frac{\partial \bar{\psi}}{\partial x_\mu} \gamma_\mu \psi: \right] - m:\bar{\psi}\psi:,$$

where the relationship between $\bar{\psi}(x)$ and $\psi^\dagger(x)$ is of the same form as that given in eq. (7.62). The Lagrangian L is the functional of the field operators given by

(8.5a) $$L\{\phi, \psi\} = \int d^3x \; \mathscr{L}[\phi(x), \psi(x)].$$

Therefore the free field Lagrangian in each of these cases is

(8.5b) $$L_0\{\phi\} = \int d^3x \; \mathscr{L}_0[\phi(x)],$$

and

(8.5c) $$L_0\{\psi\} = \int d^3x \; \mathscr{L}_0[\psi(x)].$$

The general requirement for covariance of the Euler-Lagrangian equations is that the action $\int L\, dt$ be invariant.

Not only are these forms of $\int L_0\, dt$ manifestly invariant under the proper Lorentz transformations, but they are also invariant under transformations of the form P, T, and C as defined in table 7.1. Thus, for example, for both the scalar and pseudoscalar free fields[5]

(8.6a) $$\mathbf{P}L_0\{\phi_0\}\mathbf{P}^{-1} = \int d^3x\, \mathscr{L}_0[\pm\phi_0(\bar{x})] = L_0\{\phi_0\},$$

because $\vec{x}$ may be replaced by $-\vec{x}$ as the integration variable. Also

$$\mathbf{P}L_0\{\psi_0\}\mathbf{P}^{-1} = \int d^3x\, \mathscr{L}_0[\gamma_4\psi_0(\bar{x})] = L_0\{\psi_0\},$$

a result obtained by making use of the equations

(8.6b) $$[\gamma_4\psi_0^P(x)]^\dagger\gamma_4\gamma_4\psi_0(\bar{x}) = \bar{\psi}_0(\bar{x})\psi_0(\bar{x})$$

and

(8.6c) $$[\gamma_4\psi_0(\bar{x})]^\dagger\gamma_4\gamma_\mu\gamma_4\frac{\partial}{\partial x_\mu}\psi_0(\bar{x}) = -\varepsilon_\mu\psi_0^\dagger(\bar{x})\gamma_\mu\gamma_4\frac{\partial}{\partial\bar{x}}\psi_0(\bar{x})$$

(8.6d) $$= \bar{\psi}_0(\bar{x})\gamma_4\frac{\partial}{\partial\bar{x}}\psi_0(\bar{x}),$$

as well as by replacing $\vec{x}$ by $-\vec{x}$ as the integration variable.

For the case of charge conjugation of the scalar fields, we need only note that

(8.7a) $$\mathbf{C}{:}\phi_0^\dagger(x)\phi_0(x){:}\mathbf{C}^{-1} = {:}\phi_0(x)\phi_0^\dagger(x){:} = {:}\phi_0^\dagger(x)\phi_0(x){:}$$

according to eq. (8.2), since the $\phi(x)$ satisfy commutation relations. For the Dirac fields the invariance can be demonstrated by considering the first

[5] The behavior of any given term in the Lagrangian under a kinematically admissible transformation can be determined only by expressing it in terms of the free fields $\phi_{\mu,0}$, ψ_0 with $\tau = 0$, which serve as the starting point for an iteration procedure. If *all* terms in the Lagrangian are then found to be invariant in form, the iteration of the effects of interactions starting from this zero-order Lagrangian can be used to show that the interacting fields are subject to the same transformation as the free fields, as we saw in chapters 6 and 7. The connection between the source terms in the integral equations of motion that we used as a basis for this argument in chapters 6 and 7 and the interaction term in the Lagrangian is obtained explicitly from the Euler-Lagrange equations associated with the particular Lagrangian in question.

term in eq. (8.4):

(8.7b)
$$\mathbf{C}{:}\bar{\psi}_0\gamma_\mu\frac{\partial\psi_0}{\partial x_\mu}{:}\mathbf{C}^{-1} = {:}[\gamma_2\tilde{\psi}_0^\dagger]^\dagger\gamma_4\gamma_\mu\frac{\partial}{\partial x_\mu}\gamma_2\tilde{\psi}_0^\dagger{:}$$
$$= {:}\tilde{\psi}_0\gamma_2\gamma_4\gamma_\mu\frac{\partial}{\partial x_\mu}\gamma_2\tilde{\psi}_0^\dagger{:} = -{:}\frac{\partial\bar{\psi}_0}{\partial x_\mu}\gamma_\mu\psi_0{:},$$

which is identical with the second term in eq. (8.4), thereby establishing the C invariance of the combination of the first two terms. The final equality of eq. (8.7b) is obtained by reversing the order, transposing all spinorial factors and using eq. (6.41). Also, eq. (8.2) has been used along with the fact that the spinor fields satisfy anticommutation relations. A similar procedure establishes the C invariance of the last term in eq. (8.4).

In order to establish the T invariance of $\int L_0\,dt$ we demonstrate that

(8.8a)
$$\mathbf{T}L_0\{\phi_0\}_t\mathbf{T}^{-1} = \int d^3x\;\mathbf{T}\mathscr{L}_0[\phi_0(\vec{x},t)]\mathbf{T}^{-1} = L_0\{\phi_0\}_{-t}$$

and

(8.8b)
$$\mathbf{T}L_0\{\psi_0\}_t\mathbf{T}^{-1} = \int d^3x\;\mathbf{T}\mathscr{L}_0[\psi_0(\vec{x},t)]\mathbf{T}^{-1} = L_0\{\psi_0\}_{-t},$$

which leads to the invariance of the action because the integration over t may be replaced by integration over $-t$. Eq. (8.8a) follows immediately from eq. (7.81). To establish eq. (8.8b), we first consider the last term in eq. (8.4). According to eq. (7.85),

(8.8c)
$$\mathbf{T}{:}\psi_0^\dagger(x)\gamma_4\psi_0(x){:}\mathbf{T}^{-1} = {:}\psi_0^\dagger(x')\sigma_2\gamma_4^*\sigma_2\psi_0(x'){:} = {:}\bar{\psi}_0(x')\psi_0(x'){:},$$

by virtue of eqs. (6.40) and (6.44a) and the antiunitary property of $\mathbf{T}$. Also

(8.8d)
$$\mathbf{T}{:}\bar{\psi}_0(x)\gamma_\mu\frac{\partial\psi_0(x)}{\partial x_\mu}{:}\mathbf{T}^{-1} = {:}\psi_0^\dagger(x')\sigma_2\gamma_4^*\gamma_\mu^*\sigma_2\frac{\partial\psi_0(x')}{\partial x'_\mu}{:}$$
$$= {:}\bar{\psi}_0(x')\gamma_\mu\frac{\partial\psi_0(x')}{\partial x'_\mu}{:},$$

and a corresponding result holds for the other term in $\mathscr{L}_0[\psi_0]$. Thus

(8.8e)
$$\mathbf{T}\mathscr{L}_0[\psi_0(x)]\mathbf{T}^{-1} = \mathscr{L}_0[\psi_0(x')],$$

and eq. (8.8b) follows immediately.

This demonstration of the invariance of $\int L_0\,dt$ under P, C, and T is meant to illustrate the methods whereby the invariance properties of the interaction term $L_I = \int d^3x\;\mathscr{L}_I$ may be determined. Only the interactions between tensor and spinor fields that depend bilinearly on the spinor fields (not including derivatives) will be considered here, but of course the same methods apply to other cases. The procedure is to construct the set of bilinear covariants, which are irreducible tensors formed from products of spinor fields. A term in the Lagrangian density is then formed by contracting each of these tensors into a tensor field of the same rank. This product is manifestly covariant under proper Lorentz transformations.

As the first step, then, we write down the well-known bilinear covariants of the spinor fields for kinematically independent spinor fields $\psi^{(j)}(x)$, with $j = 1, 2, 3 \ldots$. There are just five of them, and they correspond to the tensors (I) to (V) as classified in section 6.1:

(8.9a) $$S^{(j,k)} = :\bar{\psi}^{(j)}(x)\psi^{(k)}(x):, \quad \text{scalar}$$

(8.9b) $$V_\mu^{(j,k)} = i:\bar{\psi}^{(j)}(x)\gamma_\mu\psi^{(k)}(x):, \quad \text{four-vector}$$

(8.9c) $$T_{\mu\nu}^{(j,k)} = :\bar{\psi}^{(j)}(x)\sigma_{\mu\nu}\psi^{(k)}(x):,$$

antisymmetric tensor of second rank

(8.9d) $$(PV)_\mu^{(j,k)} = i:\bar{\psi}^{(j)}(x)\gamma_\mu\gamma_5\psi^{(k)}(x):, \quad \text{pseudovector}$$

(8.9e) $$P^{(j,k)} = i:\bar{\psi}^{(j)}(x)\gamma_5\psi^{(k)}(x):, \quad \text{pseudoscalar,}$$

where

(8.10a) $$\sigma_{\mu\nu} = (\gamma_\mu\gamma_\nu - \gamma_\nu\gamma_\mu)/2i$$

is a Hermitian matrix corresponding to eq. (6.44b) and

(8.10b) $$\gamma_5 = \gamma_1\gamma_2\gamma_3\gamma_4.$$

The factors of i are included so that, when $j = k$, the operators are Hermitian (or anti-Hermitian when any one index $\mu = 4$).

That the expressions eqs. (8.9) transform under proper Lorentz transformations as indicated follows directly from the transformation properties of $\psi(x)$ given by eqs. (6.26) and (6.27). It remains to be determined how tensors having the form of these bilinear covariants transform under the kinematically admissible transformations **P**, **T**, and **C**. For this purpose, we replace the $\psi^{(j)}(x)$ in eqs. (8.9) by $\psi_0^{(j)}(x)$ and apply eqs. (7.39), (7.45) (with t^0 replaced by t), and eq. (7.67). The results for **P** are then found to be:

(8.11a) $$\mathbf{P}:\bar{\psi}_0^{(j)}(x)\psi_0^{(k)}(x):\mathbf{P}^{-1} = :\bar{\psi}_0^{(j)}(\bar{x})\psi_0^{(k)}(\bar{x}):$$

(8.11b) $$\mathbf{P}:\bar{\psi}_0^{(j)}(x)\gamma_\mu\psi_0^{(k)}(x):\mathbf{P}^{-1} = -\varepsilon_\mu:\bar{\psi}_0^{(j)}(\bar{x})\gamma_\mu\psi_0^{(k)}(\bar{x}):$$

(8.11c) $$\mathbf{P}:\bar{\psi}_0^{(j)}(x)\sigma_{\mu\nu}\psi_0^{(k)}(x):\mathbf{P}^{-1} = \varepsilon_\mu\varepsilon_\nu:\bar{\psi}_0^{(j)}(\bar{x})\sigma_{\mu\nu}\psi_0^{(k)}(\bar{x}):$$

(8.11d) $$\mathbf{P}{:}\bar{\psi}_0^{(j)}(x)\gamma_\mu\gamma_5\psi_0^{(k)}(x){:}\mathbf{P}^{-1} = \varepsilon_\mu{:}\bar{\psi}_0^{(j)}(\bar{x})\gamma_\mu\gamma_5\psi_0^{(k)}(\bar{x}){:}$$

(8.11e) $$\mathbf{P}{:}\bar{\psi}_0^{(j)}(x)\gamma_5\psi_0^{(k)}(x){:}\mathbf{P}^{-1} = -{:}\bar{\psi}_0^{(j)}(\bar{x})\gamma_5\psi_0^{(k)}(\bar{x}){:},$$

where use has been made of the anticommutation of the gamma matrices. It should be noted that commutators with the generators $\mathbf{\Gamma}$, $\mathbf{\Gamma}'$ of the unitary transformations are not affected by the normal ordering because the difference between the normal-ordered and ordinary product is a c-number ("classical number") function, not an operator.

The effect of T resulting from application of eqs. (7.85) and (6.40) is

(8.12a) $$\mathbf{T}{:}\bar{\psi}_0^{(j)}(x)\psi_0^{(k)}(x){:}\mathbf{T}^{-1} = {:}\bar{\psi}_0^{(j)}(x')\psi_0^{(k)}(x'){:}$$

(8.12b) $$\mathbf{T}i{:}\bar{\psi}_0^{(j)}(x)\gamma_\mu\psi_0^{(k)}(x){:}\mathbf{T}^{-1} = -i{:}\bar{\psi}_0^{(j)}(x')\gamma_\mu\psi_0^{(k)}(x'){:}$$

(8.12c) $$\mathbf{T}{:}\bar{\psi}_0^{(j)}(x)\sigma_{\mu\nu}\psi_0^{(k)}(x){:}\mathbf{T}^{-1} = -{:}\bar{\psi}_0^{(j)}(x')\sigma_{\mu\nu}\psi_0^{(k)}(x'){:}$$

(8.12d) $$\mathbf{T}i{:}\bar{\psi}_0^{(j)}(x)\gamma_\mu\gamma_5\psi_0^{(k)}(x){:}\mathbf{T}^{-1} = -i{:}\bar{\psi}_0^{(j)}(x')\gamma_\mu\gamma_5\psi_0^{(k)}(x'){:}$$

(8.12e) $$\mathbf{T}i{:}\bar{\psi}_0^{(j)}(x)\gamma_5\psi_0^{(k)}(x){:}\mathbf{T}^{-1} = -i{:}\bar{\psi}_0^{(j)}(x')\gamma_5\psi_0^{(k)}(x'){:}.$$

From these transformations it can be seen that the bilinear covariants, eqs. (8.9), transform under T with the same phases as those assigned to the corresponding tensor fields in table 7.1, thereby establishing the correctness of that assignment as promised in connection with eq. (6.13b).[6]

Finally, the effect on those bilinear forms of charge conjugation is obtained from eq. (7.67) by use of eq. (7.66a), eq. (8.2), and

(8.13) $$\tilde{\gamma}_C = \gamma_C^{-1} = -\gamma_C,$$

where $\gamma_C = \gamma_2\gamma_4$:

(8.14a) $$\mathbf{C}{:}\bar{\psi}_0^{(j)}(x)\psi_0^{(k)}(x){:}\mathbf{C}^{-1} = {:}\bar{\psi}_0^{(k)}(x)\psi_0^{(j)}(x){:}$$

(8.14b) $$\mathbf{C}{:}\bar{\psi}_0^{(j)}(x)\gamma_\mu\psi_0^{(k)}(x){:}\mathbf{C}^{-1} = -{:}\bar{\psi}_0^{(k)}(x)\gamma_\mu\psi_0^{(j)}(x){:}$$

(8.14c) $$\mathbf{C}{:}\bar{\psi}_0^{(j)}(x)\sigma_{\mu\nu}\psi_0^{(k)}(x){:}\mathbf{C}^{-1} = -{:}\bar{\psi}_0^{(k)}(x)\sigma_{\mu\nu}\psi_0^{(j)}(x){:}$$

(8.14d) $$\mathbf{C}{:}\bar{\psi}_0^{(j)}(x)\gamma_\mu\gamma_5\psi_0^{(k)}(x){:}\mathbf{C}^{-1} = {:}\bar{\psi}_0^{(k)}(x)\gamma_\mu\gamma_5\psi_0^{(j)}(x){:}$$

(8.14e) $$\mathbf{C}{:}\bar{\psi}_0^{(j)}(x)\gamma_5\psi_0^{(k)}(x){:}\mathbf{C}^{-1} = {:}\bar{\psi}_0^{(k)}(x)\gamma_5\psi_0^{(j)}(x){:}.$$

The examples of interactions between tensor and spinor fields to be considered here are those Lagrangians that are obtained by construction of the products of these bilinear tensor operators with tensor fields of the same

[6] Note that although $\bar{\psi}\gamma_\mu\psi$ transforms like x_μ under proper Lorentz transformations, it is not possible to construct a Hermitian (anti-Hermitian for $\mu = 4$) operator transforming under T like x_μ, eq. (6.12a), because $(\bar{\psi}\gamma_\mu\psi)^\dagger = -\varepsilon_\mu\bar{\psi}\gamma_\mu\psi$.

rank. Since the L_I must be Hermitian operators, they must consist of the sum of an operator and its Hermitian conjugate. Furthermore, they are to be normal ordered, but further ordering of the product over that already indicated in eqs. (8.9) is irrelevant because at equal times (all fields appearing in $\mathscr{L}_I$ are evaluated at the same space-time point) the tensor fields commute with the spinor fields. Thus we would write for the couplings of the spinor fields with charged tensor fields:

(8.15a)
$$L_I^{\pm}(S) = \frac{1}{2}\int d^3x\{g_S^{\pm}\, \phi(x) :\bar{\psi}^{(j)}(x)\psi^{(k)}(x): + g_S^{\pm *}\, \phi^{\dagger}(x) :\bar{\psi}^{(k)}(x)\psi^{(j)}(x):\}$$

(8.15b)
$$L_I^{\pm}(V) = \frac{i}{2}\sum_{\mu}\int d^3x\{g_V^{\pm}\, \phi_{\mu}(x) :\bar{\psi}^{(j)}(x)\gamma_{\mu}\, \psi^{(k)}(x): + g_V^{\pm *}\varepsilon_{\mu}\, \phi_{\mu}^{\dagger}(x) :\bar{\psi}^{(k)}(x)\gamma_{\mu}\psi^{(j)}(x):\}$$

(8.15c)
$$L_I(T) = \frac{1}{2}\sum_{\mu,\nu}\int d^3x\{g_T\, \phi_{\mu\nu}(x) :\bar{\psi}^{(j)}(x)\sigma_{\mu\nu}\, \psi^{(k)}(x): + g_T^{*}\, \varepsilon_{\mu}\, \varepsilon_{\nu}\, \phi_{\mu\nu}^{\dagger}(x) :\bar{\psi}^{(k)}(x)\sigma_{\mu\nu}\, \psi^{(j)}(x):\}$$

(8.15d)
$$L_I^{\pm}(PV) = \frac{i}{2}\sum_{\mu}\int d^3x\{g_{PV}^{\pm}\, \phi_{\mu}(x) :\bar{\psi}^{(j)}(x)\gamma_{\mu}\gamma_5\, \psi^{(k)}(x): + g_{PV}^{\pm *}\, \varepsilon_{\mu}\, \phi_{\mu}^{\dagger}(x) :\bar{\psi}^{(k)}(x)\gamma_{\mu}\gamma_5\, \psi^{(j)}(x):\}$$

(8.15e)
$$L_I^{\pm}(P) = \frac{i}{2}\int d^3x\{g_P^{\pm}\, \phi(x) :\bar{\psi}^{(j)}(x)\gamma_5\, \psi^{(k)}(x): + g_P^{\pm *}\, \phi^{\dagger}(x) :\psi^{(k)}(x)\gamma_5\, \psi^{(j)}(x):\}.$$

The g's are coupling constants measuring the magnitude of the interactions, and the plus signs occur when $\phi(x)$ and $\phi_{\mu}(x)$ are scalar and vector fields whereas the minus signs occur when they are pseudoscalar and pseudovector fields. It should also be noted that the appearance of the factor ε_{μ} is associated with a factor i in, for example, $\phi_4 = i\phi_0$ and is therefore eliminated when the tensor fields are written in terms of "real" operators.

The charges of the fields $\psi^{(j)}$ and $\psi^{(k)}$ must differ by one unit if the charge is to be conserved by these interactions with tensor fields of unit charge.

The behavior of the interactions eqs. (8.15) under P, T, and C is to be determined by inserting the free fields into the interacton Lagrangian and applying the transformations eqs. (8.11), (8.12), and (8.14) and the corresponding transformations of the tensor fields given in table 7.1.

The effects of these transformations on the L_{I_τ} can be described most easily in terms of a substitution of the coupling constants. For example, $\mathbf{P}L_{I0}\mathbf{P}^{-1}$ has the same form as L_{I0} but with the coupling constants changed as follows:

(8.16a) $$P:\quad g_S^\pm \to \pm g_S^\pm;\quad g_V^\pm \to \pm g_V^\pm;\quad g_T \to g_T;$$
$$g_{PV}^\pm \to \mp g_{PV}^\pm;\quad g_P^\pm \to \mp g_P^\pm.$$

It follows immediately that the requirement for P invariance of the L_{I0} is

(8.16b) $$g_S^- = g_V^- = g_{PS}^+ = g_P^+ = 0.$$

Similarly, $\mathbf{C}L_{I0}\mathbf{C}^{-1}$ is equivalent to the substitutions:

(8.17a) $$C:\quad g_S^\pm \to g_S^{\pm *};\quad g_V^\pm \to \pm g_V^{\pm *};\quad g_T \to g_T^*;$$
$$g_{PV}^\pm \to \mp g_{PV}^{\pm *};\quad g_P^\pm \to g_P^{\pm *},$$

so that the conditions for C invariance are the reality conditions:

(8.17b) $$\text{Im } g_S^\pm = \text{Im } g_V^+ = \text{Re } g_V^- = \text{Im } g_T = \text{Re } g_{PV}^+ = \text{Im } g_{PV}^- = \text{Im } g_P^\pm = 0.$$

In determining $\mathbf{T}L_{I0}\mathbf{T}^{-1}$ it must be kept in mind that K operates on all constants appearing in L_{I0}. Then, the transformation under T is found to be equivalent to

(8.18a) $$T:\quad g_S^\pm \to \pm g_S^{\pm *};\ g_V^\pm \to g_V^{\pm *};\ g_T \to g_T^*;\ g_{PV}^\pm \to g_{PV}^{\pm *};\ g_P^\pm \to \mp g_P^{\pm *},$$

and the conditions for T invariance are

(8.18b) $$\text{Im } g_S^+ = \text{Re } g_S^- = \text{Im } g_V^\pm = \text{Im } g_T = \text{Im } g_{PV}^\pm = \text{Re } g_P^+ = \text{Im } g_P^- = 0.$$

The consequences of the transformation $\mathbf{CPT}L_{I0}(\mathbf{CPT})^{-1}$ where $\mathscr{L}_{I0} \equiv \mathscr{L}_I\{\phi_0, \psi_0\}$, may now be obtained by applying the substitutions eqs. (8.16a), (8.17a), and (8.18a) in succession. The result is the substitution $g \to g$ for each of the coupling constants. In fact, direct application of the transformation equations of table 7.1 yields

(8.19) $$\mathbf{CPT}\mathscr{L}_I[\phi_0(x), \psi_0(x)](\mathbf{CPT})^{-1} = \mathscr{L}_I[\phi_0(-x)\ \psi_0(-x)]$$

for each of the $\mathscr{L}_I$ given in eqs. (8.15) with any set of complex coupling constants. (Note that $\phi(x)$ symbolizes any one of the tensor fields.) Therefore the action associated with the free fields $\phi_0(x)$, $\psi_0(x)$ is invariant under CPT. An iteration of the dynamic equations starting from the free field approximation therefore leads to CPT-invariant dynamics.[7]

This demonstration of the invariance of the dynamics under CPT for a set of specific interactions serves as a set of examples corresponding to

[7] See note 5.

Pauli's general method of proof of the *CPT* theorem.[8] The assumptions underlying his proof of the theorem are those that have been made here in arriving at the form of the Lagrangian. We have assumed:

1. The validity of the basic structure of quantum field theories as outlined in section 7.1 in terms of local field operators (i.e., field operators defined at a space-time point x).
2. Invariance under proper Lorentz transformations. In particular this implies that all field operators can be analyzed in terms of normal-ordered irreducible tensor products of *anticommuting* local spinor fields at the point x. (Herein is buried the important connection between spin and statistics.)
3. The interaction Lagrangian (or Hamiltonian) is Hermitian. More generally, *all* real observables are represented by Hermitian products of field operators.

Pauli proved the theorem under these assumptions by showing that the transformation under *CPT* of any Hermitian (anti-Hermitian for an odd number of indices $\mu = 4$ in our notation) irreducible tensor product of anticommuting spinors depends only on the rank of the tensor. Therefore the behavior of each observable under *CPT* is uniquely determined and, in particular, the energy-momentum tensor is invariant.

8.2 Consequences of *CPT* Invariance and *T* Invariance for the *S* Matrix

To determine the behavior of the S matrix in the context of quantum field theory it is necessary to relate it to the field operators. Lehmann, Symanzik, and Zimmermann (1955, 1957) have developed a method (the LSZ theory) for determining the properties of the S matrix in quantum field theory that provide the desired relationships. The LSZ approach to establishing these relationships will be briefly outlined here.

We have seen that the GBS may be constructed by operating on the vacuum state with the creation operators associated with the free field, as in eq. (7.56). If the free fields used for this purpose are those we have denoted by $\phi_\tau(x)$, $\psi_\tau(x)$, etc., then the basis vectors of the GBS may be denoted by $|\alpha, \tau\rangle$, $|\beta, \tau\rangle$, etc., where α or β is a complete set of quantum numbers of all the free particles occurring in a specified state. Thus, for example, this label will specify the number of particles and antiparticles of given momentum, spin, isotopic spin, strangeness, and so on. For any given value of τ, it is assumed that these form a complete orthonormal set of states of the GBS.

To construct the S matrix, the LSZ theory introduces two complete sets

[8] See note 2.

of basis vectors,

(8.20a)
$$|\alpha, \text{in}\rangle = \lim_{\tau \to -\infty} |\alpha, \tau\rangle$$

and

(8.20b)
$$|\alpha, \text{out}\rangle = \lim_{\tau \to +\infty} |\alpha, \tau\rangle,$$

where note has been taken that both sets of states (in fact, the set associated with any given value of τ) can be labeled by the same set of quantum numbers because they correspond to free particles of the same kinds. Since either of the sets eqs. (8.20) may be used to define a basis for the GBS, they are related by a unitary transformation **S**:

(8.21a)
$$|\alpha, \text{in}\rangle = \mathbf{S}|\alpha, \text{out}\rangle$$

(8.21b)
$$= \sum_{\beta} |\beta, \text{out}\rangle \langle\beta, \text{out}|\mathbf{S}|\alpha, \text{out}\rangle.$$

In the second form of the equation the matrix element

(8.22)
$$S_{\beta\alpha} = \langle\beta, \text{out}|\mathbf{S}|\alpha, \text{out}\rangle$$

is taken with respect to the "out" basis. Of course the matrix element may also be taken with respect to the "in" basis, and by eq. (8.21a)

(8.23)
$$\langle\beta, \text{in}|\mathbf{S}|\alpha, \text{in}\rangle = \langle\beta, \text{out}|\mathbf{S}^{-1}\mathbf{S}\mathbf{S}|\alpha, \text{out}\rangle$$
$$= S_{\beta\alpha}.$$

Thus it makes no difference on which basis $S_{\beta\alpha}$ is determined.

Another form for the matrix element that follows from eq. (8.21a) is

(8.24)
$$S_{\beta\alpha} = \langle\beta, \text{out}|\alpha, \text{in}\rangle,$$

and it is evident from this form that $S_{\beta\alpha}$ is the probability amplitude for finding the system in the free particle state $|\beta\rangle$ at $t = \infty$ if at $t = -\infty$ it was in the state $|\alpha\rangle$. The presumption underlying this statement is that the states $|\alpha\rangle$ and $|\beta\rangle$ describe free particle wave packets that do not overlap in the limits $t = \pm\infty$ and therefore are not interacting in those limits. As long as the interactions have a finite effective range, those wave packets may be characterized by free particle quantum numbers (e.g., momentum) to a good approximation.

We draw the conclusion that $S_{\beta\alpha}$ as defined here is a generalized form of the S matrix of section 4.5, and its interpretation in terms of scattering and reaction cross sections may be carried over directly from the discussion of that section.

Comparing eq. (8.24) with eq. (4.71) suggests the associations

$$|\alpha, \text{in}\rangle \sim |\psi_\alpha^+\rangle, \; |\beta, \text{out}\rangle \sim |\psi_\beta^-\rangle. \tag{8.25}$$

Although these associations will be useful in making the connection with our earlier analysis, keep in mind that they are *not* equalities; the $|\psi_\alpha^\pm\rangle$ are *stationary* states, whereas $|\alpha, \text{in}\rangle$ and $|\alpha, \text{out}\rangle$ are initial and final forms of time-dependent states. Furthermore, note that there is a possibility for confusion here because this conventional language we are using associates the "in" states with the "outgoing" stationary states and the "out" states with the "incoming" stationary states, eq. (4.70), where the terms "outgoing" and "incoming" refer to the (stationary state) boundary conditions.

We now consider the consequences of CPT invariance for the matrix $S_{\beta\alpha}$. To simplify the notation we introduce the antiunitary operators

$$\mathbf{\Theta} = \mathbf{CPT} = \mathbf{U}_\theta K, \tag{8.26}$$

where $\mathbf{U}_\theta$ is unitary, and designate the CPT-transformed variables by a superscript θ:

$$CPT\colon \; x_\mu \rightarrow x_\mu^\theta = -x_\mu, \tag{8.27a}$$

$$CPT\colon \; \phi(x) \rightarrow \phi^\theta(-x) = \mathbf{\Theta}\phi(x)\mathbf{\Theta}^{-1}, \tag{8.27b}$$

$$CPT\colon \; \psi(x) \rightarrow \psi^\theta(-x) = \mathbf{\Theta}\psi(x)\mathbf{\Theta}^{-1}. \tag{8.27c}$$

Since $\mathbf{\Theta}$ is antiunitary (because of the factor $\mathbf{T}$), it satisfies the fundamental condition

$$\langle \Phi^\theta | \Psi^\theta \rangle = \langle \Psi | \Phi \rangle, \tag{8.28a}$$

where

$$|\Psi^\theta\rangle = \mathbf{\Theta}|\Psi\rangle, \; |\Phi^\theta\rangle = \mathbf{\Theta}|\Phi\rangle \tag{8.28b}$$

and $|\Psi\rangle$ and $|\Phi\rangle$ are arbitrary states of the GBS.

From the CPT theorem it follows that $\phi(x)$ and $\psi(x)$ are transformed by $\mathbf{\Theta}$ in the same way as the free fields $\phi_0(x)$ and $\psi_0(x)$. From table 7.1 these transformations are found to be[9]

$$\mathbf{\Theta}\phi(x)\mathbf{\Theta}^{-1} = \phi^\dagger(-x), \tag{8.29a}$$

[9] To obtain eqs. (8.29) we follow the procedure of section 7.5 using the occupation number representation of $\mathbf{U}_\theta$ satisfying eqs. (8.30). Then it follows that $\mathbf{\Theta}\phi_\tau(x)\mathbf{\Theta}^{-1} = \phi^C_{-\tau}(-x)$, etc., for every value of τ. But $\phi(\vec{x}, \tau) = \phi_\tau(\vec{x}, \tau)$, $\psi(\vec{x}, \tau) = \psi_\tau(\vec{x}, \tau)$, whence eqs. (8.29) follow immediately. Note that $\sigma_2 \gamma_4 \psi^C = -i\gamma_5 \tilde{\psi}^\dagger$ puts the antiparticle states into the same form as the particle states.

and

(8.29b) $$\Theta\psi(x)\Theta^{-1} = \sigma_2\gamma_4\psi^C(-x),$$

where $\phi(x)$ is either the scalar or the pseudoscalar field and $\psi^C = \gamma_2\tilde{\psi}^\dagger$ is the charge conjugate spinor field. Furthermore, the vector and pseudovector fields satisfy eq. (8.29a) with a change in sign, while the tensor field $\phi_{\mu\nu}$ satisfies the equation without the sign change.

Eqs. (8.29) may also be put in the form $\phi^\theta(x) = \phi^C(-x)$, $\psi^\theta(x) = \sigma_2\gamma_4\psi^C(-x)$, etc., and they are an expression of the way *CPT* invariance of the dynamics affects the motion. This result may be described graphically in language analogous to that used in chapter 2. The right-hand side of each of these equations represents the motion that would be observed in a "*CPT* mirror," and it would be identical to motion observed in the laboratory under *CPT*-reversed initial conditions, that is, the initial conditions as seen in the mirror. In this case the "motion" is described by the space-time dependence of the field operator and the "initial condition" is the specification of $\phi_\tau(x)$ or $\psi_\tau(x)$, the field operators that play the role of boundary values for the differential equations of motion.

To determine the behavior of $S_{\beta\alpha}$ under *CPT*, it is necessary to determine how the basis states of the GBS transform. Since these states are generated from the vacuum by the creation operators $a_\tau^\dagger(\vec{k})$, $b_\tau^\dagger(\vec{k}, s)$, etc., for $\tau = \pm\infty$, the transformation of the basis states is governed by the behavior of $a_\tau(\vec{k})$, $b_\tau(\vec{k}, s)$, etc., under *CPT*. These creation and annihilation operators are related to $\phi_\tau(x)$ and $\psi_\tau(x)$ by equations of the form eqs. (7.53a) and (7.68a).

As noted earlier, the creation and annihilation operators are real matrices that are unaffected by K, so that eqs. (8.29), (6.17), and (6.46c) lead to (note that there are two changes in the sign of $\vec{k}$, one owing to **P** and the other to **T**)

(8.30a) $$a^\theta_{-\tau}(\vec{k}) = \Theta a_\tau(\vec{k})\Theta^{-1} = \bar{a}_\tau(\vec{k})$$

(8.30b) $$b^\theta_{-\tau}(\vec{k}, s) = \Theta b_\tau(\vec{k}, s)\Theta^{-1} = i^s\bar{b}_\tau(\vec{k}, -s)$$

where $s = 2m_s$ is twice the magnetic quantum number associated with the spinor. The Wigner phase convention for the spin states has been assumed. Therefore, if we assume that the vacuum state is invariant,

(8.31) $$\Theta|0\rangle = |0\rangle,$$

and use the same set of quantum numbers α, β, etc., to label the basic vectors of the transformed GBS, we find

(8.32a) $$\Theta|\alpha, \text{in}\rangle = e^{i\phi_\alpha}|-\bar{\alpha}, \text{out}\rangle,$$

(8.32b) $$\Theta|\alpha, \text{out}\rangle = e^{i\phi_\alpha}|-\bar{\alpha}, \text{in}\rangle,$$

where the quantum numbers $-\bar{\alpha}$ are obtained from α by replacing each particle of spin $s/2$ and momentum $\vec{k}$ by its antiparticle with spin $-s/2$ and the *same* value of $\vec{k}$. The phase is

(8.32c) $$\phi_\alpha = (\pi/2)\Sigma_\alpha,$$

with Σ_α equal to the sum of all quantum numbers s included in the set α.

The consequences of CPT invariance for the S matrix follow from application of the antiunitary property eqs. (8.28) to the matrix element eq. (8.24):

(8.33) $$S_{\beta\alpha} = \langle\beta, \text{out}\,|\,\alpha, \text{in}\rangle = \langle\mathbf{\Theta}(\alpha, \text{in})\,|\,\mathbf{\Theta}(\beta, \text{out})\rangle$$
$$= e^{i(\phi_\beta - \phi_\alpha)}\langle -\bar{\alpha}, \text{out}\,|\,-\bar{\beta}, \text{in}\rangle,$$

by use of eqs. (8.32). Therefore the condition imposed by the CPT theorem is

(8.34) $$S_{\beta\alpha} = e^{i(\phi_\beta - \phi_\alpha)}S_{-\bar{\alpha},\,-\bar{\beta}},$$

from which it follows that, aside from a phase, the amplitude for a transition from α to β is equal to the amplitude of the reversed transition between antiparticle states with all spins reversed. For example, the reaction between two spin $\frac{1}{2}$ particles

$$A(\vec{k}_1, \uparrow) + B(\vec{k}_2, \uparrow) \to C(\vec{k}_3, \uparrow) + D(\vec{k}_4, \uparrow)$$

has, aside from the phase factor, the same amplitude as the reaction between antiparticles

$$\bar{C}(\vec{k}_3, \downarrow) + \bar{D}(\vec{k}_4, \downarrow) \to \bar{A}(\vec{k}_1, \downarrow) + \bar{B}(\vec{k}_2, \downarrow),$$

where the vertical arrows denote the sign of the spin-magnetic quantum number. Therefore the cross sections for these reactions are equal.

We turn now to the consequences of the *assumption* of T invariance. Under that assumption, $\phi(x)$ and $\psi(x)$ transform under T in the same way as the free fields $\phi_0(x)$ and $\psi_0(x)$:

(8.35a) $$\phi'(x') = \mathbf{T}\phi(x)\mathbf{T}^{-1} = \pm\phi(x'),$$

(8.35b) $$\psi'(x') = \mathbf{T}\psi(x)\mathbf{T}^{-1} = \sigma_2\,\psi(x'),$$

where the $\pm$ sign refers to scalar pseudoscalar fields, respectively.[10] The expansion of $\phi_\tau(x)$ and $\psi_\tau(x)$ in terms of creation and annihilation operators

[10] The second step in eqs. (8.35) follows from the agument of note 9 with $\mathbf{\Theta}$ replaced by $\mathbf{T}$ and eqs. (8.30) replaced by eqs. (8.36). Note that these results may again be interpreted in terms of the graphic language of chapter 2.

then leads to

(8.36a) $$a'_{-\tau}(\vec{k}) = \mathbf{T}a_\tau(\vec{k})\mathbf{T}^{-1} = \pm a_\tau(-\vec{k}),$$

(8.36b) $$\bar{a}'_{-\tau}(\vec{k}) = \mathbf{T}\bar{a}_\tau(\vec{k})\mathbf{T}^{-1} = \pm \bar{a}_\tau(-\vec{k}),$$

for scalar or pseudoscalar fields, respectively, and

(8.36c) $$b'_{-\tau}(\vec{k}, s) = \mathbf{T}b_\tau(\vec{k}, s)\mathbf{T}^{-1} = i^s b_\tau(-\vec{k}, -s),$$

(8.36d) $$\bar{b}'_{-\tau}(\vec{k}, s) = \mathbf{T}\bar{b}_\tau(\vec{k}, s)\mathbf{T}^{-1} = i^s \bar{b}_\tau(-\vec{k}, -s),$$

as a consequence of eqs. (8.35), (6.17), and (6.46c).

Thus we find that

(8.37a) $$\mathbf{T}|\alpha, \text{in}\rangle = \pm e^{i\phi_\alpha}|\alpha', \text{out}\rangle$$

and

(8.37b) $$\mathbf{T}|\alpha, \text{out}\rangle = \pm e^{i\phi_\alpha}|\alpha', \text{in}\rangle,$$

where again the same quantum numbers α, β, etc., are used to define the basis vectors of the transformed GBS and those denoted by α' are the motion-reversed form of the quantum numbers denoted by α; that is, every momentum vector is reversed and every quantum number s is changed to $-s$. The $\pm$ sign is determined by the number of pseudoscalar particles. In addition to the assumption of T invariance of the interactions, it has also been assumed that the vacuum state is invariant in obtaining eq. (8.37); that is,

(8.38) $$\mathbf{T}|0\rangle = |0\rangle.$$

This is of course to be expected, because our formulation of quantum field theory presumes the uniqueness of the vacuum state—that it is not a degenerate state. Some further comments on the issue of a degenerate vacuum will be made later.

Given the results eqs. (8.37), the fundamental property of antiunitary transformations may be used again to determine the consequences of T invariance for $S_{\beta\alpha}$ as in the case of CPT, eq. (8.33):

(8.39) $$\begin{aligned} S_{\beta\alpha} &= \langle\beta, \text{out}|\alpha, \text{in}\rangle = \langle\mathbf{T}(\alpha, \text{in})|\mathbf{T}(\beta, \text{out})\rangle \\ &= \pm e^{i(\phi_\beta - \phi_\alpha)}\langle\alpha', \text{out}|\beta', \text{in}\rangle \\ &= \pm e^{i(\phi_\beta - \phi_\alpha)} S_{\alpha'\beta'}. \end{aligned}$$

This result is identical with that obtained in chapter 4, eq. (4.82), and provides a generalized basis for the earlier conclusions that were derived from it.

8.3 Equality of Masses and Lifetimes of Particle and Antiparticle

That the mass of a *free* antiparticle should be equal to the mass of the free particle it is associated with is built into the assumptions underlying the definition of charge conjugation. However, that the equality of their physical masses, taking into account all interactions, should follow is not obvious because the weak interactions are known not to be C invariant. Nevertheless, the exact equality of masses does follow from the CPT theorem, as we will establish in this section.

We will also find that, for an unstable particle, the lifetime is the same as that of its associated antiparticle. Of course, the mass of an unstable particle is not sharply defined in terms of energy and momentum measurements on the decay products. Such measurements will always lead to a mass distribution having a width proportional to the decay rate. Thus the definitions of mass and lifetime (reciprocal of decay rate) are closely related, and our first task is to establish a general method for treating both in the context of quantum field theory.

The method to be used here is based on the definition of the Feynman propagator for the interacting fields, as suggested by Peierls (1955).[11] The propagator, given as a function of the four-momentum k_μ of the particle, may be expressed in the Lehmann (1954) representation as the boundary value of an analytic function $F(z)$ in the cut z plane; for example, in the case of the scalar field, we may write for the physical propagator

$$\Delta_F'(k^2) = \lim_{z \to k^2 + i\varepsilon} F(z), \tag{8.40}$$

where $k^2 = -\sum_\mu k_\mu^2$. The cut is along the positive real axis beginning at the lowest threshold for virtual decay of the particle, which is given by the square of the sum of the masses of the particles into which the particle of interest can decay.

For a stable particle the propagator has a pole at the real point $k^2 = m^2$, where m is the physical mass. For an unstable particle there is a pole in $F(z)$ at a complex value of z, $z = z_0$. This pole is below the cut, on the lower half of the z plane, so that Im z_0 is negative. Since the propagator is defined as a boundary value in the upper half plane, or on the "physical" Riemann sheet, the pole lies on the "second sheet" of $F(z)$. It can be shown[12] that if

$$z_0 = (m - \tfrac{1}{2} i\Gamma)^2, \tag{8.41}$$

[11] See Jacob and Sachs (1961) for an explicit treatment in the form used here. Also compare eq. (5.3), Sachs (1963b) and Harte and Sachs (1964).

[12] See note 11.

then Γ is the decay rate of the particle. Furthermore, when the mass of the unstable particle is measured in terms of the momenta and energies of the decay products, a distribution of masses centered on m and having a width determined by Γ will be observed. Thus we define the mass and lifetime (decay rate) of a particle in terms of the pole in $F(z)$.

For a spinor particle, $F(z)$ is a 4×4 matrix expressible in terms of the gamma matrices and the components k_μ, but Lorentz invariance requires that each matrix element have its pole at the same value of z.

To determine how the propagator transforms under CPT, we write it as a vacuum state expectation value of products of the interacting fields. Thus in the case of the charged scalar or pseudoscalar field, the physical propagator of a particle in space-time is given by[13]

(8.42a)
$$i\Delta'_F(x) = \theta(t)\langle 0 | \phi(\tfrac{1}{2}x)\phi^\dagger(-\tfrac{1}{2}x) | 0 \rangle + \theta(-t)\langle 0 | \phi^\dagger(-\tfrac{1}{2}x)\phi(\tfrac{1}{2}x) | 0 \rangle,$$

while the propagator of the antiparticle is

(8.42b)
$$i\bar{\Delta}'_F(x) = \theta(t)\langle 0 | \phi^\dagger(\tfrac{1}{2}x)\phi(-\tfrac{1}{2}x) | 0 \rangle, + \theta(-t)\langle 0 | \phi(-\tfrac{1}{2}x)\phi^\dagger(\tfrac{1}{2}x) | 0 \rangle,$$

where

(8.42c)
$$\theta(t) = 1 \quad \text{for } t > 0$$
$$\theta(t) = 0 \quad \text{for } t < 0.$$

The consequences of the CPT theorem are now found by applying the fundamental property of the antiunitary $\mathbf{\Theta}$, eq. (8.28a), to the vacuum expectation value

(8.43)
$$\langle 0 | \phi(\tfrac{1}{2}x)\phi^\dagger(-\tfrac{1}{2}x) | 0 \rangle = \langle 0 | \mathbf{\Theta}\phi(\tfrac{1}{2}x)\mathbf{\Theta}^{-1}\mathbf{\Theta}\phi^\dagger(-\tfrac{1}{2}x)\mathbf{\Theta}^{-1} | 0 \rangle^*$$

by virtue of eq. (8.31). But then from eq. (8.29a), which expresses the consequences of the CPT theorem in this case,

(8.44)
$$\langle 0 | \phi(\tfrac{1}{2}x)\phi^\dagger(-\tfrac{1}{2}x) | 0 \rangle = \langle 0 | \phi^\dagger(-\tfrac{1}{2}x)\phi(\tfrac{1}{2}x) | 0 \rangle^*.$$

Therefore it follows from eqs. (8.42) that the propagators for particle and antiparticle are related to one another by the equation

(8.45a)
$$\Delta'_F(x) = \bar{\Delta}'_F(x).$$

[13] See Bjorken and Drell (1965, pp. 137 ff.). Although the neutral (Hermitian) field is treated in this reference, the generalization to include fields with charges (either electric or hypercharge, i.e., strangeness) is straightforward when we use eq. (7.53a).

The propagator $\Delta'_F(k^2)$ in momentum representation is the four-dimensional Fourier transform of $\Delta'_F(x)$. Therefore

(8.45b) $$\Delta'_F(k^2) = \bar{\Delta}'_F(k^2)$$

or, for the analytic extension of the propagator,

(8.45c) $$F(z) = \bar{F}(z).$$

Thus the pole z_0 in $F(z)$ is identical with the pole in $\bar{F}(z)$, whence it follows that particle and antiparticle have the same (physical) mass and lifetime.

A similar treatment of the 4 × 4 matrix propagator of a spinor field using eq. (8.29b) requires some algebraic manipulation of the gamma matrices and spinor fields, but it leads to the same conclusion, that the poles in the z plane are identical for particle and antiparticle. It is evident that this proof of the equality of mass and lifetime of particle and antiparticle based on the CPT theorem can be extended to tensor fields of arbitrary rank.

8.4 Consequences of T Invariance for Electromagnetic Form Factors

The amplitudes for reactions of leptons (electrons, neutrinos, muons, etc.) with nuclear matter are governed by either the electromagnetic or the weak interactions, and they are small enough to be treated by a lowest-order perturbation approximation in most cases. For example, the scattering of a high-energy electron from a proton may be analyzed in terms of a model in which the particles exchange just one photon, as shown in figure 8.1. In this approximation, such measurements provide information about the structure of the nuclear matter from which the lepton is scattered, information that is expressed in terms of "form factors." The measurement of form factors for electron, neutrino, and muon scattering has been an important source of information concerning the structure of nuclear matter, especially the structure of the proton and neutron. It is the purpose of this section to show that the assumption of T invariance leads to a severe restriction on the electromagnetic form factors for protons and neutrons.

Let us first consider the electromagnetic form factors associated with the elastic scattering of an electron or muon from a nucleon.[14] In the approximation of one photon exchange (fig. 8.1), the scattering amplitude can be written as the product of a lepton factor of known form and the matrix

[14] The nucleon current density is the sum of contributions from all the field operators that contribute to the nucleon structure, for example, contributions from the nucleon spinor field and the pion pseudoscalar field or, alternatively, contributions of spinor quark fields. Compare eq. (8.55b) and the discussion preceding it.

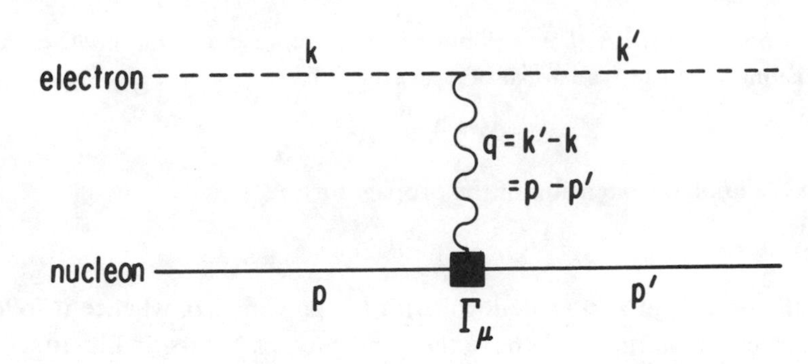

FIG. 8.1 Diagram for one-photon exchange in electron-nucleon scattering. The photon-nucleon vertex $\Gamma_\mu(p^2, p'^2, q^2)$ described by the form factors is shown as a black box. The two external nucleon lines describe a free nucleon (on the mass shell), and the internal photon line describes a virtual photon (off the mass shell).

element of the electromagnetic current density operator $j_\mu(x)$ for the nucleon. It is this matrix element that contains the information about the structure of the nucleon.

The matrix element is taken between the initial nucleon state of momentum $\vec{p}_i$ and spin magnetic quantum number $\frac{1}{2}s_i$ and a final state (after scattering of the electron or muon) of momentum $\vec{p}_f$ and spin $\frac{1}{2}s_f$. The corresponding free particle spinor states satisfying eq. (7.68) will be denoted by $u_i(\vec{p}_i)$ and $u_f(\vec{p}_f)$. The matrix element of interest is

$$\textbf{(8.46a)}\quad \langle \vec{p}_f, s_f | j_\mu(x) | \vec{p}_i, s_i\rangle = \exp(-i \sum_\nu q_\nu x_\nu)\langle \vec{p}_f, s_f | j_\mu(0) | \vec{p}_i, s_i\rangle,$$

where

$$\textbf{(8.46b)}\quad q_\nu = p_{f\nu} - p_{i\nu}$$

is the energy-momentum transferred to the nucleon. Eq. (8.46a) follows from the fact that the momentum and energy operators of the fields are also the generators of space and time translations.[15]

Since $j_\mu(0)$ transforms as a four-vector under Lorentz transformations, its matrix element must be expressible in terms of the four-vectors associated with the free nucleon. The only such four-vector in addition to $p_{i\mu}$ and $p_{f\mu}$ that can enter into a determination of the matrix element is $\bar{u}_f \gamma_\mu u_f$. Therefore the most general form of the matrix element permitted by the requirement of covariance under proper Lorentz transformations is

$$\textbf{(8.47)}\quad \langle \vec{p}_f, s_f | j_\mu(0) | \vec{p}_i, s_i\rangle = ie\bar{u}_f(\vec{p}_f)[\gamma_\mu G_1(q^2) + i(p_{f\mu} + p_{i\mu})G_2(q^2) + q_\mu G_3(q^2)]u_i(\vec{p}_i),$$

[15] See, e.g., Bjorken and Drell (1965, p. 138, eq. 16.24).

where the G_j are functions of the scalar variable $q^2 = -\sum_\mu q_\mu^2$. That only one scalar variable can be formed from the two four-momenta $p_{i\mu}$ and $p_{f\mu}$ follows from the fact that $p_i^2 = p_f^2 = M^2$, where M is the nucleon mass.

The electromagnetic current density $\vec{j}$ is an observable, therefore it is a Hermitian operator, as is the charge density $\rho = -ij_4$. (The Hermiticity of j_μ determines the factors of i appearing in eq. 8.47.) Furthermore, the classical observable $\vec{j}$, being proportional to a velocity, is odd and ρ is even under time reversal. Since $\mathbf{T}$ is defined in such a way as to preserve this classical behavior under the assumption of T invariance, we have[16]

(8.48a) $$\mathbf{T}\vec{j}(0)\mathbf{T}^{-1} = -\vec{j}(0),\ \mathbf{T}\rho(0)\mathbf{T}^{-1} = \rho(0)$$

or, since $\mathbf{T}i\rho(0)\mathbf{T}^{-1} = -i\mathbf{T}\rho(0)\mathbf{T}^{-1}$,

(8.48b) $$\mathbf{T}j_\mu(0)\mathbf{T}^{-1} = -j_\mu(0).$$

This condition will now be used to show that

(8.49) $$G_3(q^2) = 0$$

if T invariance is valid.

The result is obtained by applying the fundamental property of the antiunitary transformation:[17]

(8.50a) $$\begin{aligned}\langle \mathbf{T}(\vec{p}_f s_f) | \mathbf{T}j_\mu(0) | \vec{p}_i, s_i\rangle &= \langle \vec{p}_f, s_f | j_\mu(0) | \vec{p}_i, s_i\rangle^* \\ &= \varepsilon_\mu \langle \vec{p}_i, s_i | j_\mu(0) | \vec{p}_f, s_f\rangle,\end{aligned}$$

where the second step follows from the Hermiticity condition. But also

(8.50b) $$\begin{aligned}\langle \mathbf{T}(\vec{p}_f, s_f) | \mathbf{T}j_\mu(0) | \vec{p}_i, s_i\rangle &= \langle \mathbf{T}(\vec{p}_f, s_f) | \mathbf{T}j_\mu(0)\mathbf{T}^{-1} | \mathbf{T}(\vec{p}_i, s_i)\rangle \\ &= -i^{s_i - s_f}\langle -\vec{p}_f, -s_f | j_\mu(0) | -\vec{p}_i, -s_i\rangle,\end{aligned}$$

in accordance with eq. (8.48) and the Wigner phase convention on the spinor fields. Therefore

(8.50c) $$\begin{aligned}&\varepsilon_\mu \langle \vec{p}_i, s_i | j_\mu(0) | \vec{p}_f, s_f\rangle \\ &\quad = -i^{s_i - s_f}\langle -\vec{p}_f, -s_f | j_\mu(0) | -\vec{p}_i, -s_i\rangle\end{aligned}$$

if all relevant interactions are T invariant.

[16] The current density operator of the nucleon or quark field is proportional to $i{:}\bar{\psi}(x)\gamma_\mu\psi(x){:}$, where $\psi(x)$ is the appropriate spinor field. Since $\psi(x)$ satisfies eq. (8.35b), this current density transforms as the four-vector eq. (8.12b). The same can be said of currents associated with the pion or other tensor fields. Therefore $j_\mu(x)$ satisfies the classical conditions.

[17] Note that, since the nucleon is a free particle before and after the interaction with the photon, it is not necessary to specify whether the states are "in" states or "out" states.

We apply this condition for T invariance to eq. (8.47) by making use of eq. (7.83a), which leads to the identity

(8.51a) $$-i^{s_i-s_f}\bar{u}_{-f}(-\vec{p}_f)\gamma_\mu u_{-i}(-\vec{p}_i) \equiv -[\sigma_2 u_f^*(\vec{p}_f)]^\dagger \gamma_4 \gamma_\mu \sigma_2 u_i^*(\vec{p}_i).$$

Here the subscripts $-f$ and $-i$ are shorthand for $-s_f$ and $-s_i$, respectively. Since the quantities appearing in eq. (8.51a) are inner products of ordinary spinors and are unchanged by taking the transpose, the equation may be rewritten as

(8.51b) $$-i^{s_i-s_f}\bar{u}_{-f}(-\vec{p}_f)\gamma_\mu u_{-i}(-\vec{p}_i) \equiv \varepsilon_\mu \bar{u}_i(\vec{p}_i)\gamma_\mu u_f(\vec{p}_f),$$

where the factor $-\varepsilon_\mu$ arises from the anticommutation of the gamma matrices.

Similarly, we find that

(8.51c) $$-i^{s_i-s_f}\bar{u}_{-f}(-\vec{p}_f)u_{-i}(-\vec{p}_i) \equiv -\bar{u}_i(\vec{p}_i)u_f(\vec{p}_f).$$

Therefore the condition for time reversal eq. (8.50c) expressed in terms of the form factors, eq. (8.47), becomes

(8.51d) $$\begin{aligned}&\varepsilon_\mu \bar{u}_i(\vec{p}_i)[\gamma_\mu G_1(q^2) + i(p_{i\mu} + p_{f\mu})G_2(q^2) - q_\mu G_3(q^2)]u_f(\vec{p}_f)\\ &\equiv \varepsilon_\mu \bar{u}_i(\vec{p}_i)[\gamma_\mu G_1(q^2) + i(p_{f\mu} + p_{i\mu})G_2(q^2) + q_\mu G_3(q^2)]u_f(\vec{p}_f),\end{aligned}$$

from which eq. (8.49) follows.

Since eq. (8.49) is a direct consequence of the assumption of T invariance, it might be assumed that a test of T invariance could be obtained from a measurement of G_3. But that is not the case, because, as we shall now find, charge conservation also and independently leads to eq. (8.49).

The requirement of charge conservation can be stated as the equation of continuity

(8.52a) $$\sum_\mu \frac{\partial j_\mu(x)}{\partial x_\mu} = 0.$$

By applying this equation to eq. (8.46a), we find

(8.52b) $$\sum_\mu q_\mu \langle \vec{p}_f, s_f | j_\mu(0) | \vec{p}_i, s_i \rangle = 0$$

or, from eq. (8.47),

(8.52c) $$\bar{u}_f(\vec{p}_f)\left[\sum_\mu \gamma_\mu q_\mu G_1(q^2) - i(p_i^2 - p_f^2)G_2(q^2) - q^2 G_3(q^2)\right]u_i(\vec{p}_i) = 0.$$

That the first term vanishes follows from eq. (7.68b), and the second term vanishes because $p_i^2 = p_f^2 = M^2$. Therefore charge conservation requires that

(8.52d) $$G_3(q^2) = 0,$$

except for a possible discontinuity at $q^2 = 0$. Since in the reference frame for which $\vec{p}_f = -\vec{p}_i$ (the Breit frame) $q^2 = 0$ corresponds to $q_\mu = 0$, such a discontinuity would make no contribution to the matrix elements, eq. (8.47), and its covariance implies that it would make no contribution in any other reference frame. Thus the observation of $G_3 \neq 0$ would signify not only a violation of T invariance but also a violation of charge conservation, which is generally accepted as inviolate.

For completeness, we should also note that, by making use of eq. (7.68b), the matrix element may be expressed in terms of the two nonvanishing form factors as[18]

(8.53a) $$\langle \vec{p}_f, s_f | j_\mu(0) | \vec{p}_i, s_i \rangle = ie\bar{u}_f(\vec{p}_f)[\gamma_\mu F_1(q^2) - \sum_\nu \sigma_{\mu\nu} q_\nu F_2(q^2)] u_i(\vec{p}_i),$$

where $\sigma_{\mu\nu}$ is given by eq. (8.10a). The more conventional Dirac form factor F_1 and Pauli form factor F_2 are related to G_1 and G_2 by

(8.53b) $$G_1(q^2) = F_1(q^2) + 2MF_2(q^2),$$

(8.53c) $$G_2(q^2) = F_2(q^2).$$

Although the requirement of charge conservation rules out the possibility of testing T invariance by means of the interaction of a photon with a free nucleon, Lipschutz (1967) has observed that the connection between the conditions imposed by gauge invariance (charge conservation) and T invariance is limited to this special case. If the nucleon is in a virtual (off the mass shell) state as the result of other interactions or additional electromagnetic interactions as in figure 8.2, the interaction with a photon described by the photon-nucleon vertex functions is no longer the simple

[18] Foldy (1952) was the first to describe the structure of the matrix element. He used invariance arguments to express it in terms of an expansion in powers of the D'Alembertian. Salzman (1955) was the first to introduce eq. (8.53a) by showing that Foldy's expansion was equivalent to an expansion of the form factors in powers of q^2. The reason they obtained just two, rather than three, form factors is that Foldy's expansion was generated from a gauge-invariant generalization of the Dirac equation, eq. (7.61a). The gauge invariance guarantees charge conservation and therefore eliminates G_3. Although the possibility of the T-invariance argument was implied in the treatment of weak interaction form factors by Goldberger and Treiman (1958), the first explicit use of the argument for the electromagnetic form factor G_3 as given here was made independently by Ernst, Sachs, and Wali (1960). For a good review of the early work on electromagnetic form factors of the nucleon see Yennie, Lévy, and Ravenhall (1957).

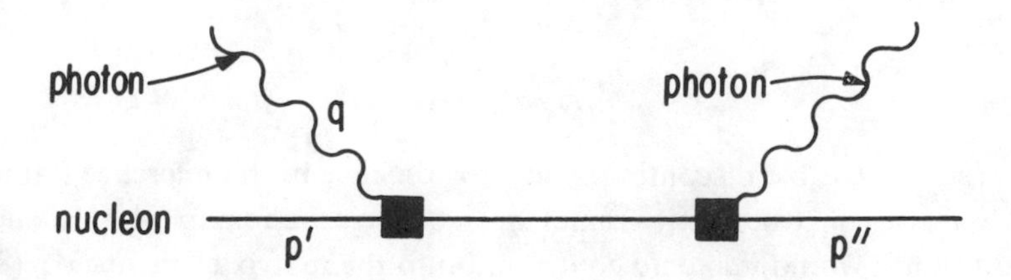

FIG. 8.2 A diagram for Compton scattering by a nucleon. The photon lines are on the mass shell, and the internal nucleon line is off the mass shell.

matrix element of the nucleon current density, eq. (8.47). It takes on a much more complicated form depending on p^2 and p'^2 as well as q^2, where p and p' are the four-momenta of the virtual nucleon before and after interaction. There are, in fact, a total of twelve parity-conserving, four-vector terms, each proportional to a scalar function of the three variables p^2, p'^2, and q^2 (Bincer 1960).

Lipschutz showed that the conditions imposed by gauge invariance do not, in general, lead to the same conditions on these generalized form factors as does T invariance. Therefore it is possible in principle to test T invariance of the interactions of photons with nucleons by means of measurements on virtual nucleons. Lipschutz considered the special example of Compton scattering of a photon by a nucleon, figure 8.2, and showed that a substantial violation of T invariance could lead to a measurable left-right asymmetry in the Compton scattering from polarized protons.

Because of the possibility of an effect for nucleons in virtual states, processes involving the scattering of photons from nucleons in nuclei (which are virtual because of scattering from other nucleons) offer an opportunity for testing T invariance. These effects would contribute to the form factors for elastic scattering of an electron from the nucleus as a whole, form factors that are defined in terms of the matrix element of the nuclear current density having a structure that depends on the nuclear spin. Of course, for a nucleus of spin $\frac{1}{2}$ this matrix element has the same form and is subject to the same conditions as that of the nucleon. Therefore that is a case in which no information about T invariance can be obtained.

It is easily seen that there is also such a limitation on the form factors for elastic scattering of an electron from nuclei of spin 0. However, for nuclei of spin greater than $\frac{1}{2}$, there is no general basis for excluding T-violating terms in the matrix element except the assumption of T invariance. For a nucleus whose angular momentum (spin) operator is $\vec{\mathbf{J}}$, there may, for example, be a term in the matrix element of the current density $\vec{j}(0)$ having the form

$$\langle m_f | \vec{\mathbf{J}} | m_i \rangle \times (\vec{p}_i + \vec{p}_f) F(q^2), \tag{8.54}$$

where m_i and m_f are the initial and final spin magnetic quantum numbers. In this case charge conservation would place no constraint on such a term, because its contribution to $\sum_\mu q_\mu \langle \vec{p}_f, m_f | j_\mu(0) | \vec{p}_i, m_i \rangle$ is proportional to $p_f^2 - p_i^2 = 0$. Since $\vec{j}$ changes sign under motion reversal (as do $\vec{\mathbf{J}}$ and $\vec{p}$) and eq. (8.54) does not, the observation of a term of the form eq. (8.54) would provide a measure of T violation after correction is made for the final state effect owing to the Coulomb interaction of electron and nucleus.

Other measures of T violation for the electromagnetic interactions of nuclei have been discussed in chapter 4, and we found there that all the experimental evidence is consistent with T invariance of nuclear and electromagnetic interactions (see eq. 4.62 et seq.).

8.5 Consequences of *T* Invariance for Weak Interaction Form Factors

Since the weak interactions of leptons with nucleons may be described in terms of nucleon current operators analogous to those describing the electromagnetic interactions (Feynman and Gell-Mann 1958; Sudarshan and Marshak 1958), their contribution to the elastic scattering of leptons by nucleons may also be described in terms of form factors analogous to the electromagnetic form factors. However, there are significant differences between the two cases that warrant some attention.

The most important of these differences is that in addition to a vector weak current density operator $J_\mu^w(x)$, the weak interactions also include a pseudovector (axial vector) current density operator $J_\mu^5(x)$ of comparable magnitude. These operators are given by the bilinear forms eqs. (8.9b) and (8.9d). Therefore there are two sets of form factors to be considered, one set for the matrix element of the vector current and the other set for the matrix element of the axial vector current (Goldberger and Treiman 1958).

Let us consider the quasi-elastic scattering of a neutrino from a proton to produce a charged lepton and a neutron, as shown in figure 8.3*a*. Since the nucleon is in a different charge state before and after the collision, it is convenient to describe both particles in terms of a single spinor field operator $\psi(x)$ having two sets of components, one corresponding to each charge state. Thus the components carry, in addition to the Dirac label $\alpha = 1, 2, 3$, or 4, an isotopic spin label $I_3 = \pm\frac{1}{2}$ (or $\tau_3 = \pm 1$). Then the current density operators for both the electromagnetic and the weak interactions may be formulated in terms of bilinear products of these isospinor operators to include both the proton and the neutron.

In terms of field operators labeled by isotopic spin, the nucleon electromagnetic current density is the sum of an isotopic scalar and the third component of an isotopic vector in the approximation that treats the neutron-proton mass difference as negligible. For example, the term associ-

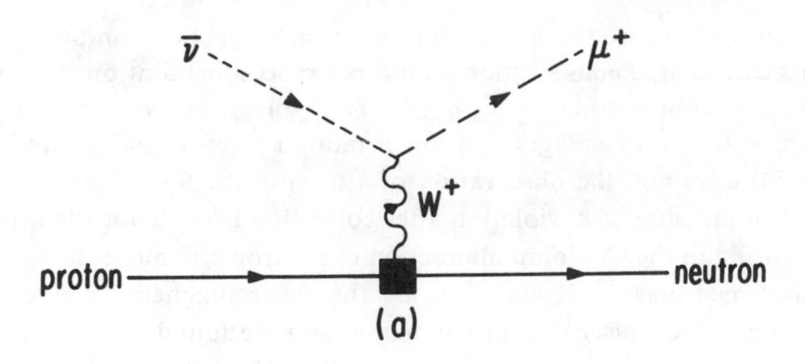

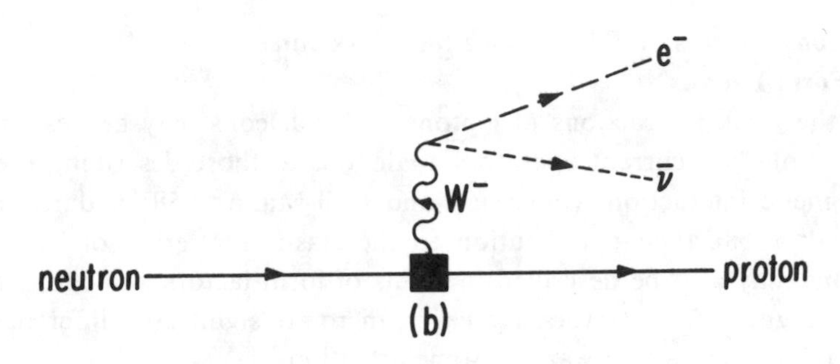

FIG. 8.3 (*a*) Quasi-elastic scattering of a neutrino by a proton described by exchange of a charged intermediate vector boson, W. (*b*) Corresponding diagram for neutron beta decay. Note that the difference between (*a*) and (*b*) is a change in sign of all four components of the neutrino momentum, converting $q^2 < 0$ to $q^2 > 0$.

ated with the isospinor field is

(8.55a) $$j_\mu(x) = \frac{ie}{2} :\bar{\psi}(x)(1 + \tau_3)\gamma_\mu \psi(x):,$$

where τ_1, τ_2, τ_3 are the 2×2 Pauli matrices for isotopic spin and have the same form as σ_x, σ_y, σ_z. Corresponding contributions to the electromagnetic current are made by other fields (the pionic term would include only an isovector). We write the total vector current operator as

(8.55b) $$J_\mu(x) = e[S_\mu(x) + V_{3\mu}(x)],$$

where S_μ is an isoscalar operator and $V_{3\mu}$ is the third component of an isovector operator.

The weak vector current density operators may be formed similarly. For example, the nucleon isospinor field contributes the term

(8.55c) $$j_\mu^\pm(x) = \frac{ig}{2} :\bar{\psi}(x)(\tau_1 \pm i\tau_2)\gamma_\mu \psi(x):,$$

since $\tau_1 \pm i\tau_2$ transform protons into neutrons and vice versa. In general, the weak vector current can be written as a linear combination of the first and second components, $j = 1, 2$, of the isovector operators $V_{j\mu}(x)$ with a weak coupling constant g as a coefficient replacing the e appearing in eq. (8.55b). The strong interactions are assumed to conserve total isotopic spin so that, if we assume we can neglect small effects of electric charge on the nucleon structure such as Coulomb effects and the neutron-proton mass difference, the relationships between the form factors that are generated by the electric current and the weak current can be expressed in terms of the isoscalar form factors given by

(8.56a)
$$\langle \vec{p}_f, s_f, I_{3f} | S_\mu(x) | \vec{p}_i, s_i, I_{3i} \rangle = i\bar{u}_f(\vec{p}_f, I_{3f}) \left[\gamma_\mu F_1^s(q^2) - \sum_\nu \sigma_{\mu\nu} q_\nu F_2^s(q^2) \right] u_i(\vec{p}_i, I_{3i}),$$

and the isovector form factors given by

(8.56b)
$$\langle \vec{p}_f, s_f, I_{3f} | V_{j\mu}(x) | \vec{p}_i, s_i, I_{3i} \rangle = i\bar{u}_f(\vec{p}_f, I_{3f}) \left[\gamma_\mu F_1^v(q^2) - \sum_\nu \sigma_{\mu\nu} q_\nu F_2^v(q^2) \right] \tau_j u_i(\vec{p}_i, I_{3i}),$$

where $u_s(\vec{p}, I_3)$ are the spinors of the plane wave nucleons having Dirac spinor labels $\alpha = 1$ to 4 and the isospin label $I_3 = \pm\frac{1}{2}$ corresponding to the physical proton and neutron, respectively.

In writing eqs. (8.56), account has already been taken of the equation of continuity that is necessarily satisfied by $V_{\pm\mu} = V_{1\mu} \pm iV_{2\mu}$ because it is satisfied by eq. (8.55b) *and* we have assumed invariance under rotations of the isospin vectors (negligible $n - p$ mass difference and charge independence of strong interactions). This result (Feynman and Gell-Mann 1958) is referred to as Conservation of the Weak Vector Current (CVC). It implies conservation of the weak vector interaction charge g in weak interaction processes just as the conservation of electric charge is implied by the equation of continuity for the electromagnetic current density.

Another consequence is that the matrix element of the weak vector current automatically satisfies the condition of T invariance because no form factor of the form $G_3(q^2)$ appears in eq. (8.56b). The demonstration that the assumption of T invariance leads to the equivalent of eq. (8.51d), and therefore of eq. (8.49), for the weak vector current eq. (8.56b) follows immediately from the fact that the current operator eq. (8.55b) behaves under T as the bilinear vector form eq. (8.9b), namely, in accordance with eq. (8.12b). The condition of T invariance for the matrix element of this current operator therefore takes the same form as eqs. (8.50). It has been emphasized that these results are valid only to the extent that the $n - p$ mass difference is negligible. We must anticipate that a term proportional to the mass difference of the form of the G_3 term will occur in the matrix element of the physical current. This term can be expected to be of the same order as electromagnetic radiative corrections to the elastic scattering or, equivalently, of the order of final state interaction effects, as will be discussed further below for the case of the axial vector weak current.

The weak axial vector current $J_\mu^5(x)$, which has no electromagnetic analogue, does not satisfy an equation of continuity,[19] and it includes only isovector components. Therefore three form factors are required to describe its matrix element in the absence of the assumption of T invariance. Hence we write (compare eq. 8.9d for the definition of a Hermitian $V_{j\mu}^5$)

(8.57)
$$\begin{aligned}\langle \vec{p}_f, s_f, I_{3f} | V_{j\mu}^5 | \vec{p}_i, s_i, I_{3i}\rangle &= i\bar{u}_f(\vec{p}_f, I_{3f})\tau_j \gamma_5[\gamma_\mu G_1^5(q^2) + (p_{f\mu} + p_{i\mu})G_2^5(q^2) \\ &\quad + iq_\mu G_3^5(q^2)]u_i(\vec{p}_i, I_{3i})\end{aligned}$$

for $j = 1$ and 2 only. The Hermitian operators appearing on the left-hand side represent the $j = 1$ and $j = 2$ components of the weak isovector, axial vector current, and J_μ^5 is a linear combination of the two.

The current $J_\mu^5(0)$ is a pseudovector that transforms under T in accordance with eq. (8.12d), that is,

(8.58)
$$\mathbf{T}J_\mu^5(0)\mathbf{T}^{-1} = -J_\mu^5(0),$$

and repetition of the steps eqs. (8.50) again leads to the same consequence if the $n - p$ mass difference is negligible:

(8.59)
$$\begin{aligned}\varepsilon_\mu \langle \vec{p}_i, s_i, I_{3i} | J_\mu^5(0) | \vec{p}_f, s_f, I_{3f}\rangle \\ = -i^{s_i - s_f}\langle -\vec{p}_f, -s_f, I_{3f} | J_\mu^5(0) | -\vec{p}_i, -s_i, I_{3i}\rangle.\end{aligned}$$

This condition for T invariance may be applied to eq. (8.57) by following the same steps that were used to obtain eq. (8.51d), with the result

(8.60)
$$\begin{aligned}&\varepsilon_\mu \bar{u}_i(\vec{p}_i)\tau_j\gamma_5[\gamma_\mu G_1^5(q^2) + (p_{i\mu} + p_{f\mu})G_2^5(q^2) - iq_\mu G_3^5(q^2)]u_f(\vec{p}_f) \\ &\equiv \varepsilon_\mu \bar{u}_i(\vec{p}_i)\tau_j\gamma_5[\gamma_\mu G_1^5(q^2) - (p_{f\mu} + p_{i\mu})G_2^5(q^2) - iq_\mu G_3^5(q^2)]u_f(\vec{p}_f).\end{aligned}$$

Therefore the consequence of the assumption of T invariance is (Goldberger and Treiman 1958)

(8.61)
$$G_2^5(q^2) \equiv 0.$$

The difference between the relative signs in eq. (8.60) and those in eq. (8.51d) results from a reversal of sign between eq. (8.51c) and the corresponding

[19] See Marshak, Riazuddin, and Ryan (1969, p. 272).

equation for $\bar{u}_i(\vec{p}_i)\gamma_5 u_f(\vec{p}_f)$ that, in turn, is due to the anticommutation of γ_4 and γ_5.

Since eq. (8.61) is a requirement of T invariance, it provides a possibility for a direct test of the time reversal symmetry. For example, measurements of the quasi-elastic scattering of neutrinos from protons might be used to determine $G_2^5(q^2)$, or at least an upper limit on $G_2^5(q^2)$, that would place an upper limit on the magnitude of direct T violation in the weak interactions of leptons and nucleons. This would be a measurement of $G_2^5(q^2)$ for space-like values of q_μ, that is, for $\vec{q}^2 > q_0^2$ ($q^2 < 0$ in our metric).

A related process determined by the same matrix element of the weak current, but for $q^2 > 0$, is the beta decay of the neutron, as can be seen from figure 8.3*b*. This case has already been discussed in chapter 5 (see figs. 5.1 and 5.2) as an example of a test of motion reversal invariance, and the measurements that were referred to there are in effect determinations of the upper limit on $G_2^5(q^2)$ for $q^2 > 0$.

The final state interactions, which correspond to higher-order corrections to the diagrams in figure 8.3, for example, to the additional exchange of a photon between the electron and proton (or to their Coulomb interaction) in figure 8.3*b*, can mimic T violation, as we found in sections 5.2 and 5.3. They lead to additional terms in the scattering amplitude, including one having the same form as the G_2^5 term. In the case of neutrino scattering from protons these effects (the interactions between a charged lepton and a neutron) would be very small, but they do place a lower limit on a measurement of T violation.[20]

Existing evidence for violation of T invariance, which will be discussed in the next chapter, is limited to the weak interactions of strange particles, and even then the effect is extremely small. Therefore it is not surprising that no additional information concerning T-violating interactions has ėmerged from experiments on the interactions of the nucleons with neutrinos. However, we shall see that larger effects might be expected in the very high energy domain, where quarks of high mass participate in the process. Therefore the study of neutrino scattering at very high energies may hold some promise for revealing new information about the nature of the T-violating weak interactions.

[20] Weak currents that give rise to form factors of the type G_3 or G_2^5 are usually referred to as "second-class currents" (Marshak, Riazuddin, and Ryan 1969, pp. 107, 314, 322). Their existence for the $n - p$ currents would be difficult to separate from the effect of T violation (and violation of the conserved weak vector current [CVC] in the case of G_3) because of the small $n - p$ mass difference. However, the T-invariance arguments used here do not apply to the strangeness changing weak currents because of the relatively large differences in the masses of particles having different strangeness. Therefore a considerable experimental effort has gone into the search for a second-class term in the strangeness changing axial vector weak current (see Hsueh et al. 1985).

9 | *CP* Violation and *T* Violation in the Decay of Neutral *K* Mesons

The only existing experimental evidence for T violation arises from observations on the decay properties of the neutral K mesons. The evidence is not direct, and there are very subtle questions associated with its interpretation. These subtleties rest on questions relating to the CPT theorem, and we must examine the *experimental* basis for that theorem in attempting to answer the questions.

The neutral K mesons have very special properties, first noted by Gell-Mann and Pais (1955), that make them a rich source of extremely interesting phenomena. It is because of these properties, in particular because they serve as an interferometer for weak interactions, that it has been possible to establish without question the existence of a tiny violation of the combined parity and charge conjugation, or CP symmetry, that in turn provides evidence for T violation. We shall first give attention to the relevant properties of the K^0 mesons and the quantum mechanics of the kinds of phenomena that result from these properties. This will provide a basis for an understanding of the connections with CP, T, and CPT invariance.

Although there is no satisfactory fundamental theory of the origin of the CP violation, in the next chapter we shall discuss some models that at least push the theory to a deeper level than that of purely phenomenological description.

9.1 Phenomenological Theory of Neutral *K* Meson Decays

Neutral K mesons play a special role in the physics of time reversal because though they are electrically neutral, they carry a different kind of "charge" called the "strangeness" quantum number or the "hypercharge."[1] Recall that the unusual behavior of the strange particles, K mesons and hyperons, may be described by invoking simple properties of the additive quantum

[1] For a review of the properties of strange particles see Gasiorowicz (1966).

number $S = 0, \pm 1, \pm 2$, etc., or equivalently, the hypercharge $Y = S + B$, where B is the baryon number ($B = 1$ for nucleon or hyperon, $B = -1$ for antinucleon or antihyperon, $B = 0$ for mesons).

The simple properties required of S are those that account for the copious production of strange particles in energetic collisions between nuclear particles and, at the same time, account for the relative stability of the strange particles once they are produced. These properties are that the total S for a system of hadrons (strongly interacting particles) is conserved in the production of strange particles by strong interactions while S changes by one unit in their decay, which is due to weak interactions. The first condition leads to the associated production of pairs of strange particles (one with $S = -1$, the other with $S = +1$) in collisions of ordinary ($S = 0$) particles, and the second accounts for the fact that a single strange particle can decay into ordinary particles.

The K^0 meson is assigned by convention the strangeness $S = +1$ and, like all other mesons, has $B = 0$. It is paired off in an isotopic spin doublet with the positively (electrically) charged K^+ meson. Just as a negatively charged antiparticle, the K^- meson, is associated with the K^+, there is an antiparticle $\bar{K}^0$ associated with the K^0. Clearly, the particle K^0 and antiparticle $\bar{K}^0$ are not to be distinguished by their electric charge in this case, because both are electrically neutral. They are distinguished by their hypercharge, however; the $\bar{K}^0$ (and the K^-) have $S = -1$.

Because its mass is about 3.5 times the mass of the π meson (pion), which has $S = 0$, there are a large number of decay channels accessible to the K^0 in accordance with the $\Delta S = \pm 1$ selection rule, including

(9.1a) $K^0 \rightarrow \pi^+\pi^-$ $\quad K^0 \rightarrow \pi^0\pi^0$

(9.1b) $K^0 \rightarrow \pi^+\pi^-\pi^0$ $\quad K^0 \rightarrow \pi^0\pi^0\pi^0$

(9.1c) $K^0 \rightarrow \pi^- e^+ \nu_e$

(9.1d) $K^0 \rightarrow \pi^+ e^- \bar{\nu}_e$

(9.1e) $K^0 \rightarrow \pi^- \mu^+ \nu_\mu$

(9.1f) $K^0 \rightarrow \pi^+ \mu^- \bar{\nu}_\mu$

(9.1ā) $\bar{K}^0 \rightarrow \pi^+\pi^-$ $\quad \bar{K}^0 \rightarrow \pi^0\pi^0$

(9.1b̄) $\bar{K}^0 \rightarrow \pi^+\pi^-\pi^0$ $\quad \bar{K}^0 \rightarrow \pi^0\pi^0\pi^0$

(9.1c̄) $\bar{K}^0 \rightarrow \pi^+ e^- \bar{\nu}_e$

(9.1$\bar{\text{d}}$) $$\bar{K}^0 \rightarrow \pi^- e^+ \nu_e$$

(9.1$\bar{\text{e}}$) $$\bar{K}^0 \rightarrow \pi^+ \mu^- \bar{\nu}_\mu$$

(9.1$\bar{\text{f}}$) $$\bar{K}^0 \rightarrow \pi^- \mu^+ \nu_\mu .$$

A variety of other very rare modes are also observed (PDG 1986, p. 14). The symbols ν_e and ν_μ refer to the neutrinos associated with the electron and muon, respectively, and the bar denotes the corresponding antineutrino. The "lepton number," l_e or $l_\mu = 1$ for e^-, μ^- and ν_e, ν_μ, respectively, and l_e or $l_\mu = -1$ for e^+, μ^+ and $\bar{\nu}_e$, $\bar{\nu}_\mu$, is additive and conserved. Application of the $\Delta S = \pm 1$ selection rule assigns strangeness only to the hadrons; thus the leptons are treated as $S = 0$ particles.

In addition to this selection rule on S, which appears to be exact for first-order weak interactions, there are other approximate selection rules for these interactions. Among them are the $\Delta I = 1/2$ rule and the $\Delta S = \Delta Q$ rule. The first, which is certainly not exact (because $K^+ \rightarrow \pi^+\pi^0$ would be forbidden and is not), asserts that $\Delta S = \pm 1$ transitions between hadron states differing by one-half a unit of *total* isotopic spin are much more probable than those having $|\Delta I| \geq \frac{3}{2}$.

The $\Delta S = \Delta Q$ rule asserts that in the decay of a hadron to hadrons plus leptons ("semileptonic decay"), the change ΔQ in the hadron electric charge (in units of the magnitude of the electron charge) must be equal to the change, ΔS, in the strangeness. Thus $\Delta S = -\Delta Q$ transitions like those described by eqs. (9.1d), (9.1f), (9.1$\bar{\text{d}}$), and (9.1$\bar{\text{f}}$) are forbidden. There is experimental evidence[2] that the ratio x of $\Delta S = -\Delta Q$ to $\Delta S = \Delta Q$ transitions is less than 1 percent, and it is generally believed at this time that the selection rule is exact, so henceforth we shall exclude the transitions (real *or* virtual) eqs. (9.1d and $\bar{\text{d}}$) and (9.1f and $\bar{\text{f}}$) from our considerations unless otherwise stated.

It appears to be merely a matter of definition of what we mean by

[2] Over a period of years a very substantial experimental effort has gone into verifying the $\Delta S = \Delta Q$ rule, the latest work being that of Niebergall et al. (1974). Their results are the basis for the upper limit of 1 percent quoted here, and their paper provides references to the earlier work. Violation of the selection rule would be of particular importance for tests of T, CPT, and CP invariance (Sachs and Treiman 1962; Sachs 1963a,b,c, 1964; Kenny and Sachs 1965). During the early 1960s there was some controversy over the experimental results on the $\Delta S = \Delta Q$ rule, which White and Sullivan (1979) use in a *Physics Today* article having the title "Social Currents in Weak Interactions" as an example of behavioral patterns of physicists; they apply the methods of behavioral science to the sociology of physics. In their attempt to identify the motivations of both theorists and experimentalists in pressing this issue, they completely overlook the invariance tests, which were among the principal original motivations. They also overlook the need for the measurement of x to interpret the charge asymmetry in semileptonic decay. See section 9.2.

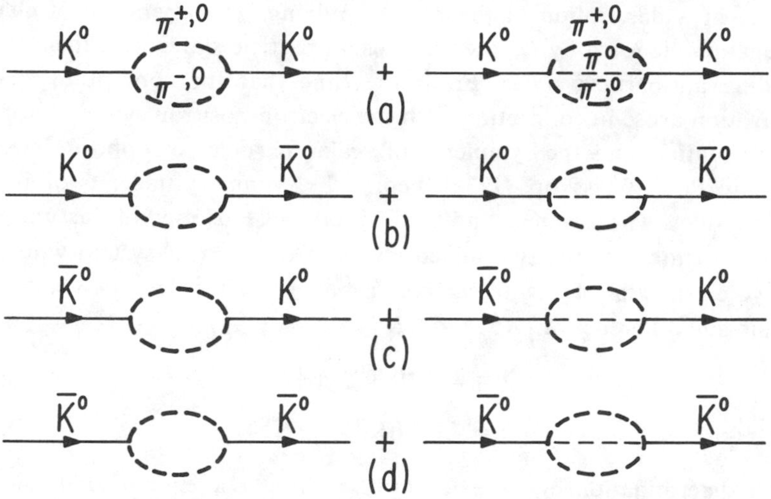

FIG. 9.1 Second-order weak interaction self-energy diagrams for the K^0, $\bar{K}^0$ system due to 2π and 3π intermediate states.

"antiparticle" that a particle and its associated antiparticle have the same mass. Although we found in section 8.3 that this is another case in which appearances are deceiving and that there are important assumptions about physics underlying this "definition," we shall take the equality of masses of the K^0 and $\bar{K}^0$ as a starting point for discussion. Then the states $|K^0\rangle$ and $|\bar{K}^0\rangle$ of these two spin 0 mesons are degenerate, that is, they have the same energy at the same value of the momentum, and the degeneracy is associated with the two values of the strangeness quantum number $S = \pm 1$.

Gell-Mann and Pais (1955) made the important observation that the existence of this degeneracy has remarkable implications for the decay properties of the K^0 mesons because the two states have the common modes of decay eqs. (9.1a and ā) and (9.1b and b̄). The existence of these common modes means that the degenerate states must be coupled by second-order weak interactions. For example, the emission of a virtual pair of pions by a K^0 can be followed by the *inverse* virtual process, formation of a $\bar{K}^0$ by annihilation of the virtual pair of pions, as indicated by diagram (*b*) in figure 9.1. This coupling means that the states $|K^0\rangle$ and $|\bar{K}^0\rangle$ are not eigenstates of the complete Hamiltonian including weak interactions. The states will be mixed and, because they are degenerate, the coefficients (mixing coefficients) of $|K^0\rangle$ and $|\bar{K}^0\rangle$ in the linear combinations forming each of the two eigenstates are expected to be comparable in magnitude.

Gell-Mann and Pais made use of the concept of charge conjugation in arriving at a description of the K^0, $\bar{K}^0$ mixing. The operation of charge conjugation, denoted by C, interchanges all particle states with their corresponding antiparticle states. From the time that the concept of charge conjugation arose in connection with the electron-positron symmetry of the Dirac equation and the symmetry of scalar particles of opposite electric charge in Pauli-Weisskopf (1934) theory, the symmetry under C of all dynamic systems was assumed, and Gell-Mann and Pais used this assumption to demonstrate that the two mixed states of the K^0, $\bar{K}^0$ system would be the two eigenstates of C with eigenvalues of ± 1. Thus these two states are the sum and difference of $|K^0\rangle$ and $|\bar{K}^0\rangle$:

(9.2a) $$|+\rangle = 2^{-1/2}(|K^0\rangle + |\bar{K}^0\rangle),$$

(9.2b) $$|-\rangle = 2^{-1/2}(|K^0\rangle - |\bar{K}^0\rangle).$$

This determination of the eigenstates without reference to dynamic details depends on the existence of a high degree of symmetry, in this case C invariance. Although the usefulness of the assumption of C invariance was short-lived, it was quickly replaced by the notion of CP invariance, which served the same purpose.

The demise of C invariance of the weak interactions was associated with violation of parity conservation. When Lee and Yang (1956) suggested that the "$\tau - \theta$ puzzle"[3] might be a consequence of parity violation in the weak interactions, Landau (1957) proposed that the basic notion of inversion symmetry would be preserved if C invariance is also violated in such a way that the weak interactions are CP invariant—that is, invariant under the combined transformation of charge conjugation and inversion.

The CP symmetry can be visualized in terms of reflection in a "CP mirror" that not only converts the right-handed system to a left-handed system but also converts every particle into its antiparticle. Graphic conceptual tests of the CP invariance of a dynamic system can then be provided by using such a mirror in the manner described in chapter 2 for ordinary reflections.

Confirmation of CP invariance would have a direct bearing on the issue of T invariance because of the CPT theorem. Since this theorem asserts

[3] That both the 2π and 3π decay modes were associated with the same particle was not understood at the time of the Gell-Mann and Pais work. In fact, the two modes (of both charged and neutral mesons) were ascribed to different mesons, the θ and the τ, respectively. Since the 2π and 3π states have opposite parity, that the θ and τ seemed to have exactly the same mass and spin was puzzling—the "τ-θ puzzle"—and the resolution of this puzzle by Lee and Yang (1956) led to the discovery of parity nonconservation. Gell-Mann and Pais treated the θ^0 case, assuming that the rare modes were semileptonic. Snow (1956) extended their ideas to the case of τ^0 decay, treating it as an independent particle.

that *all* interactions must be invariant under the combined transformation CPT, CP invariance would appear to imply T invariance. Even more important, CP violation would imply T violation to the same degree, and since there are tests of CP violation that are independent of final state interactions, they could serve to remove the shadow of T violation mimicry from the usual tests of T. The conditional phraseology here has the purpose of emphasizing that the assumptions underlying the CPT theorem, although very general and widely accepted, must themselves be subject to experimental test before a final conclusion about T can be drawn from tests of CP.

The experiment of Wu and others (1957) that establshed P violation in beta decay also established C violation (Lee, Oehme, and Yang 1957). Its results were consistent with Landau's proposal of CP invariance, as have been the results of all other experiments on beta decay since that time.

Repetition of the Gell-Mann and Pais treatment under the assumption of CP rather than C invariance leads to the conclusion that the two mixed states of the K^0, $\bar{K}^0$ system are eigenstates of CP with eigenvalues ± 1. Although the selection rules now refer to CP instead of C, the original conclusion that only the even state (i.e., eigenvalue $+1$) can decay into two pions is still valid because the 2π state of angular momentum zero is even under CP.[4] That there is a state (the odd state) that cannot decay into two pions is then a direct consequence of the assumption of CP invariance. Of course, it can still decay into the other modes of eq. (9.1).

The branching ratio for decay of the K^0 into the 2π mode is much (about 500 times) greater than that for all the other modes, therefore Gell-Mann and Pais predicted that, in addition to the short-lived particle (K_S) that decays into two pions, there must exist a long-lived ($\sim 5 \times 10^{-8}$ sec) particle (K_L) that decays into other modes. This long-lived particle was soon detected by Lande and others (1956) and Fry, Schneps, and Swami (1956).

To the extent that one accepts this model, the K^0, $\bar{K}^0$ system provides a very direct and sensitive test of CP invariance: a K_L beam, which is what remains of a K^0 (or $\bar{K}^0$) beam at some distance (many lifetimes of the K_S) from the source, must not decay into the two pion modes. It was in 1964 that Christenson and others found that the K_L *does* decay into the 2π mode, suggesting that CP invariance is violated.

[4] The K mesons are pseudoscalar; that is, they have zero spin and odd intrinsic parity. The application of CP to the state $|K^0\rangle$ therefore produces a $\bar{K}^0$ state with the opposite sign from that produced by C alone. Since, as we shall see shortly, the relative phases of the $|K^0\rangle$ and $|\bar{K}^0\rangle$ states are otherwise undetermined, we shall *define* the $\bar{K}^0$ states by $|\bar{K}^0\rangle = CP|K^0\rangle$. Compare eq. (9.25). Note that the opposite convention was used in writing eqs. (9.2) and in chapters 7 and 8. See the comments concerning η_C in the paragraph following eq. (7.58b).

However, the conclusion is model dependent and does *not* constitute a *proof* of CP violation, because the evidence that the observed 2π mode originated from a K_L decay is based on the identity of the energy and momenta of the two pions with what is to be expected from the decay of a particle having the K^0 mass. On the basis of this evidence *alone*, it is not possible to exclude a model that includes, in addition to the K^0 and $\bar{K}^0$ mesons, another neutral meson of the same mass having properties that permit the slow decay into the 2π mode. We can see very clearly that the observation is not directly relevant to CP invariance by comparing the results of the experiment with the results that would be obtained by viewing the experiment in a CP mirror. The question is whether the $K_L \rightarrow 2\pi$ rate depends on variables that appear to change when viewed in the CP mirror. The only variable entering into the rate for the 2π decay of a single particle of spin 0 is the magnitude of the relative momentum of the two pions, and that is fixed by conservation of energy and momentum. It takes on the same value on both sides of the mirror, so the comparison does not provide a test of CP invariance.

We shall find that the conclusive test requires not only that the $K_L \rightarrow 2\pi$ amplitude be different from zero, but also that it be coherent with (i.e., interfere with) the $K_S \rightarrow 2\pi$ amplitude. To demonstrate this point and discuss further the question of T invariance, we require a more general treatment of the K^0, $\bar{K}^0$ eigenstate problem, a dynamic treatment in which no assumption is made about CP, T, or for that matter CPT invariance. Such is the purpose of the remainder of this section.[5]

Before we enter into the specifics of the analysis, a few general remarks

[5] A variety of equivalent treatments have been presented in the literature, the earliest making use of the Weisskopf and Wigner (1930) modified, time-dependent perturbation theory: Treiman and Sachs (1956) and Lee, Oehme, and Yang (1957). For a complete discussion of this method applied to the K^0, $\bar{K}^0$ system see Kabir (1968a, app. A, pp. 99 ff.). The method of colliding wave packets (the "propagator method"; Jacob and Sachs 1961) described in sections 5.1 and 8.3 is used in a general treatment of the dynamics of the K^0, $\bar{K}^0$ system by Sachs (1963b). The propagator method arises naturally in the context of quantum field theory and is therefore better adapted to questions concerning fundamental theories of weak interactions. On the other hand, the generalized Weisskopf-Wigner method is expressed more directly in terms of the measured quantities and is therefore more transparent from a phenomenological viewpoint. That the two methods are consistent with one another has been demonstrated by Lipschutz (1966). The approach used here will follow the more phenomenological method.

There are many published summaries of the phenomenological theory. They include, in addition to the general treatment based on the propagator method (Sachs 1963b), Wu and Yang (1964), Enz and Lewis (1965), Bell and Steinberger (1966), Lee and Wu (1966), Okun' (1966), Gourdin (1967), Kabir (1968a), and Wolfenstein (1968, 1969a,b). The Wolfenstein (1968) article is one of an interesting collection of papers presented in a seminar on CP violation: *Uspekhi Fiz. Nauk.* 95:402–525 (translation 1969, *Soviet Phys. Uspekhi* 11:461–533). See in particular Okun' (1968).

about phase conventions may be helpful because, as we found earlier, the resolution of *T*-invariance questions inevitably comes down to an experimental determination of a relative phase between states. The states in question are the eigenstates of strong and electromagnetic interactions, which are the zero-order states in the approximation that the weak interactions are very small perturbations. The relative phases of these sets of eigenstates (sectors of the Hilbert space) between which there is no coupling in zero order are arbitrary—that is, not measurable in this approximation.

Because of the $\Delta S = 0$ selection rule for strong and electromagnetic interactions, the relative phases of the $S = 0$, $S = +1$, $S = -1$, etc., sectors may be defined arbitrarily; they are a matter of convention. Therefore a convention for fixing the relative phases between the $S = \pm 1$ sectors[6] and another for fixing that between the $S = 0$ and $S = 1$ sectors will be required. Once these conventions are fixed, physically meaningful absolute phases can be assigned to the matrix elements of the perturbations connecting states of different *S*. Although a change in the convention will yield a corresponding change in these "absolute" phases, the consequences of any physical measurement will not be altered.

The way this convention is set for the matrix elements describing decay of the neutral *K* mesons will be described later. These matrix elements may be expressed in terms of the weak interaction H_w by eq. (5.18), with $W \equiv H_w$. There are two possible initial states corresponding either to the K^0 or to the $\bar{K}^0$, and for this special case the initial state will be indicated by indices *i*, *j*, *k*, ..., each of which may be assigned the values 1 and 2 for the K^0 and $\bar{K}^0$, respectively. The decaying *K* meson will be assumed to be at rest, that is, to have zero momentum. Since the spin of the meson is also zero, no variables other than *j* are required to specify the initial state. The final states *f* are the substates of the various decay modes eq. (9.1) corresponding to the momenta and, where appropriate, spin states of the specified particles. Of course both the total momentum and angular momentum of the final states are zero. The two-particle decays in this center-of-mass (CM) reference frame are collinear, and three-particle decays are coplanar.

Since the mass of the K^0 is about 3.5 times the mass of the pion, the energies of the decay products are relativistic, and each of the energy variables used here and later will be given in units of inverse length by the relativistic expression $E = c(m^2 + p^2/c^2)^{1/2}/\hbar$, where *m* is the rest mass and *p* the magnitude of the momentum of the particle under discussion. When the K^0 is at rest, as we are assuming, the initial energy is $E_i = m_K$, where m_K is now also given in units of inverse length (the inverse of the Compton wave

[6] See note 4.

length of the K meson). This choice of units is equivalent to the choice $\hbar = c = 1$ introduced in note 2 of chapter 6.

The matrix element appearing in the definition of W_{fi}, eq. (5.18), is evaluated at $E_f = m_K$ because energy is conserved in the actual decay process. However, we shall be considering second-order processes and must therefore deal with the virtual transitions to an intermediate state of arbitrary energy, E. These intermediate states will include not only all possible decay modes but also every channel that is coupled to the K^0 or $\bar{K}^0$ state by H_w, whether or not it is energetically accessible (for example, the one-pion, 4π, 5π, etc., states). Therefore we define a virtual decay amplitude to a channel c normalized in the same way as the real decay amplitude eq. (5.25), but in the LSZ notation (see eqs. 8.25 and 5.18):

(9.3a)
$$A_{cj}(E_c) = \left(\frac{2\pi}{\hbar} d_c\right)^{1/2} \langle c, \text{out} | H_w | j \rangle$$

for $E_c > \sum_\gamma m_\gamma$, and

(9.3b)
$$A_{cj}(E_c) = 0$$

for $E_c < \sum_\gamma m_\gamma$. Here $j = 1, 2$ refers to the state $|K^0\rangle$, $|\bar{K}^0\rangle$, respectively, and the m_γ are the masses of the particles in channel c, γ labeling each of the particles.

The real decay amplitude into any final channel f that is open (i.e., for which $\sum_\gamma m_\gamma < m_K$) is, in accordance with eqs. (5.18) and (5.25),

(9.4)
$$A_{fj} = A_{fj}(E_f = m_K),$$

where m_K is the mass of the K^0 and the $\bar{K}^0$, *before corrections owing to the effects of weak interactions are taken into account.* The equality of the mass of particle and antiparticle follows from our implicit assumption that CPT invariance is valid for all interactions except, perhaps, the weak interactions. These nonweak interactions will, in fact, also be assumed to be invariant under C, P, and T separately.

When A_{fj} appears here henceforth without an explicit energy variable, it is to be interpreted as the actual decay amplitude. It will also be found convenient to use another notation for the decay amplitudes interchangeably with the A_{fj}, depending on the context. When we deal separately with the K^0 and $\bar{K}^0$ amplitudes we shall use

(9.5a)
$$A_f \equiv A_{f1}$$

for decay of the K^0 and

(9.5b) $$\bar{A}_f \equiv A_{f2}$$

for decay of the $\bar{K}^0$.

To describe the time dependence of the decay of a K^0 or $\bar{K}^0$ in the general case, it is necessary to take into account the mixing of the K^0 and $\bar{K}^0$ states owing to the second-order effects of the weak interactions as shown in figure 9.1. The results of such an analysis (Treiman and Sachs 1956; Lee, Oehme, and Yang 1957; Sachs 1963b) show that the exponentially decaying states are the eigenstates of the 2×2 matrix of the total Hamiltonian of the K^0, $\bar{K}^0$ system, each of the four matrix elements corresponding to one of the self-energy diagrams in figure 9.1. Since the decay rates are usually defined in the rest frame of the decaying particle, this Hamiltonian is usually called the "mass matrix" and will be denoted by **M**, the units again being units of inverse length, mass $\times$ $c/\hbar$.

To second order in H_w, the matrix **M** may be expressed in terms of the amplitudes $A_{cj}(E)$, where E is an arbitrary energy of the intermediate channel c. It is also important to include the possibility that W contains, in addition to H_w, a $\Delta S = \pm 2$ term H_{ww} that would give a direct coupling between $|K^0\rangle$ and $|\bar{K}^0\rangle$. Such a term arises quite naturally in the generally accepted quark theories of weak interactions, as we shall see later. In earlier years, Wolfenstein (1964) suggested the possible existence of a "superweak" interaction of this type to account for *CP* violation.

When these contributions are included, **M** is found to have the form (Kabir 1968a, app. A, pp. 99 ff.):

(9.6) $$\mathbf{M} = M - \tfrac{1}{2}i\Gamma,$$

with

(9.7a) $$\langle j|\Gamma|k\rangle = \sum_f A_{fj}^* A_{fk}$$

and

(9.7b) $$\langle j|M|k\rangle = m_K \delta_{jk} + \langle j|H_{ww}|k\rangle - \frac{1}{2\pi}\mathbf{P}\int \frac{dE}{E - m_K} \sum_c A_{cj}^*(E) A_{ck}(E).$$

The symbol **P** in front of the integral denotes the Cauchy principal value. This integral is referred to as the "dispersive" term, while Γ is referred to as the "absorptive" term. The only assumptions made in arriving at this result are that the dynamics are determined by the time-dependent Schroedinger equation, that higher-order terms in H_w and H_{ww} may be neglected, and that H_w is Hermitian.

Note that, since H_w and H_{ww} are presumed to be Hermitian, both M and Γ are Hermitian matrices:

(9.8) $$M^\dagger = M, \quad \Gamma^\dagger = \Gamma,$$

but, clearly, **M** is not. Therefore we must distinguish between the right-hand and left-hand eigenstates corresponding to multiplication by **M** from the right and from the left. These states are not, as in the case of a Hermitian operator, simply conjugate to one another. We define (Sachs 1963b) the states of the short-lived K_S and the long-lived K_L as the left-hand eigenstates

(9.9) $$\mathbf{M}|\alpha\rangle = \lambda_\alpha|\alpha\rangle,$$

where the Greek labels $\alpha = S$ or L correspond to the eigenstates $|K_S\rangle$ or $|K_L\rangle$, in contrast to the Latin labels $j = 1$ or 2 used earlier to correspond to $|K^0\rangle$ or $|\bar{K}^0\rangle$. Because **M** is not Hermitian the eigenvalues λ_α are complex, and the real and imaginary parts are denoted by m_α and $-\frac{1}{2}\Gamma_\alpha$, respectively:

(9.10) $$\lambda_\alpha = m_\alpha - \tfrac{1}{2}i\Gamma_\alpha.$$

We continue to use the standard Dirac notation

(9.11) $$|\alpha\rangle^\dagger = \langle\alpha|$$

so that $\langle S|$ and $\langle L|$ are *not* right-hand eigenstates.

The states are normalized in the standard way:

(9.12) $$\langle\alpha|\alpha\rangle = 1,$$

but the states $|S\rangle$ and $|L\rangle$ are not orthogonal in general. The usual argument for the orthogonality of eigenstates leads to a different result. It may be paraphrased as follows: eq. (9.9) may be rewritten as

(9.13a) $$(M - \tfrac{1}{2}i\Gamma)|\alpha\rangle = \lambda_\alpha|\alpha\rangle,$$

and its Hermitian conjugate becomes

(9.13b) $$\langle\alpha|(M + \tfrac{1}{2}i\Gamma) = \lambda_\alpha^*\langle\alpha|,$$

since M and Γ are Hermitian. Thus, from eq. (9.13a),

(9.14a) $$\langle\beta|(M - \tfrac{1}{2}i\Gamma)|\alpha\rangle = \lambda_\alpha\langle\beta|\alpha\rangle.$$

But by making use of eq. (9.13b) with α replaced by β, we also find

(9.14b) $$\langle\beta|(M + \tfrac{1}{2}i\Gamma)|\alpha\rangle = \lambda_\beta^*\langle\beta|\alpha\rangle.$$

The difference between eqs. (9.14b) and (9.14a) is

$$i\langle\beta|\Gamma|\alpha\rangle = (\lambda_\beta^* - \lambda_\alpha)\langle\beta|\alpha\rangle. \tag{9.15}$$

Thus, if we set $\alpha \equiv S$ and $\beta \equiv L$, instead of an orthogonality condition we find

$$[m_L - m_S + \tfrac{1}{2}i(\Gamma_L + \Gamma_S)]\langle L|S\rangle = i\langle L|\Gamma|S\rangle. \tag{9.16}$$

Also, when we set $\beta = \alpha$ and use the normalization condition $\langle\alpha|\alpha\rangle = 1$, the diagonal elements of Γ are found to be given by

$$\langle\alpha|\Gamma|\alpha\rangle = \Gamma_\alpha, \tag{9.17}$$

where $-\frac{1}{2}\Gamma_\alpha$ is the imaginary part of λ_α, eq. (9.10).

The eigenstates $|\alpha\rangle$ are linear combinations of the K^0 and $\bar{K}^0$ states:

$$|\alpha\rangle = \sum_{j=1}^{2} |j\rangle C_{j\alpha}, \tag{9.18}$$

and the matrix elements therefore are given by

$$\langle\beta|\Gamma|\alpha\rangle = \sum_{j,k} C_{j\beta}^*\langle j|\Gamma|k\rangle C_{k\alpha}. \tag{9.19}$$

From the definition of Γ, eq. (9.7a), it then follows that

$$\langle\beta|\Gamma|\alpha\rangle = \sum_f \left(\sum_j C_{j\beta} A_{fj}\right)^* \left(\sum_k C_{k\alpha} A_{fk}\right). \tag{9.20}$$

Since the decay amplitude of the state $|\alpha\rangle$ is given, according to eqs. (9.18) and (9.3a), by

$$A_{f\alpha} = \sum_j C_{j\alpha} A_{fj}, \tag{9.21}$$

eq. (9.20) is simply

$$\langle\beta|\Gamma|\alpha\rangle = \sum_f A_{f\beta}^* A_{f\alpha}. \tag{9.22}$$

Therefore the condition eq. (9.16) may be written as the "Bell-Steinberger relation"

$$[\tfrac{1}{2}(\Gamma_L + \Gamma_S) - i(m_L - m_S)]\langle L|S\rangle = \sum_f A_{fL}^* A_{fS}, \tag{9.23}$$

where A_{fL} and A_{fS} are the decay amplitudes into final state f of the K_L and K_S mesons, respectively. This relationship was first derived by Bell and Steinberger (1966), who showed that it followed from the requirement that probability be conserved (unitarity). The more direct method used here implicitly contains the assumption of unitarity because it depends on the condition that the interactions be Hermitian.[7] As is well known, the requirement that the Hamiltonian be Hermitian arises from its identification as the generator of time changes of the state of the system, eq. (3.5), and the condition that those changes be unitary.

The other condition, eq. (9.17), may be interpreted even more directly, since from eq. (9.22) it becomes:

(9.24) $$\Gamma_\alpha = \sum_f \Gamma_{f\alpha},$$

where $\Gamma_{f\alpha} = |A_{f\alpha}|^2$ is the partial decay rate of the state $|\alpha\rangle$ into the channel f. Thus Γ_S and Γ_L are the total decay rates of the K_S and K_L, respectively.

We find, then, the usual[8] connection between the complex mass of an unstable particle and its decay rate: in this case the complex mass is the eigenvalue λ_α of the mass matrix. The real part of λ_α, m_S or m_L, is what we call "the mass" of the K_S and K_L, respectively. These masses now include corrections owing to weak interactions, as can be seen from eq. (9.7b).

The coefficients $C_{j\alpha}$ by means of which the eigenstates $|\alpha\rangle$ are expressed in terms of the states $|K^0\rangle$ and $|\bar{K}^0\rangle$ may easily be obtained by standard methods of solution of eq. (9.13a). They would then be given as simple functions of the matrix elements of **M**. However, we can take advantage of the knowledge, based on experimental results, that the eigenstates are almost the same as those expected for a CP-invariant weak interaction, namely, the CP-even and CP-odd states, and treat the difference as a small correction.

A convention for the relative phases of the states $|K^0\rangle$ and $|\bar{K}^0\rangle$ must first be established. We choose to define the antiparticle state by the relationship[9]

(9.25) $$|\bar{K}^0\rangle = \mathbf{CP}|K^0\rangle,$$

which serves to fix the relative phase. Then the CP-even and CP-odd states are the states $|+\rangle$ and $|-\rangle$ given by eqs. (9.2).

The short-lived state is approximately equal to the even state, and we can write it as

(9.26a) $$|S\rangle = |+\rangle + (\varepsilon - \bar{\varepsilon})|-\rangle$$
$$= 2^{-1/2}[(1 + \varepsilon - \bar{\varepsilon})|K^0\rangle + (1 - \varepsilon + \bar{\varepsilon})|\bar{K}^0\rangle],$$

[7] The case in which the interaction is not Hermitian will be considered in section 9.4 in connection with the possibility of a violation of the CPT theorem. See Kenny and Sachs (1973) and Sachs (1986).

[8] See eq. (5.3) and section 8.3.

[9] See note 4.

while we write for the long-lived state

(9.26b) $$|L\rangle = |-\rangle + (\varepsilon + \bar{\varepsilon})|+\rangle$$
$$= 2^{-1/2}[(1 + \varepsilon + \bar{\varepsilon})|K^0\rangle - (1 - \varepsilon - \bar{\varepsilon})|\bar{K}^0\rangle],$$

where ε and $\bar{\varepsilon}$ are the two small parameters measuring the deviation from the *CP*-invariant form.

Eqs. (9.26) can be inserted into eq. (9.13a) and solved for ε and $\bar{\varepsilon}$ in first order with the results

(9.27a) $$\varepsilon = (\langle K^0|\mathbf{M}|\bar{K}^0\rangle - \langle \bar{K}^0|\mathbf{M}|K^0\rangle)/2(\lambda_S - \lambda_L)$$

and

(9.27b) $$\bar{\varepsilon} = (\langle \bar{K}^0|\mathbf{M}|\bar{K}^0\rangle - \langle K^0|\mathbf{M}|K^0\rangle)/2(\lambda_S - \lambda_L).$$

The states $|K^0\rangle$ and $|\bar{K}^0\rangle$ are orthonormal, so that we find from eqs. (9.26)

(9.28) $$\langle L|S\rangle = (\varepsilon + \bar{\varepsilon})^* + (\varepsilon - \bar{\varepsilon}) = 2 \text{ Re } \varepsilon - 2i \text{ Im } \bar{\varepsilon}.$$

The Bell-Steinberger relation, eq. (9.23), then takes the form

(9.29) $$[(\Gamma_L + \Gamma_S) - 2i(m_L - m_S)][\text{Re } \varepsilon - i \text{ Im } \bar{\varepsilon}] = \sum_f A_{fL}^* A_{fS}.$$

It will be noted that all the quantities appearing in eq. (9.29) except perhaps $[\text{Re } \varepsilon - i \text{ Im } \bar{\varepsilon}]$ are, at least in principle, directly measurable.

The dynamic behavior of the isolated K^0, $\bar{K}^0$ system is expressed in terms of the time-dependent behavior of the eigenstates $|S\rangle$ and $|L\rangle$, which has the usual form e^{-iEt} where, in the *CM* system, the "energy" E is given by the mass eigenvalues λ_S and λ_L, respectively. Note that, in keeping with our use of the unit of inverse length for measuring mass, we are using units of length for measuring time: $t = c \times$ time where c is the speed of light.

If at time $t = 0$ the state is described by a linear combination

(9.30) $$|0\rangle = |S\rangle C_S + |L\rangle C_L,$$

then at later time t it is given by

(9.31) $$|t\rangle = |S\rangle C_S e^{-im_S t} e^{-\Gamma_S t/2} + |L\rangle C_L e^{-im_L t} e^{-\Gamma_L t/2}.$$

Therefore the decay amplitude into mode f at time t for this state is given by

(9.32) $$A_f(t) = C_S A_{fS} e^{-im_S t} e^{-\Gamma_S t/2} + C_L A_{fL} e^{-im_L t} e^{-\Gamma_L t/2},$$

and the decay rate $|A_f(t)|^2$, that is, the rate of production of mode f at time t, is

$$R_f(t) = |C_S A_{fS}|^2 \left\{ e^{-\Gamma_S t} + 2 \operatorname{Re}\left[\frac{C_L}{C_S}\eta_f e^{-i\Delta m t}\right] e^{-(\Gamma_S+\Gamma_L)t/2} + \left|\frac{C_L}{C_S}\right|^2 |n_f|^2 e^{-\Gamma_L t} \right\}, \tag{9.33}$$

where $A_{f\alpha}$ is defined in eq. (9.21),

$$\Delta m = m_L - m_S, \tag{9.34}$$

and

$$\eta_f = A_{fL}/A_{fS}. \tag{9.35}$$

Some remarks concerning the determination of the coefficients C_S and C_L in eq. (9.30) are in order. These coefficients are specified by the initial condition, that is, by determining the state of the system at $t = 0$. If the instant of production of the neutral K meson is taken to be the initial time, the coefficients are determined by the choice of a production process.

When K mesons are produced by the collision of ordinary (i.e., $S = 0$) hadrons—for example, energetic protons or pions striking a nuclear target—the conservation of strangeness for strong interactions leads to production of a K^0 in association with an $S = -1$ particle or a $\bar{K}^0$ in association with an $S = 1$ particle. In the former case the associated particle may be a hyperon (Λ^0, $\Sigma^{0,\pm}$) or a $\bar{K}^0$, in the latter it may be a K^0 or, if the energy is high enough, an antihyperon. The high energy is required for antihyperon production because conservation of baryon number requires that baryons (nucleons, hyperons) be produced in antiparticle-particle pairs, in this case antihyperons paired off with nucleons (as well as the associated $\bar{K}^0$).

Since the threshold for K^0, $\bar{K}^0$ pair production is also higher than that for (K^0, Λ^0) or (K^0, Σ) production, it is possible to produce a pure K^0 state by restricting the energy of bombardment to low enough values. Then the coefficients C_S and C_L may be found by inverting eqs. (9.26) to obtain $|K^0\rangle$ in term of $|S\rangle$ and $|L\rangle$. The result, to first order in ε and $\bar{\varepsilon}$, is

$$|K^0\rangle = 2^{-(1/2)}[(1 - \varepsilon - \bar{\varepsilon})|S\rangle + (1 - \varepsilon + \bar{\varepsilon})|L\rangle]. \tag{9.36a}$$

For completeness, we also write down the corresponding form of the state

$|\bar{K}^0\rangle$:

(9.36b) $$|\bar{K}^0\rangle = 2^{-(1/2)}[(1 + \varepsilon + \bar{\varepsilon})|S\rangle - (1 + \varepsilon - \bar{\varepsilon})|L\rangle].$$

At higher production energies, the initial state may be a mixture of the two pure states eq. (9.36a) and eq. (9.36b). This more general case may be treated by introducing the appropriate 2 × 2 density matrix for the K^0, $\bar{K}^0$ system. In the usual situation the K^0 and $\bar{K}^0$ states are incoherent, and the initial density matrix is in diagonal form, the one diagonal element being proportional to the number of K^0's and the other being proportional to the number of $\bar{K}^0$'s at production ($t = 0$).

It was pointed out by Pais and Piccioni (1955) and by M. L. Good (1957) that a coherent mixture of $|K^0\rangle$ and $|\bar{K}^0\rangle$ could be obtained by making use of the fact that the amplitudes for scattering and absorption of the K^0 by nuclei are quite different from those of the $\bar{K}^0$. The reason is again the conservation of strangeness and baryon number (conservation of hypercharge) for the reaction process. The reaction amplitudes for the K^0 and $\bar{K}^0$ are quite different because a slow $\bar{K}^0$ can be absorbed by nuclei whereas a slow K^0 cannot. Therefore a beam that consists purely of K_L, after passing through a plate of material having a high-reaction cross section, will contain a different mixture of $|K^0\rangle$ and $|\bar{K}^0\rangle$ than that in the original beam, given by eq. (9.26b). If, in this new mixture, eqs. (9.36) are substituted for $|K^0\rangle$ and $|\bar{K}^0\rangle$, it will be found to be a mixture of $|S\rangle$ and $|L\rangle$ rather than the pure $|L\rangle$ that was incident on the plate. It turns out that the amplitudes C_S and C_L appearing in this mixture depend on the ratio of the sum and difference of the reaction amplitudes for the K^0 and $\bar{K}^0$ (Good et al. 1961), if the *CPT* theorem is valid (Sachs 1963b, eq. 86). Thus, if the initial time is taken to be the instant of emergence of the beam from the plate, it is possible in principle to predetermine the initial coefficients C_S and C_L for an experiment by selection of the physical properties of the plate. The first successful experimental use of this method was made by R. H. Good and others (1961) to determine the value of the $K_L - K_S$ mass difference Δm, eq. (9.34). The high degree of sensitivity of K^0, $\bar{K}^0$ interferometry is indicated by the measured values of Δm, as shown in table 9.1, which are of the order of $7 \times 10^{-15}\ m_K$. Table 9.1 also gives data on the basic decay parameters entering into the determination of the decay rate eq. (9.33). The rate for $K^\pm$ decay is included for comparison and future reference.

If the observed phenomenon of a long-lived 2π decay mode is due to $K_L \rightarrow 2\pi$ decay, the relative magnitude of its decay rate and that of the K_S gives $|\eta_{\pi\pi}|^2$, where $\eta_{\pi\pi}$ is the ratio eq. (9.35) of 2π amplitudes. However, we have seen that the observation of 2π decay of a long-lived species does not

TABLE 9.1 Decay Parameters for the K Mesons

$\Gamma_S(\text{sec}^{-1})$*	$\Gamma_L(\text{sec}^{-1})$*	$\Delta m(\text{sec}^{-1})$*	$\Gamma(K^{\pm}) = \Gamma_{\pm}(\text{sec}^{-1})$*
$(1.121 \pm 0.003) \times 10^{10}$	$(1.929 \pm 0.015) \times 10^{7}$	$(0.5349 \pm 0.0022) \times 10^{10}$	$(0.808 \pm 0.002) \times 10^{8}$

Source: PDG (1986).
*The units of mass, energy, and decay rate (width in energy) in the text are inverse lengths. To make it explicit that the rates concern time dependence, in this table they are given in units of inverse time, corresponding to a multiplication by c.

in itself confirm its assignment to the K_L and therefore does not establish CP violation unambiguously. For the latter purpose, it is necessary to demonstrate the coherence between the long-lived and short-lived modes that would be the consequence of our assumption that K_S and K_L are two states of the same system.

The important feature of eq. (9.33) is the occurrence of the interference term[10] that is required to establish coherence. That the rate is *not* simply the sum of two exponentials corresponding to a mixture of two independent mesons would provide the required evidence. The existence of interference also makes possible extremely sensitive measurements on the weak interactions of the K mesons, measurements that not only demonstrate the existence of CP violation but also provide some quantitative information about the details of the CP-violating interaction, for example, the phase of η_f.

K^0 meson interferometry also provides a very sensitive test of the CPT theorem as applied to K meson phenomena, thereby providing a precise measure of T violation based on the measurement of CP-violating parameters. These tests and measurements are treated in the following sections.

9.2 *CP* Violation

The first indication that the strangeness changing weak interaction violates CP invariance was the discovery (Christenson et al. 1964) of the decay of the K_L into two pions, a process that, as we have seen, is forbidden by CP invariance *if the Gell-Mann and Pais model is the correct one.* We have already found in the previous section that to establish CP violation unequivocally it is necessary to carry out an experiment showing that the decay rate depends on a variable that is changed by the operator **CP**. Therefore we now turn to consideration of the behavior of the decay amplitudes under the CP transformation.

The 2π decay is the mode of interest, and its rate is governed by the amplitudes $A_{2\pi,S}$ and $A_{2\pi,L}$ appearing in eq. (9.32). These amplitudes may be

[10] The possible occurrence of the interference term in K^0, $\bar{K}^0$ decay follows from the Weisskopf-Wigner perturbation method (Treiman and Sachs 1956).

expressed in terms of $A_{2\pi}$ and $\bar{A}_{2\pi}$ by making use of the relationship eq. (9.21) with coefficients $C_{j\alpha}$ obtained by means of eqs. (9.26):

(9.37a) $$A_{2\pi,S} = 2^{-(1/2)}[(1 + \varepsilon - \bar{\varepsilon})A_{2\pi} + (1 - \varepsilon + \bar{\varepsilon})\bar{A}_{2\pi}]$$

and

(9.37b) $$A_{2\pi,L} = 2^{-(1/2)}[(1 + \varepsilon + \bar{\varepsilon})A_{2\pi} - (1 - \varepsilon - \bar{\varepsilon})\bar{A}_{2\pi}].$$

The effect of the transformation *CP* on $A_{2\pi,S}$ and $A_{2\pi,L}$ is determined by the effect on $A_{2\pi}$, $\bar{A}_{2\pi}$, ε, and $\bar{\varepsilon}$, which in turn is determined by the behavior of the general amplitude $A_{cj}(E)$ given by eq. (9.3a). Now *CP* transforms $|K^0\rangle$ into $|\bar{K}^0\rangle$ and vice versa, and

(9.38) $$\mathbf{CP}|c, \text{out}\rangle = |\bar{c}, \text{out}\rangle,$$

where $\bar{c}$ is the channel charge conjugate to c with all quantum numbers (including parity) transformed by *P*. Therefore the effect of the transformation is the substitution

(9.39) $$A_c(E) \leftrightarrow \bar{A}_{\bar{c}}(E).$$

The 2π channel has orbital angular momentum zero (the *K* meson has spin 0 and is at rest) and the intrinsic parity of the *two* pseudoscalar pions is even. Therefore the state $|2\pi\rangle$ has even parity. Furthermore, a neutral pion state is even, and π^+ and π^- are interchanged under charge conjugation so the zero angular momentum, neutral, 2π state is also even under *C*, whence $\bar{c} \equiv c$ for this mode. Thus eq. (9.39) becomes

(9.40) $$A_{2\pi}(E) \leftrightarrow \bar{A}_{2\pi}(E).$$

To determine the effect of the transformation *CP* on ε and $\bar{\varepsilon}$, it suffices to note that the transformation is equivalent to the substitution $K^0 \leftrightarrow \bar{K}^0$ of the labels in the mass matrix elements eqs. (9.7), because the Σ_c or Σ_f appearing in those equations includes sums over *all* states, both c and $\bar{c}$ (or f and $\bar{f}$). The substitution $K^0 \leftrightarrow \bar{K}^0$, when applied to eqs. (9.27), yields

(9.41a) $$\varepsilon \rightarrow -\varepsilon$$

and

(9.41b) $$\bar{\varepsilon} \rightarrow -\bar{\varepsilon}.$$

Application of eqs. (9.40) and (9.41) to eqs. (9.37) then provides the desired information about the behavior of the K_S and K_L amplitudes for 2π decay under *CP*:

(9.42) $$A_{2\pi,S} \rightarrow A_{2\pi,S}, \quad A_{2\pi,L} \rightarrow -A_{2\pi,L}.$$

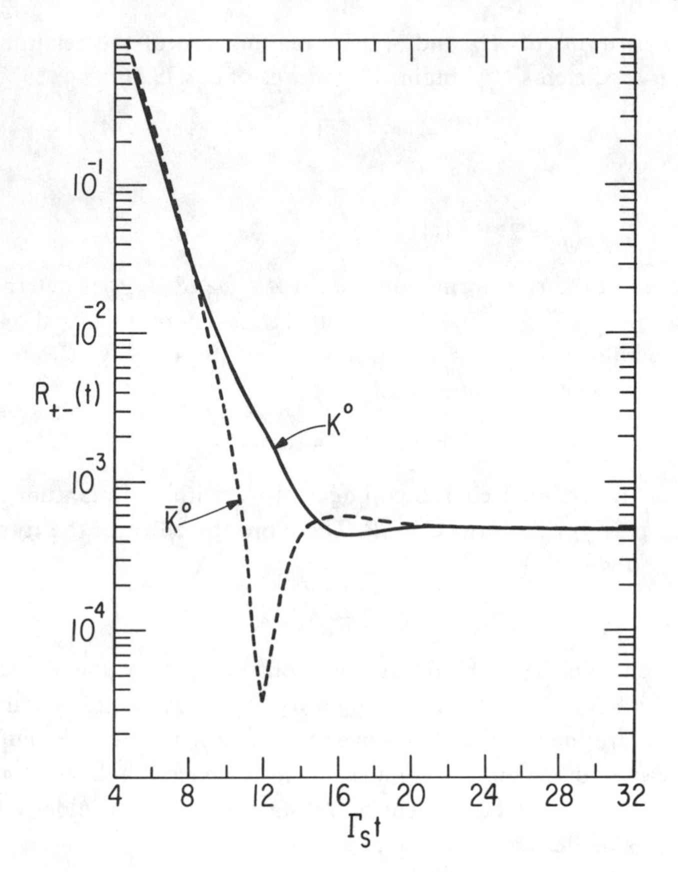

FIG. 9.2 Decay rates (arbitrary units) for the $\pi^+\pi^-$ mode when the intial state ($t = 0$) is a K^0 (*solid curve*) and when it is a $\bar{K}^0$ (*dashed curve*) if $\eta_{+-} \neq 0$, showing that the existence of the interference term is a manifestation of CP violation. The values of the parameters used in eq. (9.33) to obtain these curves are given in table 9.1 and in eqs. (9.47).

This is the result required to show that the nonzero value of $A_{2\pi,L}$ implies CP violation because it shows that the interference between the 2π decay of the K_S and K_L changes sign under CP.

Consider the 2π decay rate $R_{2\pi}(t)$ given as a function of time by eq. (9.33). It will manifest an interference term proportional to

(9.43)
$$\eta_{\pi\pi} = A_{2\pi,L}/A_{2\pi,S}\,.$$

The 2π decay curve for the case in which the initial state is purely $|K^0\rangle$ has the form shown as the solid curve in figure 9.2.

To test CP invariance we may imagine using a CP mirror to carry out a "classical" reflection experiment of the kind described in chapter 2. The decay curve is observed in the *CP* mirror and, because the number of particles observed in the mirror in a given time interval is the same as the original, this mirror image of $R_{2\pi}(t)$ is the same as the original function. But the initial conditions as viewed in the mirror are different from those obtaining in the original experiment; all particles are replaced by their *CP* conjugates. Thus, if in accordance with our standard test procedure the experiment were carried out with initial conditions corresponding to those seen in the mirror, the decay rate would be $\bar{R}_{2\pi}(t)$, where $\bar{R}_{2\pi}(t)$ is obtained from eq. (9.33) with $\eta_{\pi\pi}$ transformed in accordance with eq. (9.42) or

(9.44) $$\eta_{\pi\pi} \rightarrow -\eta_{\pi\pi}.$$

The dashed curve in figure 9.2 shows $\bar{R}_{2\pi}(t)$, that is, the rate when the initial state is $|\bar{K}^0\rangle$. *CP* invariance would imply that $\bar{R}_{2\pi}(t)$ is the same as the decay curve seen in the mirror, namely, that

(9.45) $$\bar{R}_{2\pi}(t) \equiv R_{2\pi}(t),$$

a condition that will be met if and only if

(9.46) $$\eta_{\pi\pi} = 0.$$

Thus the very existence of the interference term in the 2π decay curve provides the required evidence for *CP* violation.[11]

The key to an unambiguous test of *CP* is therefore the coherence of the $|K_S\rangle$ and $|K_L\rangle$ states, that is, the evidence that the K_S and K_L "particles" represent two states of the same system. Total absence of coherence could, as we have seen, indicate that the observed long-lifetime 2π mode is due to a particle that is independent of the short-lived K^0.

The first demonstration of coherence between the K_L and K_S mesons responsible for the 2π modes was carried out by Fitch and others (1965), who showed that a K_S beam regenerated from a K_L beam by passage through matter was coherent with it, thereby establishing that *CP* invariance is violated. Coherence by direct measurement of the shape of $R(t)$ (sometimes called "vacuum regeneration") was first demonstrated with sufficient accuracy to measure $\eta_{\pi\pi}$ (actually η_{+-}, for the $\pi^+\pi^-$ mode) by Jensen and others (1969). The *CP*-violating parameter $\eta_{\pi\pi}$ is a complex number whose phase $\phi_{\pi\pi}$ can be determined from this type of interference

[11] The occurrence of interference (i.e., "nonexponential decay") in the 2π modes as a result of the existence of a $K_L \rightarrow 2\pi$ mode was predicted by Sachs (1964). The interference term in the $K^0 \rightarrow 2\pi$ rate was first discussed in detail by Lyuboshitz (1965), who pointed out that it would have the opposite sign in the $\bar{K}^0 \rightarrow 2\pi$ decay rate. Sakurai and Wattenberg (1967) emphasized that this "absolute" distinction between the behavior of the K^0 and the $\bar{K}^0$ would be a manifest experimental demonstration of *CP* violation.

measurement only if the $K_L - K_S$ mass difference, Δm, is known. Experiments making use of interference between a K_L beam and a K_S beam regenerated from it have been used to measure $\eta_{\pi\pi}$, Δm, and other parameters in variations on a method first proposed by M. L. Good (1958). Measurements of the magnitude of $\eta_{\pi\pi}$, have been made by determining the $K_L \to 2\pi$ decay rate or branching ratio. The value of $|\eta_{+-}|$ (i.e., $|\eta_{\pi\pi}|$ for the $\pi^+\pi^-$ mode), $|\eta_{+-}| \approx 2 \times 10^{-3}$, was obtained in this way in the historic experiment of Christenson and others (1964).

The latest values that will be used here are[12]

(9.47a) $$|\eta_{+-}| = (2.27 \pm 0.02) \times 10^{-3},$$

(9.47b) $$\phi_{+-} = 44.7° \pm 1.2°$$

for the $\pi^+\pi^-$ mode, and

(9.48a) $$|\eta_{00}|/|\eta_{+-}| = 1.0138 \pm 0.0174,$$

(9.48b) $$\phi_{00} - \phi_{+-} = 10° \pm 6°$$

for the $\pi^0\pi^0$ mode.

The parameters η_{+-} and η_{00} provide a measure of *CP* violation, but their relationship with more fundamental parameters, for example, the *CP*-violating interactions, remains to be determined. For that purpose we must rewrite the definition of $\eta_{\pi\pi}$ given by eq. (9.43) in terms of the decay amplitudes $A_{2\pi}$ and $\bar{A}_{2\pi}$ that are directly related to the fundamental interactions. This rewriting is accomplished by making use of the relationships between the decay amplitudes of K^0, $\bar{K}^0$ and K_S, K_L given by eqs. (9.37). To first order in the small parameters ε, $\bar{\varepsilon}$, and $\eta_{\pi\pi}$, we find

(9.49) $$\eta_{\pi\pi} = \varepsilon + \bar{\varepsilon} + \frac{A_{2\pi} - \bar{A}_{2\pi}}{A_{2\pi} + \bar{A}_{2\pi}}.$$

Eq. (9.40) shows that for a *CP*-invariant system

(9.50) $$\bar{A}_{2\pi} = A_{2\pi},$$

[12] These data, except for eq. (9.48a), are taken from Cronin's Nobel lecture (1981), which provides references to some of the latest sources. See also PDG (1986, p. 18). Eq. (9.48a) is reconstructed by means of eq. (9.71a) from the recent measurement eq. (9.72), of $|\varepsilon'/\eta_{+-}|$ by the Chicago-Saclay group (Bernstein et al. 1985). The point is that although the result eq. (9.72) is obtained by implicit use of the measurement of $|\eta_{00}|/|\eta_{+-}|$, an explicit value was not extracted from the data. The excellent review by Kleinknecht (1976) provides an extensive bibliography of the original experimental papers and a good summary of the experimental methods. Note, however, that in his treatment of the phenomenological theory there are differences in notation from mine, although the approach is similar, with the exception that he uses ab initio the assumption of *CPT* invariance.

so that the quantity

(9.51)
$$\varepsilon_{\pi\pi} = \frac{A_{2\pi} - \bar{A}_{2\pi}}{A_{2\pi} + \bar{A}_{2\pi}}$$

vanishes in that case. Therefore $\varepsilon_{\pi\pi}$ may be taken as a measure of *direct CP* violation in the 2π decay interaction. The measured parameter

(9.52)
$$\eta_{\pi\pi} = \varepsilon + \bar{\varepsilon} + \varepsilon_{\pi\pi}$$

is then made up of the sum of the direct contribution and a contribution owing to the parameters ε and $\bar{\varepsilon}$, which are also a measure of CP violation, as can be seen from eq. (9.41). However, from the definitions eq. (9.27), the latter contribution is seen to be an *indirect* measure depending on CP violation in all modes of decay, real or virtual. If there exists a $|\Delta S| = 2$ interaction (which can lead only to virtual decay, because there are no known $S = \pm 1$ states with mass $< m_K$), its contribution to CP violation is also included in the indirect terms.

The 2π decay amplitudes that are measured consist of either the $\pi^+\pi^-$ amplitude or the $\pi^0\pi^0$ amplitude. In order to take account of the effect of final state interactions on the amplitudes, it is shown in section 5.2 that the final states should be chosen to be eigenstates of the strong interactions between the two pions. The strong interactions conserve isotopic spin, I, hence the eigenstates are states of total isotopic spin. Since the isotopic spin of a pion is $I = 1$, corresponding to the three states π^+, π^0, π^-, the possible total isotopic spin values for the two pions are $I = 0$, 1, 2. However, the pions satisfy Bose statistics, so that the wave function must be symmetric for interchange of the pions. Conservation of angular momentum limits the space function to the symmetric $J = 0$ state so that only the symmetric isotopic spin states $I = 0$ and $I = 2$ are allowed as final states for the two pions. Therefore we must express the amplitudes $A_{2\pi}$ and $\bar{A}_{2\pi}$ in terms of amplitudes for decay into states of given I, A_I and $\bar{A}_I$ with $I = 0$ and $I = 2$.

The relationships given by the standard Clebsch-Gordan reduction are

(9.53a)
$$A_{+-} = \frac{1}{\sqrt{3}}(\sqrt{2}A_0 + A_2)$$

for the $\pi^+\pi^-$ amplitude and

(9.53b)
$$A_{00} = \frac{1}{\sqrt{3}}(A_0 - \sqrt{2}A_2)$$

for the $\pi^0\pi^0$ amplitude, where we have adopted the same convention as Wu and Yang (1964) for the relative phases of coefficients. The relationships for $\bar{A}$, A_S, and A_L are identical.

The effect of final state interactions may now be brought out by making explicit the dependence of the 2π decay amplitudes on the $\pi - \pi$ scattering phase shifts δ_I for the eigenstates $I = 0$ and $I = 2$ by writing

(9.54a) $$A_I = a_I e^{i\delta_I}, \quad \bar{A}_I = \bar{a}_I e^{i\delta_I}$$

and

(9.54b) $$A_{I,S} = a_{I,S} e^{i\delta_I}, \quad A_{I,L} = a_{I,L} e^{i\delta_I},$$

as in eq. (5.28). Then by combining eqs. (9.53) and (9.54a) we find that eq. (9.51) takes the form

(9.55a) $$\varepsilon_{+-} = [(a_0 - \bar{a}_0) + 2^{-(1/2)}(a_2 - \bar{a}_2)e^{i\delta}] \cdot [(a_0 + \bar{a}_0) + 2^{-(1/2)}(a_2 + \bar{a}_2)e^{i\delta}]^{-1}$$

for the $\pi^+\pi^-$ mode and

(9.55b) $$\varepsilon_{00} = [(a_0 - \bar{a}_0) - 2^{1/2}(a_2 - \bar{a}_2)e^{i\delta}] \cdot [(a_0 + \bar{a}_0) - 2^{1/2}(a_2 + \bar{a}_2)e^{i\delta}]^{-1}$$

for the $\pi^0\pi^0$ mode, where

(9.56) $$\delta = \delta_2 - \delta_0$$

is the (measurable) difference between the two π-π scattering phase shifts.

The consequences of the CPT theorem have not been taken into account up to this point because later we shall want to consider the possibility of a failure of the theorem. However, let us now assume its validity and consider its consequences for the 2π decay modes.

Because it involves the antiunitary transformation T, CPT is also antiunitary. Therefore we may apply the fundamental property eq. (3.32) (or eq. 8.28a) of the inner product of two state vectors when subjected to an antiunitary transformation:

(9.57) $$\langle \mathbf{CPT}\phi \,|\, \mathbf{CPT}\psi \rangle = \langle \phi \,|\, \psi \rangle^*.$$

Under the assumption that the weak interaction is CPT invariant,

(9.58) $$\mathbf{CPT}\; H_w (\mathbf{CPT})^{-1} = H_w,$$

the argument leading to eq. (5.27) can be paraphrased by making use of eqs. (8.32b) and (8.24) to show that the amplitude A_c, eq. (9.3), for decay into any eigenchannel c has the property

(9.59) $$A_c^* \sim \langle \mathbf{CPT}(c, \text{out}) | \mathbf{CPT}\, H_w (\mathbf{CPT})^{-1} | \mathbf{CPT}\, K^0 \rangle$$
$$= e^{-2i\delta_c} \langle \bar{c}', \text{out} | H_w | \bar{K}^0 \rangle,$$

or

(9.60) $$A_c^* = e^{-2i\delta_c} \bar{A}_{\bar{c}'}$$

where $\bar{c}$ is the *CP* conjugate channel to c and c' is obtained from c by motion reversal of all quantum numbers. Since the K^0 is taken to be at rest and has spin 0, the state $| K^0 \rangle$ is unaffected by time reversal, and use of the Wigner phase convention eliminates any phase change of the final state. In terms of the reduced amplitudes for the 2π modes this result becomes

(9.61) $$\bar{a}_I = a_I^*,$$

because the 2π channel is unaffected by either the *CP* transformation or motion reversal except for compensating reversals of the direction of the relative momentum of the pions, one due to P and the other to T.

Before this result is applied to eqs. (9.55), it is desirable to fix the remaining arbitrary phase in the amplitudes. Because neither the strong nor the electromagnetic interactions connect the $S = \pm 1$ sector of the Hilbert space of state vectors with the $S = 0$ sector, the relative phase of these two sectors is undetermined. This phase may be adjusted so as to make the decay amplitude into any given state have phase zero, that is, to make that *one* amplitude real. The phases of all other amplitudes are then fixed by this choice.

We shall follow the convention established by Wu and Yang (1964) and choose the $I = 0$ reduced amplitude a_0 to be real. Then

(9.62) $$a_0^* = a_0 = \bar{a}_0,$$

according to eq. (9.61), and eqs. (9.55) are simplified to read

(9.63a) $$\varepsilon_{+-} = \varepsilon'/(1 + \omega)$$

and

(9.63b) $$\varepsilon_{00} = -2\varepsilon'/(1 - 2\omega),$$

with

(9.64) $$\varepsilon' = 2^{-(1/2)} i e^{i\delta} \operatorname{Im} a_2/a_0$$

and

(9.65) $$\omega = 2^{-(1/2)} e^{i\delta} \operatorname{Re} a_2/a_0 .$$

That there is a direct CP-violating 2π amplitude only if a_2 is not real is another manifestation of our repeated observation that phase differences between states are introduced by T violation, which is equivalent here to CP violation because eqs. (9.63) are based on CPT invariance. The relationship of the phase of a_2 to T violation will be made explicit in the next section.

The direct and indirect contributions to eq. (9.52) may be separated by comparing η_{+-} and η_{00}. Since the approximate validity of the $\Delta I = 1/2$ rule implies that ω is small,[13]

(9.66)
$$|\omega| \gtrsim 0.032,$$

and we will find that $\varepsilon' \ll |\eta_{+-}|$, it is a good approximation to write

(9.67)
$$\eta_{00} = \eta_{+-} - 3\varepsilon'.$$

From eq. (9.64) we see that the phase of ε' is $\delta + \frac{\pi}{2}$. The most recent estimate[14] of the π-π phase shifts gives $\delta + \frac{\pi}{2} = (45 \pm 10)°$. Purely as a matter of chance, this phase is nearly equal to the phase ϕ_{+-} of η_{+-}, given by eq. (9.47b). Cronin (1981) has emphasized that the coincidence greatly simplifies the determination of $|\varepsilon'|$. If eq. (9.67) is written in the form

(9.68)
$$\frac{3\varepsilon'}{\eta_{+-}} = 1 - \frac{\eta_{00}}{\eta_{+-}},$$

the phase of the left-hand side is either μ or $\mu + \pi$, where

(9.69)
$$\mu = \delta + \tfrac{\pi}{2} - \phi_{+-} = 0.3° \pm 10°.$$

Then eq. (9.68) may be rewritten in the form

(9.70)
$$\pm 3\frac{|\varepsilon'|}{|\eta_{+-}|}\, e^{i\mu} = 1 - \frac{|\eta_{00}|}{|\eta_{+-}|}\, e^{i(\phi_{00} - \phi_{+-})},$$

[13] The K mesons have isotopic spin $I = \frac{1}{2}$ corresponding to the doublets K^+, K^0 or $\bar{K}^0, K^-$, and the $\Delta I = \frac{1}{2}$ rule for the K^0 decay would exclude the $I = 2$ state (for which $\Delta I = 3/2$). However, from the decay of the $K^\pm$ into $\pi^\pm\pi^0$, which cannot belong to the $I = 0$ state, it follows that the $\Delta I = 3/2$ transition does occur, but with an amplitude suppressed by a factor of about 20 compared with the $\Delta I = 1/2$ transition. The ratio a_2/a_0 may be determined directly from the ratio $C = \Gamma_S(\pi^+\pi^-)/\Gamma_S(\pi^0\pi^0)$ by use of eqs. (9.53). However, uncertainties in the measured π-π phase shifts and the radiative corrections introduce some ambiguity in the result obtained this way. The generally accepted alternative is to assume that there is no $\Delta I = 5/2$ contribution to the transition, and then the magnitude of the $\Delta I = 3/2$ amplitude a_2 can be obtained directly from $\Gamma_+(\pi^\pm\pi^0) = B_{2\pi,\pm}\Gamma_\pm = 3|a_2|^2/2$, since a_0 does not contribute to $K^\pm \to \pi^\pm\pi^0$. See Kenny (1967) and Marshak, Riazuddin, and Ryan (1969, pp. 543–49). From the values of the branching ratio for $K^\pm \to 2\pi$, $B_{2\pi,\pm} = 0.21$ (PDG 1986, p. 13), $\Gamma_\pm$ and $\Gamma_S = 2(|a_0|^2 + |a_2|^2)$ given in table 9.1 it follows that $|a_2/a_0| \approx 0.045$, a result that leads to a value of C that is consistent with its direct measurement. That $\mathrm{Re}(a_2/a_0)$ is positive follows from the fact that $C > 2$ (eq. 9.143), $\mathrm{Im}(a_2/a_0) \ll \mathrm{Re}(a_2/a_0)$, and $\delta \approx -\pi/4$ (see note 12).

[14] See note 12.

where use has been made of eq. (9.48b) to fix the relative sign of η_{00} and η_{+-}.

Since $\mu \ll 1$ radian, the imaginary terms in eq. (9.70) must be small so that both phase factors may safely be expanded to first order to obtain

(9.71a)
$$\pm 3 \frac{|\varepsilon'|}{|\eta_{+-}|} = 1 - \frac{|\eta_{00}|}{|\eta_{+-}|}$$

and

(9.71b)
$$\phi_{00} - \phi_{+-} = \left[1 - \frac{|\eta_{+-}|}{|\eta_{00}|}\right]\mu$$

by equating the real and imaginary parts in the resulting equation.

The good fortune is that eq. (9.71a) makes it possible to determine $|\varepsilon'|$ from the measurement of $|\eta_{00}|$; the very much more difficult direct determination of the phase ϕ_{00} is *not* required. From eq. (9.48a) it follows that[15]

(9.72)
$$\varepsilon'/\eta_{+-} = -0.0046 \pm 0.0058.$$

Since μ is given with higher precision than the directly measured ϕ_{00}, we may use eq. (9.71b) to obtain

(9.73)
$$\phi_{00} - \phi_{+-} = 0.006° \pm 0.61°,$$

which is not inconsistent with (within less than two standard deviations of) eq. (9.48b). Although less direct, eq. (9.73) appears to be the more precise measure of $\phi_{00} - \phi_{+-}$. The precision depends, of course, on the assumption of *CPT* invariance and on the precision of the measurement of $\delta_2 - \delta_0$, which is not so high as in the direct measurement of $\phi_{00} - \phi_{+-}$ ($\pm 10°$ as compared with $\pm 6°$). However, the precision obtained from eq. (9.71b) is far greater because the error in $\delta_2 - \delta_0$ is suppressed by the very small factor $1 - |\eta_{+-}|/|\eta_{00}|$.

In view of the small value of ε', and therefore of $\varepsilon_{\pi\pi}$, given by eqs. (9.63), it is clear from eq. (9.52) that the dominant contribution to $\eta_{\pi\pi}$ arises from the term ε or $\bar{\varepsilon}$. However, it will now be demonstrated that the *CPT* theorem implies $\bar{\varepsilon} = 0$, so that eq. (9.52) becomes

(9.74a)
$$\eta_{+-} = \varepsilon + \varepsilon'$$

or

(9.74b)
$$\eta_{00} = \varepsilon - 2\varepsilon',$$

if the very small corrections to $\eta_{\pi\pi}$ owing to ω in eqs. (9.63) are neglected.

[15] See note 12.

To demonstrate the consequence of *CPT* invariance for ε and $\bar{\varepsilon}$, which are given by eqs. (9.27), we must first consider the consequences for the mass matrix, eqs. (9.6) and (9.7). The *CPT* theorem leads to the condition eq. (9.60) for the decay amplitudes, real or virtual. When this result is applied to the amplitudes appearing in eq. (9.7a) for $\langle j|\Gamma|k\rangle$ it leads immediately to the result

(9.75) $$\langle K^0|\Gamma|K^0\rangle = \langle \bar{K}^0|\Gamma|\bar{K}^0\rangle,$$

while it leads to a trivial identity for the two off-diagonal elements. A similar result holds for the dispersive term in eq. (9.7b), and as a consequence of *CPT* invariance of the strong interactions, the masses of the K^0 and $\bar{K}^0$ represented by the leading term are also equal.

The fundamental eq. (9.57) may be applied to the only remaining term, that due to the $\Delta S = 2$ interaction H_{ww}, which is assumed to be Hermitian and to satisfy

(9.76) $$\mathbf{CPT}H_{ww}(\mathbf{CPT})^{-1} = H_{ww}.$$

Thus

(9.77) $$\langle j|H_{ww}|k\rangle^* = \langle \mathbf{CPT}j|\mathbf{CPT}H_{ww}(\mathbf{CPT})^{-1}|\mathbf{CPT}k\rangle = \langle \bar{j}'|H_{ww}|\bar{k}'\rangle,$$

in the notation of eq. (9.59). But since H_{ww} is Hermitian,

(9.78) $$\langle j|H_{ww}|k\rangle^* = \langle k|H_{ww}|j\rangle,$$

and eq. (9.77) leads to

(9.79a) $$\langle K^0|H_{ww}|K^0\rangle = \langle \bar{K}^0|H_{ww}|\bar{K}^0\rangle$$

and

(9.79b) $$\langle K^0|H_{ww}|\bar{K}^0\rangle = \langle K^0|H_{ww}|\bar{K}^0\rangle,$$

the second equation again being a trivial identity. Thus we find that the consequence of the *CPT* theorem is that the *diagonal elements of the mass matrix are equal:*

(9.80) $$\langle K^0|\mathbf{M}|K^0\rangle = \langle \bar{K}^0|\mathbf{M}|\bar{K}^0\rangle.$$

This is a simple generalization of the principle that, as a consequence of the *CPT* theorem, the mass of a particle is equal to the mass of its antiparticle.

Since $\bar{\varepsilon}$ is proportional to the difference between the diagonal elements of the mass matrix (eq. 9.27b), we have the anticipated result

(9.81) $$\bar{\varepsilon} = 0$$

as a consequence of *CPT* invariance. Under this assumption, then, the *CP* violation in the 2π mode is due essentially to the mixing parameter ε. In fact, because the observed direct term ε', eq. (9.72), is so small, it is a good approximation for most purposes to write

(9.82) $$\eta_{+-} = \eta_{00} = \varepsilon.$$

The further interpretation of the measurements may now be made on the basis of the definition of ε, eq. (9.27a), which can be rewritten in the form

(9.83) $$\varepsilon = \frac{\operatorname{Im} M_{12} - \frac{1}{2}i \operatorname{Im} \Gamma_{12}}{i\Delta m - \frac{1}{2}(\Gamma_S - \Gamma_L)},$$

where Δm is given by eq. (9.34), $M_{12} = \langle K^0 | M | \bar{K}^0 \rangle$ and

(9.84) $$\Gamma_{12} = \langle K^0 | \Gamma | \bar{K}^0 \rangle = \sum_f a_f^* \bar{a}_f,$$

in accordance with eq. (9.7a), if the channels f are eigenchannels of the S matrix and the reduced amplitudes eq. (5.28) are introduced. In order to obtain an indication of the important contributions to ε, we consider the absorptive term Γ_{12}, which is made up of measurable decay amplitudes, and we simplify the analysis by neglecting the very small (or vanishing) $\Delta S = -\Delta Q$ terms. Then the only important contributions to Γ_{12} arise from the 2π and 3π modes. Of these, the 2π mode is dominant by a factor of about 580 (PDG 1986). However, when the condition eq. (9.62) of *CPT* invariance is taken into account we find that the $I = 0$ term

(9.85) $$a_0^* \bar{a}_0 = a_0^2$$

is real and therefore makes no contribution to $\operatorname{Im} \Gamma_{12}$.

On the other hand, we find from eq. (9.61) that for $I = 2$

(9.86) $$a_2^* \bar{a}_2 = (a_2^*)^2.$$

Therefore the 2π contribution to $\operatorname{Im} \Gamma_{12}$ is

(9.87) $$\operatorname{Im} \Gamma_{12} \approx -2(\operatorname{Re} a_2)(\operatorname{Im} a_2).$$

The real and imaginary parts of a_2 may be expressed in terms of ω and ε'. The former is measured by the $\Delta I = 3/2$ contribution to the 2π decay, which has an amplitude of about 1/20 of the $\Delta I = 1/2$ amplitude a_0.[16] But the other factor in eq. (9.87) is a measure of the direct *CP* violation in the

[16] See note 13.

2π amplitude ε', given by eq. (9.64). Use of eqs. (9.64) and (9.65) leads to

(9.88)
$$|\mathrm{Im}\,\Gamma_{12}| \approx 4a_0^2|\omega||\varepsilon'| \lesssim 2\times 10^{-3}a_0^2|\varepsilon|$$

on the basis of the experimental values of ω and ε', eqs. (9.66) and (9.72).

Since $2a_0^2 \approx \Gamma_S$ and the denominator of eq. (9.83) is of the same order of magnitude, it is clear that $\mathrm{Im}\,\Gamma_{12}$ makes a very small contribution to ε. Therefore it is a good approximation to write

(9.89)
$$\varepsilon = \frac{\mathrm{Im}\,M_{12}}{i\Delta m - \frac{1}{2}\Gamma_S},$$

where we have also used (see table 9.1)

(9.90)
$$\Gamma_S \approx 580\,\Gamma_L \gg \Gamma_L.$$

The phase of ε is given by eq. (9.89) and may be determined from the measured values of $\Delta m/\Gamma_S$, which is (table 9.1)

(9.91)
$$\Delta m/\Gamma_S = 0.477 \pm 0.002.$$

The phase ϕ_0 of ε is then

(9.92)
$$\phi_0 = \arctan\frac{2\Delta m}{\Gamma_S} = 43.7°,$$

in good agreement with the measured phase ϕ_{+-} of η_{+-}, as would be expected on the basis of eq. (9.82).

An independent experimental confirmation of the determination of ε is provided by the measurement of the charge asymmetry in the semileptonic decay of the K_L, which also has the virtue of providing the observation of a manifest violation of CP. The charge asymmetry δ_L is defined by

(9.93)
$$\delta_L = \frac{\Gamma_L(\pi^-l^+) - \Gamma_L(\pi^+l^-)}{\Gamma_L(\pi^-l^+) + \Gamma_L(\pi^+l^-)},$$

where $\Gamma(\pi^\pm l^\mp)$ is the partial rate for $K_L \to \pi^\pm l^\mp \left\{ {\bar\nu_l \atop \nu_l} \right\}$ and l stands for either charged lepton: e or μ. It should be clear from an application of the standard test of comparing the decay of a K_L as seen in a CP mirror (a $\bar{K}^0$ at $t = 0$) with the actual decay (a K^0 at $t = 0$) that an experiment leading to $\delta_L \neq 0$ shows that CP invariance is violated. The change in initial conditions makes no difference, of course, in the way the K_L behaves in the laboratory.

The calculation of δ_L in terms of the CP-violating parameters may be accomplished by using the analogue of eq. (9.37b) for the semileptonic modes. If the parameters measuring the small (if any) deviation from the $\Delta S = \Delta Q$ rule are defined as

(9.94a) $$x = \bar{A}(\pi^- l^+)/A(\pi^- l^+)$$

and

(9.94b) $$\bar{x} = A(\pi^+ l^-)/\bar{A}(\pi^+ l^-),$$

the analogue of eq. (9.37b) for each of the modes is

(9.95a) $$A_L(\pi^- l^+) = 2^{-1/2} A(\pi^- l^+)[(1 + \varepsilon + \bar{\varepsilon}) - (1 - \varepsilon - \bar{\varepsilon})x]$$

and

(9.95b) $$A_L(\pi^+ l^-) = -2^{-1/2} \bar{A}(\pi^+ l^-)[(1 - \varepsilon - \bar{\varepsilon}) - (1 + \varepsilon + \bar{\varepsilon})\bar{x}].$$

By substituting for $A(\pi^- l^+)$ and $\bar{A}(\pi^+ l^-)$ and for x and $\bar{x}$ suitable averages over the kinematic variables of the decay channels, we find

(9.96) $$\begin{aligned}\delta_L = [&|1 - x|^2 |A(\pi^- l^+)|^2 - |1 - \bar{x}|^2 |\bar{A}(\pi^+ l^-)|^2 \\ &+ 2\,\mathrm{Re}\{(\varepsilon + \bar{\varepsilon})(1 - x)^*(1 + x)\} |A(\pi^- l^+)|^2 \\ &+ 2\,\mathrm{Re}\{(\varepsilon + \bar{\varepsilon})(1 - \bar{x})^*(1 + \bar{x})\} |\bar{A}(\pi^+ l^-)|^2] \\ \times [&|1 - x|^2 |A(\pi^- l^+)|^2 + |1 - \bar{x}|^2 |\bar{A}(\pi^+ l^-)|^2 \\ &+ 2\,\mathrm{Re}\{(\varepsilon + \bar{\varepsilon})(1 - x)^*(1 + x)\} |A(\pi^- l^+)|^2 \\ &- 2\,\mathrm{Re}\{(\varepsilon + \bar{\varepsilon})(1 - \bar{x})^*(1 + \bar{x})\} |\bar{A}(\pi^+ l^-)|^2]^{-1}\end{aligned}$$

to first order in ε and $\bar{\varepsilon}$.

This expression is greatly simplified by the *CPT* theorem, which, in addition to $\bar{\varepsilon} = 0$, yields on the basis of eq. (9.60)

(9.97) $$\bar{x} = x^*$$

and

(9.98) $$|A(\pi^- l^+)|^2 = |\bar{A}(\pi^+ l^-)|^2$$

after the appropriate averaging is carried out. Therefore, under the assumption of *CPT* invariance, eq. (9.96) reduces to

(9.99) $$\delta_L = 2\,\frac{1 - |x|^2}{|1 - x|^2}\,\mathrm{Re}\,\varepsilon,$$

to lowest order in ε. Thus a measurement of δ_L provides a direct measure of

Re ε. The measurements on the 2π mode led to eq. (9.82), from which it follows that, to a good approximation,

(9.100) $$\text{Re } \varepsilon = |\eta_{+-}| \cos \phi_{+-},$$

so that the value of δ_L predicted on the basis of eqs. (9.47) is

(9.101) $$\delta_L \approx 3.23 \times 10^{-3},$$

since $x \approx 0$ (the $\Delta S = \Delta Q$ rule).

The most recent direct experimental value of the asymmetry for both electrons and muons is (PDG 1986)

(9.102) $$\delta_L = (3.30 \pm 0.12) \times 10^{-3},$$

in good agreement with eq (9.101). Therefore we can have confidence not only in the existence of CP violation but also in the mutual consistency of the direct measurements of the CP-violating parameters under the assumption of CPT invariance. However, this consistency in itself does not provide a conclusive demonstration of CPT invariance for the K^0, $\bar{K}^0$ system. That question will be addressed further in the next section in the context of information provided by other data on the K^0, $\bar{K}^0$ system.

9.3 *T* Violation in the K^0, $\bar{K}^0$ System

Given the existence of a CP-violating term in the weak interactions responsible for neutral meson decay, the CPT theorem would imply the existence of T violation by the same interaction and to exactly the same degree as the CP violation. On the other hand, if the CPT theorem is not valid, the interaction may be made up of two parts, one T invariant, the other T violating, and the degree of T violation would then be different from the degree of CP violation. The T-invariant part of the interaction would be CPT violating, and the T-violating part would be CPT invariant (unless there is a non-Hermitian contribution to the interactions).[17] Thus there are two distinct types of measurements that are of interest, one that measures directly the amount of CPT violation, if any, and the other that measures directly the amount of T violation.

The first question is how the analysis and conclusions of section 9.2 are affected if the CPT theorem is *not* valid. Then eqs. (9.58) and (9.76) are replaced by

(9.103a) $$(\mathbf{CPT})H_w(\mathbf{CPT})^{-1} = H''_w$$

[17] See note 7.

and

(9.103b) $$(\mathbf{CPT})H_{ww}(\mathbf{CPT})^{-1} = H''_{ww},$$

where the notation reflects the definitions

(9.104a) $$\mathbf{T}H_w\mathbf{T}^{-1} = H'_w; \quad \mathbf{T}H_{ww}\mathbf{T}^{-1} = H'_{ww}$$

and

(9.104b) $$\mathbf{CP}H'_w(\mathbf{CP})^{-1} = H''_w; \quad \mathbf{CP}H'_{ww}(\mathbf{CP})^{-1} = H''_{ww}.$$

That H''_w and H''_{ww} were not equal to H'_w and H'_{ww} would be a restatement of the *CP* violation of the dynamics as demonstrated by the experiments discussed in section 9.2. Eq. (9.104a) with $H'_w \neq H_w$ or $H'_{ww} \neq H_{ww}$ would be an expression of the way the dynamics violate *T* invariance. Similarly, $H''_w \neq H_w$ or $H''_{ww} \neq H_{ww}$ would be an expression of failure of the *CPT* theorem. Because proof of the *CPT* theorem involves both dynamic and kinematic assumptions, as we shall see later, such a failure may reflect a breakdown in either the dynamics or the kinematics, but for the purpose of analyzing experiments the burden may be placed on the dynamics, as we have done here, by carrying out the analysis in terms of effective (phenomenological) interactions H_w and H_{ww} without ascribing to them further fundamental significance.

Repetition of the analysis leading from the *CPT* theorem to eq. (9.60) now leads to

(9.105) $$A_c^*(E) = e^{-2i\delta_c}\bar{A}''_{\bar{c}'}(E),$$

where A''_{cj} is defined as in eq. (9.3a) with H''_w replacing H_w:

(9.106) $$A''_{cj}(E) \sim \langle c, \text{out} \,|\, H''_w \,|\, j\rangle.$$

Eq. (9.61) for the reduced amplitudes now becomes

(9.107) $$\bar{a}_I = a''^*_I.$$

If there is any deviation from *CPT* invariance, its effect on these amplitudes would be expected to be no greater than the *CP* violation. Therefore it is reasonable to assume

(9.108a) $$a''_I = a_I - 2\alpha_I a_I$$

with

(9.108b) $$|\alpha_I| \lesssim |\eta_{\pi\pi}|,$$

and α_I is a measure of CPT violation. From eq. (9.107) it follows that

(9.109) $$\bar{a}_I = a_I^* - 2\alpha_I^* a_I^* .$$

Again there is freedom of choice of the phase of one of the amplitudes, but unlike the case of CPT invariance, the choice of the phase of a_I does not fix the phase of $\bar{a}_I$ because of the term $\alpha_I^* a_I^*$ in eq. (9.109). We choose this free phase in such a way that a_0 is real. Then, for $I = 0$, eq. (9.109) becomes

(9.110) $$\bar{a}_0 = a_0 - 2\alpha_0^* a_0 .$$

By use of this relationship, eqs. (9.55) take on a form similar to eqs. (9.63) but differing by terms in α_I. To first order in the α_I, ε' and ω (given by eqs. 9.64 and 9.65) we find

(9.111a) $$\varepsilon_{+-} = \varepsilon'' + \alpha_0^* ,$$

and

(9.111b) $$\varepsilon_{00} = -2\varepsilon'' + \alpha_0^* ,$$

where

(9.112) $$\varepsilon'' = \varepsilon' + 2^{-1/2} e^{i\delta} \alpha_2^* a_2^* / a_0 - \omega \alpha_0^* .$$

Again, to the same approximation used in eq. (9.67),

(9.113) $$\eta_{00} = \eta_{+-} - 3\varepsilon'',$$

an equation that may be used as a basis for analysis of the data on η_{+-} and η_{00}. However, the fortuitous circumstance that fixed the simplifying relationship between phases no longer applies because the phases of α_I and a_2, and therefore of ε'', remain to be determined. Clearly, eqs. (9.47) and (9.48) do not provide enough information to determine all the parameters. Only when the assumption of CPT invariance is replaced by some other restriction on the parameters can they be determined by these measurements. However, an upper limit can be placed on Re α_2 on the assumption that $\alpha_2 \approx \alpha_2^{\pm} = (a_{2+} - a_{2-})/2a_{2\pm}$, where $a_{2\pm}$ are the $I = 2$, 2π decay amplitudes of the $K^{\pm}$ decay (see table 9.3), which gives

(9.114) $$[|a_{2+}|^2 - |a_{2-}|^2] = (8.0 \pm 12.0) \times 10^{-3} |a_{2\pm}|^2.$$

Then since

(9.115) $$|a_{2+}|^2 - |a_{2-}|^2 = 4|a_{2\pm}|^2 \text{ Re } \alpha_2^{\pm} ,$$

we find

(9.116) $$\text{Re}\,\alpha_2 \approx \text{Re}\,\alpha_2^{\pm} \approx (2.0 \pm 3.0) \times 10^{-3},$$

which is consistent with eq. (9.108b).

Although measurements limited to the 2π channels are not able to separate the measured value of η_{+-} or η_{00} into the part arising from *CPT* violation and the part arising from *T* violation, it is possible to use data on the amplitudes of decay channels other than the 2π channel to do so. It will be found that the totality of the data can place limits on the amount of *CPT* violation and that the results tend to confirm *CPT* invariance for the K^0, $\bar{K}^0$ system. Therefore the extreme example of *CPT* violation permitted by the analysis of the 2π mode alone appears to be ruled out. But again, some caution is advisable in drawing a final conclusion from the analysis that is about to be presented, because it depends on the assumption of unitarity.[18] An extension of the analysis following relaxation of this assumption will be discussed in the next section.

The parameters $\eta_{\pi\pi}$, which appear linearly in the 2π decay rate, provide a measure of *CP* violation *because they are odd under the transformation CP*, as shown by eq. (9.44). This suggests that a similar parameter—for example, any measurable linear combination of η_{+-} and η_{00}—that is odd under the transformation *T* would serve as a measure of *T* violation, and any such combination that is *even* under *T* but still odd under *CP* would serve as a measure of *CPT* violation. However, the existence of observables that transform linearly under *T* is not a foregone conclusion because *T* is itself not a linear transformation.

The nonlinearity of *T* is associated with the operation of complex conjugation. We have seen that this operation transforms incoming to outgoing states and vice versa, as in eq. (8.37), and that decay processes are thereby reversed. Thus the direct application of the transformation *T* clearly does not lead to the required simple transformation of a decay curve.

This awkward situation arises from the implied reversal of initial conditions when the transformation is applied directly to the states of the system. To circumvent the difficulty, we must determine the way the dynamics alone transform under *T* and then consider the behavior of states satisfying the original initial conditions under the transformed dynamic equations. Therefore it is not the effect of a formal *transformation of the states* that is to be considered but, instead, the effect of the *substitution*

(9.117) $$H_w \to H'_w, \quad H_{ww} \to H'_{ww},$$

[18] See note 7.

which replaces the Hamiltonian by the time-reversed Hamiltonian, eq. (9.104a). We shall find that, although T is a nonlinear transformation, the substitution eq. (9.117) transforms $\eta_{\pi\pi}$ linearly. Therefore $\eta_{\pi\pi}$ may be separated into even and odd terms,

$$\eta_{\pi\pi} = \eta^e_{\pi\pi} + \eta^o_{\pi\pi}, \tag{9.118}$$

where

$$\eta^e_{\pi\pi} \rightarrow \eta^e_{\pi\pi} \quad \text{and} \quad \eta^o_{\pi\pi} \rightarrow -\eta^o_{\pi\pi} \tag{9.119}$$

under the substitution. Thus $\eta^o_{\pi\pi}$ is the T-violating part of $\eta_{\pi\pi}$ and $\eta^e_{\pi\pi}$ is the CPT-violating part.

The validity of this conclusion may be visualized by again considering a thought experiment based on the decay rate curve eq. (9.33) for the $\pi^+\pi^-$ mode shown as the solid curve in figure 9.2. The interference term is proportional to η_{+-}. Only the part of the interference term associated with η^o_{+-} would change sign if the same decay process were observed in a time reversed world. Furthermore, if $\eta^e_{+-} \neq 0$, the curve measured in the time-reversed world would be different from that measured in the CP mirror, shown as the dashed curve in figure 9.2, implying violation of CPT invariance.

If the interaction H_w is replaced by H'_w, the calculated decay amplitude $A_{cj}(E)$ is replaced by $A'_{cj}(E)$ where, according to eq. (9.3a),

$$A'_{cj}(E) \sim \langle c, \text{out} | H'_w | j \rangle, \tag{9.120}$$

the proportionality constant being omitted because it is unchanged by the substitution. $A'_{cj}(E)$ may be related to $A^*_{cj}(E)$ by making use of the procedure leading to eq. (5.27):

$$\begin{aligned} A^*_{cj}(E) &\sim \langle \mathbf{T}(c, \text{out}) | \mathbf{T} H_w \mathbf{T}^{-1} | \mathbf{T} j \rangle \\ &= e^{-2i\delta c} \langle c', \text{out} | H'_w | j \rangle, \end{aligned} \tag{9.121}$$

since $\phi_c = \phi_j = 0$ and $j' = j$ because the K^0 is spinless and assumed to be at rest. Therefore

$$A'_{cj}(E) = e^{2i\delta_c} A^*_{c'j}(E). \tag{9.122}$$

The same procedure leads to the result

$$\langle j | H_{ww} | k \rangle \rightarrow \langle j | H'_{ww} | k \rangle = \langle j | H_{ww} | k \rangle^* \tag{9.123}$$

under the substitution eq. (9.117). Since H_{ww} is Hermitian,[19] the matrix

[19] See note 7.

element of H'_{ww} may be rewritten as

(9.124) $$\langle j | H'_{ww} | k \rangle = \langle k | H_{ww} | j \rangle.$$

The consequences of the substitution for the mass matrix **M** follow directly from the definition, eqs. (9.6) and (9.7), of **M** and from eqs. (9.122) and (9.124), which yield

(9.125) $$\mathbf{M} \rightarrow \mathbf{M}',$$

where

(9.126) $$\langle j | \mathbf{M}' | k \rangle = \langle k | \mathbf{M} | j \rangle.$$

Again use has been made of the fact that all channels are included in the sums over intermediate states, so that the substitution of the summation variable c' or f' for c or f leaves the sums unchanged.

From the expressions eqs. (9.27) for ε and $\bar{\varepsilon}$ it follows immediately that

(9.127a) $$\varepsilon \rightarrow -\varepsilon$$

and

(9.127b) $$\bar{\varepsilon} \rightarrow \bar{\varepsilon}$$

under the substitution eq. (9.117), which represents the action of time reversal for our purposes. Therefore, when $\eta_{\pi\pi}$ is expressed in terms of ε and $\bar{\varepsilon}$ as in eq. (9.49), ε contributes only to the odd term of $\eta_{\pi\pi}$ and $\bar{\varepsilon}$ to the even term.

If we could also separate $\varepsilon_{\pi\pi}$, and therefore $\eta_{\pi\pi}$, into even and odd parts, the result would not in itself be very useful unless $\eta^o_{\pi\pi}$ and $\eta^e_{\pi\pi}$ were measurable quantities. Because their definitions involve the dispersive terms in the mass matrix that depend on all virtual as well as actual decay amplitudes, ε and $\bar{\varepsilon}$ are not measured. Furthermore, to express $\varepsilon_{\pi\pi}$ and $\varepsilon'_{\pi\pi}$ in terms of the measurable amplitudes $A_{2\pi,S}$ and $A_{2\pi,L}$, it is necessary to know ε and $\bar{\varepsilon}$, as can be seen from eqs. (9.37). Therefore it does not appear to be possible to measure directly the four independent quantities η^e_{+-}, η^o_{+-}, η^e_{00}, and η^o_{00}. Fortunately, it is possible to determine the linear combinations

(9.128) $$Z^e = (\eta^e_{00} + C\eta^e_{+-})/(1 + C)$$

and

(9.129) $$Z^o = (\eta^o_{00} + C\eta^o_{+-})/(1 + C),$$

where

(9.130) $$C = |A_{+-,S}|^2/|A_{00,S}|^2$$

is the (measured) ratio of the $\pi^+\pi^-$ decay rate to the $\pi^0\pi^0$ decay rate of the K_S. We note that the combination of the observed values of η_{+-} and η_{00} that was introduced by Ashkin and Kabir (1970),

(9.131) $$Z = (\eta_{00} + C\eta_{+-})/(1 + C),$$

can be written as

(9.132) $$Z = Z^e + Z^o$$

and it is a known complex quantity to the extent that both the magnitudes and the phases of η_{+-} and η_{00} have been measured.

The Bell-Steinberger relation, eq. (9.29), may be expressed directly in terms of Z:

(9.133a) $$(\Gamma_S + 2i\Delta m)(\text{Re }\varepsilon + i \text{ Im } \bar{\varepsilon}) = \Gamma_S Z + \Sigma'_f A^*_{fS} A_{fL},$$

where

(9.133b) $$\Gamma_S = |A_{+-,S}|^2 + |A_{00,S}|^2,$$

Σ'_f denotes the sum over all modes *other than* the 2π modes and very small quantities of the order of Γ_L/Γ_S have been neglected.

All the amplitudes contributing to Σ'_f are small compared with the 2π amplitudes. Therefore corrections of order ε and $\bar{\varepsilon}$ to each term in the sum may be neglected, and we can write

(9.134) $$A^*_{fS} A_{fL} = \tfrac{1}{2}(A_f + \bar{A}_f)(A_f - \bar{A}_f)$$

on the basis of eqs. (9.26). Then by eq. (9.122) the result of the substitution eq. (9.117) is found to be

(9.135) $$\Sigma'_f A'^*_{fS} A'_{fL} = \Sigma'_f A_{fS} A^*_{fL},$$

so that the even and odd parts, Σ^e and Σ^o, of the sum

(9.136) $$\Sigma = (2\Lambda)^{-1}\Sigma'_f A_{fS} A^*_{fL} = \Sigma^e + \Sigma^o$$

can be identified as

(9.137) $$\Sigma^e = \text{Re } \Sigma, \quad \Sigma^o = i \text{ Im } \Sigma.$$

It is convenient to rewrite eq. (9.133a) in the form

(9.138) $$2\Lambda e^{i\phi_0}(\text{Re }\varepsilon + i \text{ Im } \bar{\varepsilon}) = \Gamma_S Z + 2\Lambda\Sigma^*,$$

where

(9.139) $$\Lambda = [(\Delta m)^2 + \tfrac{1}{4}\Gamma_S^2]^{1/2}$$

and ϕ_0 is given by eq. (9.92). We may separate eq. (9.138) into its even and odd parts by making use of eqs. (9.127):

(9.140a) $$e^{-i\phi_0}[\cos \phi_0 Z^e + \Sigma^{e*}] = i \text{ Im } \bar{\varepsilon}$$

(9.140b) $$e^{-i\phi_0}[\cos \phi_0 Z^o + \Sigma^{o*}] = \text{Re } \varepsilon,$$

since $\cos \phi_0 = \Gamma_S/2\Lambda$.

Eq. (9.138) may also be expressed in terms of its real and imaginary parts as follows:

(9.141a) $$\text{Re } \varepsilon = \text{Re } \{e^{-i\phi_0}[\cos \phi_0 Z + \Sigma^*],$$

(9.141b) $$\text{Im } \bar{\varepsilon} = \text{Im } \{e^{-i\phi_0}[\cos \phi_0 Z + \Sigma^*].$$

By inserting these expressions on the right-hand side of eqs. (9.140) and making use of eq. (9.137), we find

(9.142a) $$Z^e = e^{i\phi_0}[i \text{ Im } \{e^{-i\phi_0} Z\} - \Sigma],$$

(9.142b) $$Z^o = e^{i\phi_0}[\text{Re } \{e^{-i\phi_0} Z\} + \Sigma].$$

These final expressions, which are represented as components of Z in figure 9.3, involve only the measurable quantities ϕ_0, Z, A_{fL}, A_{fS}, where the f include all modes other than the 2π mode. Therefore they provide the direct measures of *CPT* and *T* violation that are implicit in the measurements of η_{+-} and η_{00}. In principle the procedure for determining Z^o and Z^e delineated by figure 9.3 is to plot the vector Z determined from eq. (9.131) in the complex plane and rotate it clockwise by the angle ϕ_0. Then its projection on the real axis is to be constructed, and the vector Σ given by eq. (9.136) is to be added as shown. The resultant is rotated counterclockwise by ϕ_0 to obtain Z^o, and Z^e is obtained by closing the triangle. Although the available data are not sufficiently complete or precise to execute this process, the existing data on the decay modes do provide enough information to yield an estimate of Z^o and an upper limit on $|Z^e|$.

The data on the 2π mode, eqs. (9.47) and (9.48), combined with the value (PDG 1986, p. 14)

(9.143) $$C = \Gamma_S(\pi^+\pi^-)/\Gamma_S(\pi^0\pi^0) = 2.186$$

and eq. (9.92), yield for eq. (9.131)

(9.144) $$e^{-i\phi_0}Z = [2.27 \pm 0.04 + i(0.17 \pm 0.12)] \times 10^{-3}.$$

Note that the real part is essentially $|\eta_{+-}|$ with an increase in assigned error because of the relatively poor precision of η_{00}. The imaginary part is almost entirely due to the phase of η_{00}, which is difficult to measure.

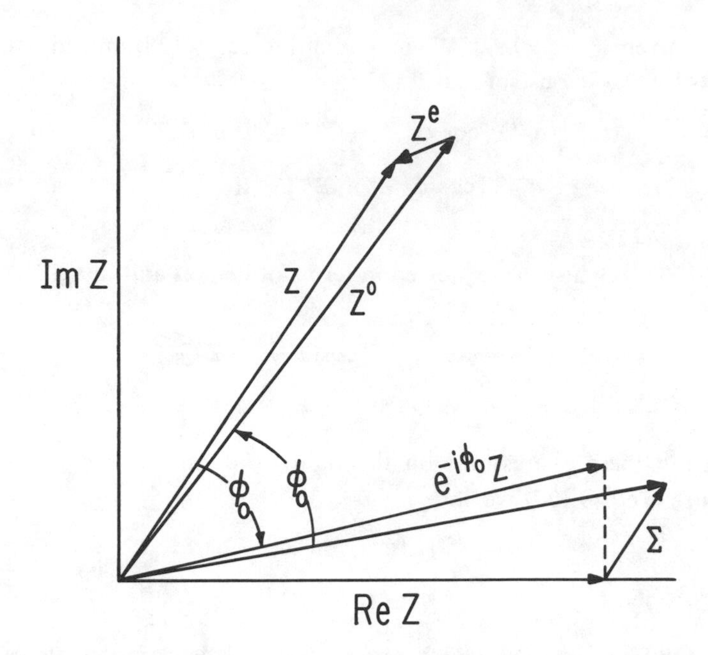

FIG. 9.3 Diagram of the Ashkin-Kabir parameter Z in the complex plane illustrating the T-violating component Z^o and the CPT-violating component Z^e in terms of Σ, eq. (9.136), which is a measure of the absorptive contributions to the mass matrix. The magnitude of Σ is exaggerated for clarity.

The quantity $\Sigma'_f A^*_{fL} A_{fS}$ is dominated by the 3π and the semileptonic modes, for which the K_L branching ratios are given in table 9.2. For the K_S the (CP-violating)[20] relative amplitudes for the 3π mode are measured by $\eta_{3\pi} = A_{3\pi,S}/A_{3\pi,L}$ and (PDG 1986, p. 18)

$$|\eta_{+-0}|^2 < 0.12, \quad |\eta_{000}|^2 < 0.1. \tag{9.145}$$

Therefore we find

$$\left|\sum_{3\pi} A^*_{3\pi,L} A_{3\pi,S}\right| < 0.11\Gamma_L. \tag{9.146}$$

In the case of the semileptonic modes, eqs. (9.95) and (9.96) lead to the

[20] Because the 3π state has opposite parity, and therefore opposite CP signature from the 2π state, the amplitudes $a_{3\pi}$ and $\bar{a}_{3\pi}$ have opposite signs. Therefore the roles of the K_S and K_L are reversed from the case for 2π decay, and it is $K_S \rightarrow 3\pi$ that violates CP. Thus the conventional definitions of η_{+-0} and η_{000}, which are used here, are the inverse of η_f defined in eq. (9.35).

TABLE 9.2 Branching Ratios for K_L

Mode f	B_{fL}
$\pi^0\pi^0\pi^0$	$(21.5 \pm 1.0)\%$
$\pi^+\pi^-\pi^0$	$(12.39 \pm 0.20)\%$
$\pi^\pm\mu^\mp\nu$	$(27.1 \pm 0.4)\%$
$\pi^\pm e^\mp\nu$	$(38.7 \pm 0.5)\%$.

Source: PDG (1986, p. 14).

result

(9.147) $$\sum A_L^*(\pi l)A_S(\pi l) = 2|A_L(\pi^- l^+)|^2(x - \bar{x} + \delta_L)$$

to first order in x, $\bar{x}$ and δ_L, where the sum is over the two modes $\pi^- l^+$ and $\pi^+ l^-$. The quantity $x - \bar{x}$, which is a measure of the difference between the violation of the $\Delta S = \Delta Q$ rule in the oppositely charged modes, is usually reported as $2i$ Im x because the *CPT*-invariance condition eq. (9.97) is assumed. The reported value (PDG 1986, p. 18) is

(9.148) $$(x - \bar{x}) = -2i(0.004 \pm 0.026),$$

while δ_L is given by eq. (9.102). By combining these results with the branching ratios above, we find

(9.149) $$\sum_{l=e,\mu} \sum A_L^*(\pi l)A_S(\pi l) = [0.00 - i(0.01 \pm 0.034)]\Gamma_L$$

if we assume that the same parameters x, $\bar{x}$ and δ_L apply to both the electron and muon modes.

Since all other modes have much smaller branching ratios, we conclude that

(9.150) $$|\Sigma'_f A_{fL}^* A_{fS}| < 0.12\Gamma_L.$$

Then from eqs. (9.142a), (9.139), (9.90), and (9.91) it follows that

(9.151) $$|Z^e| < 0.33 \times 10^{-3},$$

while, from eq. (9.142b)

(9.152) $$|Z^o| = (2.27 \pm 0.55) \times 10^{-3}.$$

Since $Z^0 \neq 0$, we draw the conclusion that *T invariance is violated in the $K_L \to 2\pi$ decay.* On the other hand, although eq. (9.151) is consistent with *CPT* invariance, it is a weak constraint *permitting as much as 15 percent of* *the CP-violating amplitude to result from CPT violation.*[21] To reduce this ambiguity we must increase the precision in the measurements of η_{00}, especially its phase, and of Im x.[22]

The reasoning leading to this conclusion depends very sensitively on one feature of the experimental data, that the measured phase of Z is, within the limits of its precision, equal to the "natural" phase ϕ_0. The contribution from the 2π mode to the CPT-violating term $\text{Im}\{e^{-i\phi_0}Z\}$ in eq. (9.142a) is due almost entirely to the uncertain value of $\phi_{00} - \phi_0$. Our present knowledge of contributions from the other modes is only good enough to place a limit on the other term in Z^e, a limit that provides a definitive statement about T violation because it is considerably smaller than $\text{Re}\{e^{-i\phi_0}Z\}$.

Determining Z^e or Z^o provides an average of the contributions of CPT violation or T violation, respectively, to the 2π mode. The contributions to the $\pi^+\pi^-$ and $\pi^0\pi^0$ modes may now be separated by use of the experimental value of the linear combination

(9.153a) $$X = \eta_{+-} - \eta_{00}.$$

To show that the even and odd parts of X,

(9.153b) $$X^e = \eta^e_{+-} - \eta^e_{00},$$

(9.153c) $$X^o = \eta^o_{+-} - \eta^o_{00},$$

may be determined separately we note that, according to eq. (9.49),

(9.154) $$X = \varepsilon_{+-} - \varepsilon_{00},$$

with $\varepsilon_{\pi\pi}$ given by eq. (9.51).

[21] The separation of the question of T invariance from that of CP invariance of the decay amplitudes has been considered by Gourdin (1967), Okun' (1968), Kabir (1968b), Casella (1968, 1969), Wolfenstein (1969b), Achiman (1969), Ashkin and Kabir (1970), Schubert et al. (1970), Dass and Kabir (1972), and Winter (1972). In these cases the method has been based on showing by use of the Bell-Steinberger relation, eq. (9.29), that one or more of the various parameters that are odd under T do not vanish. Also the CPT question has been addressed by placing upper limits on various CPT-odd parameters. However, to obtain a well-defined measure of the degree of T violation or CPT violation, we obtain here those combinations of the parameters that would offer a direct manifestation of the violation in a hypothetical comparison of experiments between the real world and the time-reversed world or CP-reversed world. Since η_{+-} is not by itself separated into the two parts, the rate curve to be observed for this purpose is, instead of figure 9.2, the curve $R(t) = [R_{00}(t) + CR_{+-}(t)]$, for which the interference term, according to eq. (9.131), is proportional to Z.

A different method for measuring Reε and $\bar{\varepsilon}$, one that does not involve the decay process, has been suggested by Aharony (1970) and by Kabir (1970). Although the experiment would be very difficult, Tanner and Dalitz (1986) have suggested it might be feasible in the near future. This method is essentially a direct test of CPT invariance, since it entails a comparison of the time dependences of the K^0 and $\bar{K}^0$ states. See also Sachs (1963a, eq. 22).

[22] See note 2.

The even and odd parts of $\varepsilon_{\pi\pi}$ may be expressed in terms of the even and odd parts of A_{Ij}. It follows immediately from eq. (9.122) that

(9.155a) $$A_{Ij}^{e} = e^{i\delta_I}\,\mathrm{Re}(e^{-i\delta_I}A_{Ij})$$

and

(9.155b) $$A_{Ij}^{o} = ie^{i\delta_I}\,\mathrm{Im}(e^{-i\delta_I}A_{Ij}).$$

Since the $I = 0$ amplitude is much larger than the $I = 2$ amplitude (see eq. 9.66) and the CP-violating (and CPT-violating) fraction of the amplitude is very small, we may neglect all but the even amplitude $A_0^e = \bar{A}_0^e$ in the denominator of eq. (9.51). It follows that the separation of $\varepsilon_{\pi\pi}$ into even and odd parts is obtained in terms of the even and odd parts of the numerator. Therefore

(9.156a) $$X^e = e^{i\delta}\,\mathrm{Re}(e^{-i\delta}X)$$

(9.156b) $$X^o = ie^{i\delta}\,\mathrm{Im}(e^{-i\delta}X),$$

with $\delta = \delta_2 - \delta_0$. These relationships are most easily visualized by presenting $\eta_{\pi\pi}$ as a vector in the complex plane, figure 9.4.

The procedure suggested by this figure is to construct the vector X as shown and swing it counterclockwise through the (negative) angle $|\delta|$ to obtain $Xe^{-i\delta}$, represented by the dashed line. Then the real component of $Xe^{-i\delta}$ is projected to the indicated position, and this projection is swung clockwise by angle $|\delta|$ to obtain X^e. The vector X^o closes the X, X^e triangle.

X may be expressed in terms of the parameters $|\eta_{\pi\pi}|$, $\phi_{\pi\pi}$ and δ to obtain finally

(9.157a) $$\eta_{+-}^e - \eta_{00}^e = e^{i\delta}[|\eta_{00}|\Delta - (|\eta_{00}| - |\eta_{+-}|)\mu],$$

(9.157b) $$\eta_{+-}^o - \eta_{00}^o = -ie^{i\delta}[|\eta_{00}| - |\eta_{+-}| + |\eta_{00}|\mu\Delta],$$

where we have included only the lowest-order terms in the small quantities

(9.158) $$\Delta = \phi_{+-} - \phi_{00},$$

given by eq. (9.48b), and μ, given by eq. (9.69). The magnitude of $|\eta_{00}| - |\eta_{+-}|$ is, according to eqs. (9.47a) and (9.48a), of the order of $(3 \pm 4) \times 10^{-5}$. Therefore

(9.159a) $$|\eta_{+-}^e - \eta_{00}^e| \approx (0.39 \pm 0.25) \times 10^{-3}$$

and

(9.159b) $$|\eta_{+-}^o - \eta_{00}^o| \approx (0 \pm 0.11) \times 10^{-3}.$$

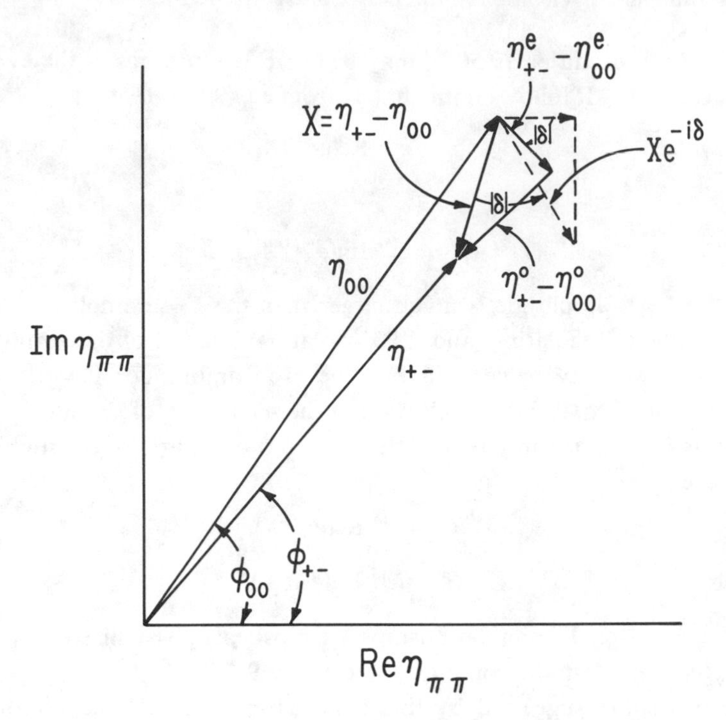

FIG. 9.4 The *CP*-violating complex "vectors" $\eta_{\pi\pi}$. Their difference $X = \eta_{+-} - \eta_{00}$ is resolved into a *T*-violating component X^o and a *CPT*-violating component X^e. The differences in phases, $\phi_{00} - \phi_{+-}$, and magnitude are exaggerated for illustration.

Comparing eq. (9.159b) with $|Z^o|$ given by eq. (9.152) shows that *both* η_{+-} and η_{00} are dominantly odd; that is, they are *T*-violating. Comparing eq. (9.159a) with $|Z^e|$ given by eq. (9.151) confirms our earlier conclusion that to place a smaller limit on *CPT* violation in the 2π mode will require a better determination both of ϕ_{00} *and* of the contribution to Z owing to the 3π and semileptonic decay modes.

Before we draw a final conclusion from these considerations, some attention must be given to a flaw in the logical edifice we have constructed to establish *T* violation (Kenny and Sachs 1973).

9.4 Supplementary Evidence for *T* Violation

The flaw in the analysis has to do with the assumptions underlying the *CPT* theorem that are outlined at the end of section 8.1. When we consider the possibility of a violation of *CPT* invariance, we imply that one or another of these assumptions may be incorrect. Therefore in constructing the analytical tools to be used for testing *CPT* invariance, we must not make explicit use of any one of the assumptions underlying the *CPT* theorem. But that is exactly what we have done.

What must be questioned is the assumption that the Hamiltonian operator, including all interactions, is Hermitian, which is fundamental to the proof of the *CPT* theorem. That assumption has been made throughout this chapter, in particular in arriving at eqs. (9.7), which are implicit in eqs. (9.142). Since the Hermiticity of the Hamiltonian is required to guarantee the unitarity of the dynamics, the validity of the assumption is generally agreed to be sacrosanct. However, the possibility that there exist in nature systems that apparently deviate from unitarity by an amount of the order of the very weak interaction responsible for *CP* violation is not easily excluded.

We must bear in mind that the terms in the weak interaction Hamiltonian on which the Hermiticity condition has been imposed, namely H_w and H_{ww}, are part of an *effective* or "phenomenological" Hamiltonian and are not to be construed as the actual fundamental interactions at, say, the quark level (see chap. 10). Now it is well known that an *effective* interaction need not be Hermitian even when the theory is fully unitary. For example, in the analysis of elastic potential scattering problems, the effect of absorption is often taken into account by adding an imaginary term to the potential, which is simply a device for including the effect of channels other than the scattering channel. And there are other, more subtle ways small non-Hermitian terms might be introduced into the phenomenological interactions when very high energy effects or effects associated with quantum relativity (Wald 1980) are taken into account (Sachs 1986).

From these considerations it follows that a logical analysis of the *T*-invariance question would avoid the explicit assumption that H_w and H_{ww} are Hermitian. But then the form of the mass matrix, eqs. (9.7), is changed significantly. Kenny and Sachs (1973) have treated the effect of this change in general and have shown, in particular, that the measurable amplitudes appearing in the Bell-Steinberger relation eq. (9.29) are in part replaced by quantities that may not be measurable. The same can be said of our eqs. (9.142). Therefore the quantitative separation of the *T*-violating and *CPT*-violating parts of the *CP* violation would no longer be possible.

The interesting question raised by this situation is whether it might be possible to "rescue" *T* invariance by assuming that the observed *CP* violation arises from the introduction of a non-Hermitian contribution to an effective weak interaction. Imposing *T* invariance constrains the number of parameters in such a way that it is possible to obtain an experimental answer to this question. This point has been demonstrated (Sachs 1986) by repeating some of the foregoing analysis under the assumptions that *all* interactions are T invariant and that there is a non-Hermitian term in H_w.

Although the effect of the non-Hermitian term is to foreclose the possibility of separating the T-violating and CPT-violating parts of $\eta_{\pi\pi}$, the conditions imposed on $\eta_{\pi\pi}$ by assuming T invariance *do* provide a test of the assumption. The test is a consequence of the fact that $\eta_{\pi\pi}$ can still be expressed in terms of other quantities that have been measured or on which limits can be placed. Then T invariance can be valid under the assumption of non-Hermitian interactions only if the limits placed on $\eta_{\pi\pi}$ in this way are consistent with the directly measured value of $\eta_{\pi\pi}$.

To carry out this analysis, we must assume that the non-Hermitian weak interactions satisfy the same isotopic spin selection rules as the Hermitian weak interactions, so that measurements on $K^{\pm}$ decay amplitudes may be used to obtain estimates of some of the parameters determining K^0, $\bar{K}^0$ amplitudes. In particular, the observed limits on CP and CPT violation in $K^{\pm}$ decay, as shown in table 9.3, limit the size of the "predicted" value of η_{+-} that would be consistent with the assumption of T invariance.

The resulting limit is (Sachs 1986)

$$|\eta_{+-}| < 3.4 \times 10^{-4}, \tag{9.160}$$

which is a clear contradiction by at least an order of magnitude of the directly measured $|\eta_{+-}|$, eq. (9.47a).

We can conclude from this contradiction that our assumption of T invariance is invalid *even if anti-Hermitian interactions occur up to the limit permitted by the data on unitarity.* When this argument is combined with that given in section 9.3, the existence of T violation appears to be well established.

A more direct test of the assumption of unitarity underlying the Bell-Steinberger relation would be made possible by the direct experimental determination of Re ε and Im $\bar{\varepsilon}$ using the method[23] proposed by Aharony (1970) and by Kabir (1970). Unitarity would require that these measurements be consistent with the results obtained by inserting experimental values of Z and Σ into eqs. (9.141).

Although the analysis leading to the conclusion that T invariance is violated is convincing, it has the disadvantage of indirection; the connection with the directly established CP violation invokes an elaborate theoretical structure and the combination of many different measurements. Clearly, a direct demonstration of T violation is desirable. A violation of invariance

[23] See note 21.

TABLE 9.3 Tests of *CP* Invariance and *CPT* Invariance from Decay Modes of $K^{\pm}$

Test of	Mode	Measurement	Reference
CP	$K^{\pm} \to \pi^{\pm}\pi^{0}$	$\alpha^{\pm} = \dfrac{B_{2\pi,+} - B_{2\pi,-}}{4B_{2\pi,-}} = (0.2 \pm 0.3) \times 10^{-2}$	Herzo et al. (1969)
CP	$K^{\pm} \to \mu^{\pm}\nu$	$\alpha^{\pm}_{\mu 2} = \dfrac{B_{\mu 2,+} - B_{\mu 2,-}}{4B_{\mu 2,-}} = -(1.4 \pm 1.0) \times 10^{-3}$	Ford et al. (1967)
CP	$K^{\pm} \to \pi^{\pm}\pi^{\pm}\pi^{\mp}$	$\alpha^{\pm}_{\tau} = \dfrac{B_{\tau+} - B_{\tau-}}{4B_{\tau-}} = (0.2 \pm 0.3) \times 10^{-3}$	Ford et al. (1970)
CP	$K^{\pm} \to \pi^{\pm}\pi^{0}\pi^{0}$	$\alpha^{\pm}_{\tau'} = \dfrac{B_{\tau'+} - B_{\tau'-}}{4B_{\tau'-}} = (0.2 \pm 1.5) \times 10^{-3}$	Smith et al. (1973)
CPT	$K^{\pm}$ lifetime	$\dfrac{\Gamma_{+} - \Gamma_{-}}{\Gamma_{+}} = (1.1 \pm 0.9) \times 10^{-3}$	Lobkowicz et al. (1969)

Note: $B_{f,\pm}$ is branching ratio into mode *f*. The definition of $\alpha^{\pm}_{f}$ is $\alpha^{\pm}_{f} = (a_{f+} - a_{f-})/2a_{f-}$, which is real under the assumption of *T* invariance.

under motion reversal that is inconsistent with measured final state interactions, a nonzero value of the static electric dipole moment of a particle or an atom, or the splitting of a Kramers degeneracy by an electric field is the kind of evidence that would be required.

Because the observed *T*-violating phenomenon in the K^0, $\bar{K}^0$ system is a very small effect and is primarily due to indirect causes associated with the mass matrix, the chances for observing a motion reversal violation in that system seem remote. That an effect is seen at all is due entirely to the remarkable sensitivity associated with mixing of the degenerate K^0, $\bar{K}^0$ states. Convincing evidence that $\varepsilon_{\pi\pi}$ (or ε'), which is a direct measure of the *CP*-violating interaction, is nonvanishing would be helpful but still would not satisfy the desire for evidence based on motion reversal. The best hope is that such evidence might be forthcoming in experiments on particles of much higher mass than the *K* mesons because, as we shall see in the next chapter, the origin of the phenomenon in the K^0, $\bar{K}^0$ mass matrix may be due to large *T* violation in states of high mass.

10 | Quark Models and Tests of *CP* Violation and *T* Violation

Having established the existence of *CP* and *T* violation, we come now to the fundamental question: How does this violation originate in a natural way in the theory of weak interactions? Although it is an easy matter to introduce a *T*-violating phase into most theories of the weak interactions (an exception is the four-quark model, as we will find later), this is strictly an ad hoc approach. A fundamental theory would be expected to have the characteristic that the permissible values of the *T*-violating phase would be determined by general features of the theory. It seems possible that the development of such a theory will have profound implications for the structure of theoretical physics at the deepest levels, just as the development of the concept of chiral invariance[1] in response to the discovery of parity nonconservation has had such implications.

Such a fundamental insight into the origins of *CP* violation has still to emerge. In fact, there may be reason to assume that the explanation will not lie in that direction, that the value of the *T*-violating phase is in some sense an accident of nature owing to a fluctuation in the early universe (see chap. 11). In these circumstances, there might not be any simple connection between *T* violation and the basic structure of the theory.

An understanding of the origins of *CP* violation can be expected to emerge only from a step-by-step process taking into account the implications of new developments in both theory and experiment, especially experiments that can add quantitative information on *CP* and *T* going beyond the available data on *K* mesons.

In this chapter we consider some small steps that have been taken in that

[1] Early work on chiral symmetry is reviewed by Adler and Dashen (1968), Bernstein (1968), Renner (1968), Treiman, Jackiw, and Gross (1971), and Lee (1972). Chiral symmetry underlies the unification of electroweak interactions and grand unification theories. See the Nobel lectures of Glashow (1980), Salam (1980), and Weinberg (1980); also see Georgi and Glashow (1980).

direction. We examine the phenomenological implications of quark models and some of the possibilities for *CP* experiments arising from the discovery of the existence of heavy quarks. As far as possible the discussion will be based on general physical arguments. The generic features of the widely accepted Kobayashi-Maskawa (KM) model will be combined with highly simplified models of the decay process to estimate orders of magnitude of *CP*-violation effects to be expected in some heavy meson experiments.

10.1 Phenomenological Constraints on the Theory

The analogy between the implications of the discovery of *CP* violation and of *P* violation is obscured by the quantitative difference between them. Within the context of weak interactions, *P* violation is a large effect showing that the *P*-odd interaction is "maximal," that is, it is comparable to the *P*-even interaction. And it is ubiquitous; whenever weak interactions show themselves, *P*-violating effects occur. In contrast, *CP* or *T* violation has been seen only in the case of K^0, $\bar{K}^0$ decay, apparently because the effect is so small that only the highly sensitive interferometry of the K^0, $\bar{K}^0$ system can reveal it. The *CP*-violating $K_L \rightarrow 2\pi$ rate is about 10^{-6} of the $K_S \rightarrow 2\pi$ rate. If expressed in terms of a $K_L \rightarrow 2\pi$ *CP*-odd interaction, this would mean that that interaction is 10^{-3} times the weak interaction of the K_S. Thus the *CP*-odd interaction would be "milliweak," a far cry from the maximal *T* violation, for example, equality of *CP*-odd and *CP*-even interactions that might be expected by analogy with *P* violation.[2]

However, we have already seen that this interpretation in terms of a direct $K_L \rightarrow 2\pi$ interaction is not correct because $\eta_{\pi\pi}$ includes, in addition to

[2] This notion of "maximal" *CP* violation or maximal *T* violation was introduced by Sachs (1964; see also Wolfenstein 1966) based on an observation by Sachs and Treiman (1962, n. 8) that a difference by $\frac{1}{2}\pi$ between the phases of equal $\Delta S = -\Delta Q$ and $\Delta S = \Delta Q$ semileptonic interactions of the *K* mesons would lead to a maximum value of the *CP*-violation parameter ε. The subsequent evidence that $\Delta S = -\Delta Q$ interactions are very small or zero rendered that observation uninteresting. This concept of maximal violation is based on the notion that there are two distinct and equally important contributions to the interaction that have opposite signature under the symmetry operation. In the case of *P* violation those contributions arise from vector and axial vector currents of equal strength. In the case of *CP* violation, the meaning of "maximal" is not at all clear because there is no such natural pairing off of equal weak interactions (except in the failed $\Delta S = \pm\Delta Q$ case). There have been several recent attempts to define maximal *CP* violation in the Kobayashi-Maskawa (KM) model. See Stech (1983), Hochberg and Sachs (1983), Wolfenstein (1984a,b, 1986a), Gronau and Schechter (1985), Jarlskog (1985a,b), and Dunietz, Greenberg, and Wu (1985). Because of the variety of *CP*-violating interactions permitted by the KM model, it is not clear how to define the maximal interaction in such a way as to maximize all *CP*-violating effects. The first three of the references cited above define it in such a way as to maximize the effect between selected pairs of heavy and light meson currents, and the rest of them deal with the generic question of an overall maximization.

the direct effect measured by $\varepsilon_{\pi\pi}$, the indirect contribution of K^0, $\bar{K}^0$ mixing measured by ε. Since $\varepsilon_{\pi\pi}$ is found to be very small, possibly zero, the observed effect does not provide a direct measure of the CP-odd interaction. In fact there may be no such CP-odd weak interaction between the K meson and the pions, the observed phenomenon being primarily due to a CP-violating effect in K^0, $\bar{K}^0$ mixing. It is quite possible that this mixing effect arises from other interactions that violate T maximally in some sense.

At present there exists no generally accepted fundamental theory of the weak interactions that leads naturally to the desired mixing effect. However, there are ad hoc models of CP violation at the quark level, and to the extent that these models can be confirmed by experiments, they introduce constraints that must be satisfied by the ultimate theory. In addition there is Wolfenstein's (1964) original "superweak" model that places all the responsibility for CP violation on a $\Delta S = 2$ interaction term H_{ww} that is much weaker than milliweak and not directly related to the usual weak interactions H_w.

H_{ww} was taken to be T odd in the original form of Wolfenstein's model, so that the matrix element $\langle K^0 | H_{ww} | \bar{K}^0 \rangle$ is purely imaginary relative to dispersive contributions to the mass matrix, since H_w is assumed to be T invariant. Thus all of the CP violation in the mixing is due to the contribution of the matrix element of H_{ww} to M_{12}, eq. (9.7*b*). Then eqs. (9.27*a*) and (9.7*b*) yield

$$\varepsilon = (i\Lambda)^{-1} e^{i\phi_0} \langle \bar{K}^0 | H_{ww} | K^0 \rangle. \tag{10.1}$$

Since CPT invariance implies $\bar{\varepsilon} = 0$ and the assumed T invariance of H_w implies $\varepsilon_{\pi\pi} = 0$, it is found from eq. (9.52) that

$$\eta_{\pi\pi} = \varepsilon. \tag{10.2}$$

It has been assumed here (and will be for the remainder of the chapter) that all interactions are Hermitian and that CPT is valid.

It is immediately apparent that this model is consistent with the data on K mesons; in particular, η_{+-} has the natural phase ϕ_0, which is therefore referred to as the "superweak phase." Furthermore, the prediction $\varepsilon' = 0$ is consistent with the data. Neither of these data were available when Wolfenstein proposed the model. Although there are physically quite different models that lead to the same result, any such model is usually referred to in the literature as "superweak." Operationally that simply means that $\eta_{\pi\pi}$ has the phase ϕ_0 and $\varepsilon' = 0$. The superweak model provides a sufficient but not a necessary condition for meeting these superweak requirements. The necessary condition is that there be no CP-violating amplitudes for any of the actual (on-the-mass-shell) decay modes. Eq. (10.2) follows immediately because there is then no direct *CP* violation in the 2π mode, and the only question concerns the phase of ε. But the *T* invariance of the interactions leading to all decay modes implies that in eq. (9.120) $A'_{fj} = A_{fj}$, from which it follows that the absorptive contribution Γ_{12}, given by eq. (9.7a), vanishes and ε, given by eq. (9.27a), has the superweak phase.

Under this condition the *T* violation (or *CP* violation) must arise in one or more of the virtual amplitudes occurring in the dispersive term of the mass matrix at energies greater than the *K* meson mass, or in a $\Delta S = 2$ interaction, or both. It suggests that there may be intermediate states of energy much greater than m_K, possibly involving the excitation of high-mass particles, for which the weak interactions are strongly *T* violating but whose contribution to the dispersive term in the *K* mass matrix is greatly reduced because of the large energy denominator or for other reasons (such as Cabibbo suppression; see eq. 10.11).

Models that depend on effects due to states of high mass can in principle be distinguished from the original superweak model by means of predicted results of high-energy experiments on *T* invariance or *CP* invariance, if the experiments are designed to produce the massive particles responsible for the *CP*-violating virtual amplitudes. The disintegration (decay) of these unstable particles would be expected to manifest large *CP*- or *T*-violating effects.

Although the phenomenological theory based on the effective interaction H_w makes clear the need for high-energy experiments, a more fundamental approach is required to provide a complete description of the high-energy phenomena. In view of its successes, the Weinberg-Salam-Glashow-Iliopoulos-Maiani model, otherwise known as the "standard model" of weak interactions,[3] offers the appropriate starting point. Therefore we turn our attention to this model, and our first task will be to determine the form of the effective interactions H_w and H_{ww} that arise from the model in order to make the connection with the K^0, $\bar{K}^0$ phenomenological theory.

10.2 The Standard Model and *CP* Violation

As an introduction to the standard model, we consider the four-quark model that was the basis of the Glashow-Iliopoulos-Maiani (GIM) conjecture on the nature of weak interactions of the hadrons and that is capable of describing all the weak interaction properties of the strange particles except the *CP* violation. The starting point for the model is the assumption

[3] See Quigg (1983, pp. 148 ff.).

that the hadrons are composed of quarks, the mesons of a quark-antiquark pair, and the baryons of three quarks.[4]

The four kinds of quark of interest are those having the four "flavors": up, down, strange, and charm, usually denoted by u, d, s, c, respectively. The generic quark of flavor f will be denoted here by q_f. All the quarks are spin 1/2 fermions and have baryon number 1/3. The electric charge of both u and c is $2e/3$, and that of d and s is $-e/3$, where e is the magnitude of the electron charge. The quarks occur in two "families" (u, d) of strangeness $S = 0$ and a family (c, s) where the strangeness of s is $S = -1$ and c carries a new internal quantum number c analogous to S, with $c = 1$ for the c quark. The member of a family having charge 2/3 will be denoted generically by u_f and one having charge $-1/3$, by d_f, since there should be no confusion between this usage of the same index symbol for flavor and for family. The K mesons have the compositions

$$K^+ = (u\bar{s}), \quad K^0 = (d\bar{s}), \quad K^- = (\bar{u}s), \quad \bar{K}^0 = (\bar{d}s),$$

where $\bar{q}$ denotes the antiparticle to q, which not only has the opposite sign of electric charge from q but also has the opposite sign of strangeness, baryon number, and "charm," which can also be treated as an additive quantum number $c = \pm 1$, conserved in strong interactions.

The Weinberg-Salam-Glashow theory of electroweak interactions is based on the assumption that the carriers of these interactions are gauge fields[5] that manifest themselves as the photon and three intermediate vector bosons, the $W^\pm$ and Z^0. The fermions act as sources of these fields, the source current being proportional to the four-vector $\bar{\psi}_i \gamma_\mu \psi_j$, where ψ_i is a spinor field describing the fermions of type i and γ_μ, γ_5 are the Dirac gamma matrices, eqs. (6.24) and (8.10b). If the Dirac fields of the electrons or muons are denoted by l and the field of the associated neutrino by v_l, the leptonic source of the electrically charged bosons $W^\pm$ is the weak lepton current ("charged weak current")

$$j_\mu^l = \frac{1}{2} \sum_l \bar{v}_l \gamma_\mu (1 + \gamma_5) l \tag{10.3}$$

and its Hermitian conjugate, where the factor $\frac{1}{2}(1 + \gamma_5)$ guarantees that only the left-handed leptons,

$$l_L = \tfrac{1}{2}(1 + \gamma_5) l, \tag{10.4}$$

[4] For a good review of quark structure of hadrons see Close (1979), and for the gauge theory of their interactions, Quigg (1983).

[5] See note 4.

and so on, take effect. In reference to this behavior under *P*, *T*, and *C*, applying eqs. (8.11), (8.12), and (8.14) shows that these currents do not satisfy the conditions for *P* invariance and *C* invariance but do satisfy the condition eq. (6.21a) for *T* invariance.

The corresponding weak quark "charged" current (the one that transfers electric charge) takes a similar form:

(10.5) $$j^q_\mu = \frac{1}{2}\sum_f \bar{u}_f \gamma_\mu(1 + \gamma_5)\, d'_f,$$

where the sum is over the two families. The prime on d'_f indicates that it does not, in fact, refer to a specific family but instead refers to a linear combination of families:

(10.6) $$d'_f = \sum_{f'} V_{ff'}\, d_{f'}.$$

This mixing of families is needed to account for $\Delta S = \pm 1$ decays, while the requirement that $V_{ff'}$ be a unitary 2×2 matrix guarantees that the weak interactions will take the "universal" form suggested by Cabibbo (1963). The suggestion that the charged weak current be given by eq. (10.5) is due to Glashow, Iliopoulos, and Maiani (1970), who pointed out that the introduction of a fourth quark flavor ("charm," before it was discovered) in this way provided a mechanism (the "GIM mechanism") accounting for the absence of $\Delta S = \pm 1$ transitions between electrically neutral states. This requires only that the "neutral" weak currents be formed from bilinear expressions in u_f or in d'_f, where d'_f is a unitary mixture, eq. (10.6), as will be shown below.

The weak decay of a hadron into leptons occurs via the emission by one of the quarks in the hadron of a $W^\pm$ or a Z^0, which in turn is the source of a pair of leptons. Weak decay into hadrons also occurs through the intermediary of the boson. In the latter case the boson is the source of a pair of quarks that combine with the original quarks of the decaying hadron and quark pairs from the vacuum, if necessary, to form (at least) two hadrons. This process is illustrated in figure 10.1. The amplitudes for emission of the $W^\pm$ due to the weak current eq. (10.5) is proportional to an element $V_{ff'}$ of the matrix V and the amplitude for production of a hadron by $W^\pm$ to an element of the matrix $V^\dagger$, or vice versa.

On the other hand, the neutral weak current responsible for emission of the Z^0 by a hadron must be made up of a linear combination of products of quark fields, $\sum_f \bar{u}_f \gamma_\mu u_f$ or $\sum_f \bar{d}'_f \gamma_\mu d'_f$ and similar expressions with γ_5 inserted. The first of these expressions clearly has no matrix elements be-

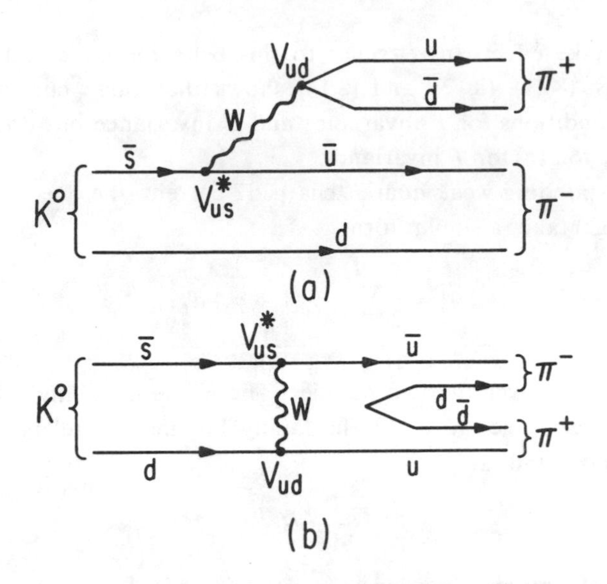

FIG. 10.1 Schematic of $K^0 \to \pi^+\pi^-$ decay. The K^0 is represented as a (bound) $d\bar{s}$ pair of quarks. In (*a*) the $\bar{s}$ emits a W^+ while becoming a $\bar{u}$ that combines with the "spectator" d to form a π^-. The W^+ decays into a $u\bar{d}$ pair that forms a π^+. The $u\bar{u}$ and $d\bar{d}$ final states could also be combined to form the $2\pi^0$ mode. In (*b*) the W^+ is emitted by the $\bar{s}$ as before but is absorbed by the d in the K^0 to become a u. A $\bar{d}d$ pair from the vacuum provides the $\bar{d}$ and d that separately combine with the u and $\bar{u}$ to form the π^+ and π^-. This process is called "hadronization" of the u and $\bar{u}$. None of the multitude of gluon exchanges between quarks that are responsible for their binding into quarks, for their hadronization, and for the final state interactions between pions are shown in the diagram.

tween states of different S, since a difference in S means a difference of flavors. The second has $\Delta S = \pm 1$ matrix elements from the term $\bar{d}s$ and $\bar{s}d$ that are proportional to

$$(V^\dagger V)_{ds} = (V^\dagger V)_{sd} = 0 \tag{10.7}$$

because V is unitary. Thus the GIM mechanism for eliminating strangeness changing neutral currents follows directly from the unitarity of V.

The important question for us is how to introduce CP or T violation in this model. Because the model assumes Hermitian local interactions between the weak currents and the bosons, CPT invariance is implicit, and we may limit attention to the question of T violation. From the form of eqs. (9.64), (9.7), and (9.27a), it is apparent that we need the freedom to introduce an appropriate phase factor into the effective interactions H_w and H_{ww} in order to produce the required imaginary parts to the amplitudes. We have already seen (fig. 10.1) that the terms in H_w leading to hadron emission are

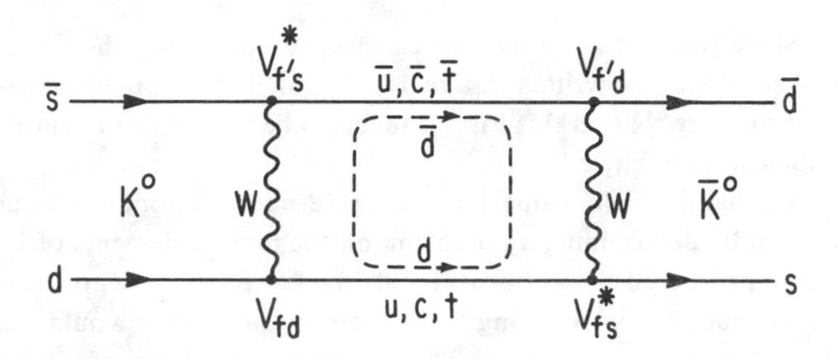

FIG. 10.2 "Box" diagram representing M_{21} for an effective $\Delta S = 2$ interaction, that is, the conversion of a K^0 into a $\bar{K}^0$. Note that by inserting a closed loop of a *u* or *d* quark (arising from the vacuum) into the box and hadronizing the *u* and $\bar{u}$ in the intermediate states, the 2π intermediate state of the self-energy diagram, figure 9.1*b* can be reproduced. However, there are many possible quark and gluon intermediate states that cannot be represented solely in terms of hadrons. The top quark *t* is included in this figure for later reference.

proportional to λ_u, where λ_f is defined by

(10.8)
$$\lambda_f = V_{fd} V^*_{fs}.$$

Furthermore, we can see from figure 10.2 that the model leads to a $\Delta S = 2$ term H_{ww} that is proportional to products of the form $\lambda_f \lambda_{f'}$, where f and f' refer to either the *u* or the *c* quarks (or to the *t* quark in the six-quark model discussed later).

Since in this model the only source of *CP*-violating phases of the effective interactions H_w and H_{ww} are the phases of the matrix elements of the unitary matrix V, the phases of V must be chosen to fit the phases of H_w and H_{ww} determined from the data on *CP* violation. The question then arises of the amount of freedom there is to choose the phases of V to fit the data. The answer is that there is no such freedom; the matrix elements of V in the four-quark model may always be taken to be real. The reason is the one discussed on page 195. The phases of independent sectors of the basic Hilbert space are arbitrary; they are not measurable and therefore are simply a matter of convention.

In this case the independent sectors are those referring to different flavors. In the four-quark model the weak current includes V as a 2×2 matrix connecting the d_f to the $u_{f'}$. There is one overall phase in V, and there are the two relative phases of $|d\rangle$, $|s\rangle$, and of $|u\rangle$, $|c\rangle$, making a total of three that may be adjusted arbitrarily by changing the phase conventions for each one of the d_f and each of the $u_{f'}$. On the other hand, four independent real numbers are required to specify an arbitrary 2×2 unitary matrix.[6] Since three of these may be canceled by adjusting the three arbitrary phases, V may be written as a real orthogonal matrix determined by a single real number that, in this case, is taken to be the angle of rotation θ_C, called "the Cabibbo angle."[7]

It follows that in the four-quark model there is no freedom to account for CP violation by determining a phase among the matrix elements of V. This point was emphasized by Kobayashi and Maskawa (1973), who then went on to point out that introducing a third family of quarks would make it possible to overcome the problem. The subsequent discovery of the b quark has led to a general acceptance of the Kobayashi-Maskawa (KM) six-quark model, although convincing direct experimental evidence for the partner of the b, called the t quark, remains to be established.

The fact that a six-quark model can take account of a CP-violating phase follows from the evaluation of the number of parameters required to determine a 3×3 matrix V that mixes the three families in accordance with eq. (10.6). In this case there are[8] $N^2 = 9$ real numbers required to specify V, but of these one is an arbitrary overall phase and $2(N-1) = 4$ are relative phases among the $N(=3) d_f$ and among the $N\ u_{f'}$ states. Thus V is determined by four real numbers, of which three may be taken to be the Euler angles of a three-dimensional rotation and the fourth is a phase factor that may be chosen to fit data on CP violation.

In the KM notation the matrix

$$V = \begin{pmatrix} V_{ud} & V_{us} & V_{ub} \\ V_{cd} & V_{cs} & V_{cb} \\ V_{td} & V_{ts} & V_{tb} \end{pmatrix}$$

is parameterized as follows:

$$\begin{aligned} &V_{ud} = c_1, && V_{us} = s_1 c_3, && V_{ub} = s_1 s_3 \\ \textbf{(10.9)}\quad &V_{cd} = -s_1 c_2, && V_{cs} = c_1 c_2 c_3 - e^{i\delta} s_2 s_3, && V_{cb} = c_1 c_2 s_3 + e^{i\delta} s_2 c_3 \\ &V_{td} = -s_1 s_2, && V_{ts} = c_1 s_2 c_3 + e^{i\delta} c_2 s_3, && V_{tb} = c_1 s_2 s_3 - e^{i\delta} c_2 c_3, \end{aligned}$$

[6] A direct way to obtain this result is to note that any $N \times N$ unitary matrix can be written in the form $U = \exp(iQ)$, where Q is a Hermitian $N \times N$ matrix. Q is specified by N real diagonal elements and $\frac{1}{2}N(N-1)$ complex off-diagonal elements, or a total of N^2 real numbers. Thus U is specified by N^2 real numbers.

[7] Cabibbo (1963) introduced the concept of a rotation θ_C of the two-dimensional space spanned by the $\Delta S = 0$ weak current and the $\Delta S = 1$ weak current as a basis for a "universal Fermi interaction." When these phenomenological currents of an earlier era are expressed in terms of the quark currents, the only change is to identify θ_C as a rotation of the two-dimensional space spanned by the d and s quark states.

[8] See note 6.

where $s_i \equiv \sin\theta_i$, $c_i \equiv \cos\theta_i$, and θ_1, θ_2, θ_3 are three independent angular variables (Euler angles). The values $\theta_2 = \theta_3 = 0$, $\theta_1 = \theta_C$ imply that there is no coupling between the first two families and the third family, which corresponds to the four-quark model and would lead to the Cabibbo form of the weak interactions.

By a change of phase conventions for the quark states, it is possible to move the phase factor around the matrix. For example, by an appropriate change in the phase of $|t\rangle$ relative to $|c\rangle$ and $|u\rangle$, V_{tb} can be converted to a real number and V_{td} becomes a complex number. Of course these changes do not affect the observables of the system.

Determination of the four parameters of V from measurements on meson decay properties requires that we calculate the phenomenological weak interactions H_w and H_{ww} of the mesons in terms of the parameters by taking appropriate matrix elements of the quark currents eq. (10.5). These matrix elements may then be used to fit the parameters to measured decay rates and other data. The form of the dependence of H_w and H_{ww} on the matrix elements of V is evident from diagrams such as those in figures 10.1 and 10.2, as we have already seen. Although the strong interaction (quantum chromodynamic = QCD) aspects of the calculation of H_w and H_{ww} are not susceptible to exact treatment, it has been possible to obtain good enough approximations to be able to estimate or place limits on the magnitudes of the three real parameters s_i in the matrix V. Strong interaction effects do not themselves introduce T-violating or CP-violating phases, since QCD is assumed to be T invariant.

Analysis[9] of available data, including information on B meson (composed of $\bar{u}b$ or $\bar{d}b$; see sec. 10.3) decay rates (Chau and Keung 1984), shows that s_2 and s_3 are of the order of s_1^2, while $s_1 c_3 \equiv \sin\theta_C$ is known from the Cabibbo fit to the data (Shrock and Wang 1978) to be small: $\sin\theta_C = 0.22 \pm 0.002$. Wolfenstein (1983) has suggested that these results may be used to simplify the parameterization of the matrix V by expanding in powers of

(10.10) $$s_C = V_{us} = \sin\theta_C .$$

[9] Ellis, Gaillard, and Nanopoulos (1976) provided the basic framework for analysis. A recent review of methods is given by Chau (1983), but the final results of any analysis are subject to some caution because of the many approximations required to obtain quantitative results. In particular, most analyses are based on the assumption that ε is to be obtained from the "box" diagram, figure 10.2, although it has been shown (Hochberg and Sachs 1983) that there are other diagrams of comparable importance. Much remains to be done before a fully satisfactory quantitative treatment of the KM model and its derivatives is available, but the qualitative features of the results to be expected from the model are clear enough and elucidate the prospects for future tests of *CP* invariance and *T* invariance.

The order of magnitude of the contributions of the quark currents to weak interaction phenomena is then governed by the lowest-order terms in each matrix element. Wolfenstein has shown that, when terms of order s_C^4 and higher are neglected in the real parts of the matrix elements of V and only the lowest-order terms in the imaginary parts are retained, V can be put in the form

$$
\textbf{(10.11)}\qquad
\begin{array}{cccc}
 & d & s & b \\
u & 1-\frac{1}{2}s_C^2 & s_C & \Omega s_C^3(\mu - i\xi) \\
c & -s_C & 1-\frac{1}{2}s_C^2 - i\xi\Omega^2 s_C^4 & \Omega s_C^2(1+i\xi s_C^2) \\
t & \Omega s_C^3(1-\mu-i\xi) & -\Omega s_C^2 & 1
\end{array}
$$

by a judicious choice of the arbitrary phases. Here Ω, μ, and ξ are on the order of or less than one.

The KM model can account for the observed CP violation in the K^0, $\bar{K}^0$ system because it includes the phase δ (or ξ) as a free parameter. We found in chapter 9 that the observed CP violation is primarily due to the indirect term ε; the direct contribution ε' to the 2π mode is much smaller, as shown by eq. (9.72). That the direct term arising from the matrix V is consistent with this result can be seen by noting that the decay amplitude associated with the diagram in figure 10.1 is proportional to $V_{ud}V_{us}^* = s_C(1-\frac{1}{2}s_C^2)$. This amplitude is real and, by eq. (9.64), contributes nothing to ε'. If the only diagrams contributing to the 2π decay were those shown in figure 10.1, the result would be $\varepsilon' = 0$, which is consistent with the data. However, Vainshtein, Zakharov, and Shifman (1975; see also Shifman, Vainshtein, and Zakharov 1977) have remarked that there is good reason to include another diagram, the "penguin" graph of figure 10.3*a*, in order to account for the $\Delta I = 1/2$ rule. Gilman and Wise (1979) then noted that because the amplitude associated with the penguin graph includes a contribution proportional to the complex matrix element V_{sc}, it contains an imaginary term and can therefore yield a nonvanishing value for ε', although the relative size of this amplitude is of order s_C^4. The calculations[10] of ε' on the basis of the KM model are subject to all the uncertainties associated with QCD corrections as well as those associated with the determination of the $V_{f'f}$, but they lead to small values of ε', well within the range of the experimental result eq. (9.72). The important point is that the KM model leads to a finite but very small direct contribution to CP violation. Experimental confirmation of

[10] Reviews of recent results are presented by Wolfenstein (1985a, 1986b). See also Gilman and Hagelin (1983), Buras, Slominski, and Steger (1984), and Chau, Cheng, and Keung (1985).

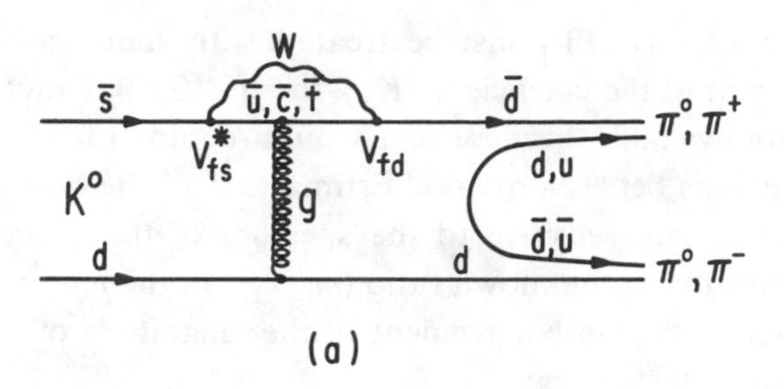

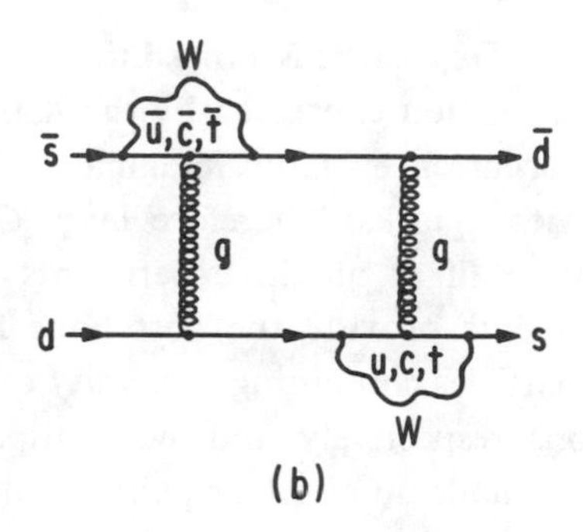

FIG. 10.3 (*a*) Penguin diagrams for decay $K^0 \to \pi^0\pi^0$ or $\pi^+\pi^-$. (*b*) Double penguin contribution to $K^0 \to \bar{K}^0$ transition.

such a term will require a substantial increase in precision over that estimated in eq. (9.72).

Calculation of ε, which is given by eq. (9.27a), requires a determination of the off-diagonal element of the mass matrix. Since the direct *CP* violation effects are very small, the contribution to ε of the absorptive term Γ_{12} may be neglected. Therefore ε is determined by Im M_{12}. From figure 10.2 and figure 10.3*b* it is evident that M_{21} has the general form

(10.12) $$M_{21} = \sum_{f,f'} K_{ff'}\lambda_f\lambda_{f'},$$

where λ_f is given by eq. (10.8) and f, f' run through the values u, c, t. As shown by Ellis, Gaillard, and Nanopoulos (1976), $K_{ff'}$ is a function of m_K^2, m_f^2, and $m_{f'}^2$. The lowest-order contributions to Im M_{12} are found from eq. (10.11) to be proportional to $\xi\Omega^2 s_C^6$, and from figure 10.1 we find that $\Gamma_S \sim \lambda_u^2 \sim s_C^2$ in lowest order. Therefore

(10.13) $$|\varepsilon| \sim \xi\Omega^2 s_C^4,$$

and the order of magnitude of $|\varepsilon|$ is accounted for by the small size of the Cabibbo angle, $s_C \approx 0.22$, from which $s_C^4 \approx 2.3 \times 10^{-3} \approx |\varepsilon|$.

The estimate eq. (10.13) must be treated with some caution because it takes no account of the coefficients $K_{ff'}$ in eq. (10.12). Exact calculation of the $K_{ff'}$ is not available, because that would require a full treatment of the strong interactions between quarks. Estimates[11] of the dependence of $K_{ff'}$ on quark masses are sensitive to the specifics of the approximation and depend strongly on the unknown ratio $(m_t/m_c)^2$ of the masses of the t and c quarks. Therefore the close agreement of the magnitude of s_C^4 with $|\varepsilon|$ may be to some extent fortuitous.

10.3 Heavy Mesons: Time-Dependent Mixing Effects

The T violation and CP violation expressed by the KM matrix eq. (10.11) arises as a result of the occurrence of the imaginary terms in V that are associated with the heaviest quarks. Therefore large CP-violating effects may be expected to appear, if at all, in experiments involving t and b quarks, and such experiments provide the best hope for confirming the model. The lowest-mass particles containing c, b, and t quarks are or should be the D, B, and T mesons, respectively, and their antiparticles, since these mesons are presumed to be made up of quark pairs as follows:[12]

(10.14a) $$D^+ = c\bar{d}, \quad D^0 = c\bar{u}, \quad D^- = \bar{c}d, \quad \bar{D}^0 = \bar{c}u,$$

(10.14b) $$B^+ = \bar{b}u, \quad B^0 = \bar{b}d, \quad B^- = b\bar{u}, \quad \bar{B}^0 = b\bar{d},$$

(10.14c) $$T^+ = t\bar{d}, \quad T^0 = t\bar{u}, \quad T^- = \bar{t}d, \quad \bar{T}^0 = \bar{t}u.$$

Other ("exotic") mesons of comparable mass include

(10.14d) $$F^+ = c\bar{s}, \quad F^- = \bar{c}s,$$

(10.14e) $$B_c^+ = \bar{b}c, \quad B_s^0 = \bar{b}s, \quad B_c^- = b\bar{c}, \quad \bar{B}_s^0 = b\bar{s},$$

(10.14f) $$T_s^+ = t\bar{s}, \quad T_b^+ = t\bar{b}, \quad T_c^0 = t\bar{c},$$

$$T_s^- = \bar{t}s, \quad T_b^- = \bar{t}b, \quad \bar{T}_c^0 = \bar{t}c.$$

The D, B, and F varieties of mesons have been observed (*PDG* 1986), but the T's have not.

The similarity of the neutral meson systems D^0, $\bar{D}^0$; B^0, $\bar{B}^0$; etc., to the K^0, $\bar{K}^0$ system suggests immediately that they should offer comparable opportunities for tests of CP violation. However, a special circumstance

[11] See note 9.

[12] My convention for the B^0 and $\bar{B}^0$ mesons is different from that usually found in the literature ($B^0 = b\bar{d}$, etc.), but it is more in keeping with the usual assignments of isotopic spin as, for example, in the case of the K mesons. It is also the convention used by the Particle Data Group (PDG 1986, p. 70).

that does not apply to the heavier mesons has made possible the successful demonstrations of *CP* violation in the K^0, $\bar{K}^0$ experiments, namely, that the K meson mass, $m_K \approx 3.5\, m_\pi$, is such that there is a very large difference in the available kinetic energy and therefore in the density of the final states between the *CP*-even 2π decay mode and the *CP*-odd 3π decay mode. Thus the decay rate for the former is much greater than the decay rate for the latter, $\Gamma_S \gg \Gamma_L$, and it is possible to separate the K_S and K_L components of a K^0 or $\bar{K}^0$ beam. The successful *CP* experiments are all based on this separation.

In contrast, the mass of the charmed meson is $m_D \approx 13\, m_\pi$, while the much heavier B meson is estimated to have $m_B \approx 38\, m_\pi$ (*PDG* 1986). Therefore the two eigenstates of the neutral heavy mesons are expected to have approximately equal decay rates. That is, if we construct a mass matrix for, say, the D^0, $\bar{D}^0$ system by means of diagrams analogous to figure 10.2 and figure 10.3*b*, and the two eigenvalues of this mass matrix are denoted by $z_\alpha = m_\alpha - \frac{1}{2}i\,\Gamma_\alpha$ and $z_\beta = m_\beta - \frac{1}{2}i\,\Gamma_\beta$, it is to be expected that

(10.15) $$\Gamma_\alpha \approx \Gamma_\beta .$$

There is no clear separation into long-lived and short-lived states as in the case of the K^0, $\bar{K}^0$. The same is true, of course, for the other neutral heavy mesons. Therefore different methods must be used to test for *CP* violation in the decay of the heavy mesons.

A generic study of *CP* tests in the heavy-meson systems has been presented by Pais and Treiman (1975).[13] Aside from methods that would demonstrate an asymmetry between the decays of oppositely charged heavy mesons (Carter and Sanda 1981; Chau and Cheng 1984), these methods depend on neutral particle-antiparticle mixing or on time-dependent interference between the $|\alpha\rangle$ and $|\beta\rangle$ eigenstates of the mass matrix $\mathbf{M}$.

The qualitative features of the mixing and interference phenomena for the heavy mesons are governed by the time dependence of the eigenstates. If the heavy mesons are denoted generically by H^0 and $\bar{H}^0$, the eigenstates are given by

(10.16a) $$|\alpha\rangle = [p\,|H^0\rangle + q\,|\bar{H}^0\rangle],$$

(10.16b) $$|\beta\rangle = [p\,|H^0\rangle - q\,|\bar{H}^0\rangle],$$

where

(10.16c) $$p^2 = \mathbf{M}_{12}/[|\mathbf{M}_{12}| + |\mathbf{M}_{21}|], \quad q^2 = \mathbf{M}_{21}/[|\mathbf{M}_{12}| + |\mathbf{M}_{21}|],$$

[13] See also Carter and Sanda (1981) and Bigi and Sanda (1981).

and the notation for the matrix elements is analogous to that used for the K^0, $\bar{K}^0$ system, eq. (9.6). The state of a particle that is initially $|H^0\rangle$ has the time dependence

(10.17)
$$|\psi(t)\rangle = (1/2p)[|\alpha\rangle e^{-im_\alpha t}e^{-\Gamma_\alpha t/2} + |\beta\rangle e^{-im_\beta t}e^{-\Gamma_\beta t/2}]$$
$$= \tfrac{1}{2}e^{-im_\alpha t}e^{-\Gamma_H t/2}(e^{\Delta\Gamma_H t/4} + e^{-i\Delta m_H t}e^{-\Delta\Gamma_H t/4})|H^0\rangle$$
$$+ (q/p)(e^{\Delta\Gamma_H t/4} - e^{-i\Delta m_H t}e^{-\Delta\Gamma_H t/4})|\bar{H}^0\rangle],$$

with

(10.18a)
$$\Gamma_H = \tfrac{1}{2}(\Gamma_\alpha + \Gamma_\beta),$$

(10.18b)
$$\Delta\Gamma_H = \Gamma_\beta - \Gamma_\alpha,$$

(10.18c)
$$\Delta m_H = m_\beta - m_\alpha.$$

The observation of time-dependent mixing effects requires that the state live long enough for $|H^0\rangle$ to convert into $|\bar{H}^0\rangle$. Therefore the coefficient of $\bar{H}^0$ must be appreciable at $t = \Gamma_H^{-1}$. Since we have already seen that $\Gamma_\alpha \approx \Gamma_\beta$ or

(10.19)
$$|\Delta\Gamma_H|/\Gamma_H \ll 1,$$

it is clear that the amplitude of the $|\bar{H}^0\rangle$ term at $t = \Gamma_H^{-1}$ will be determined by $\Delta m_H/\Gamma_H$.

Consider for example the process suggested by Okun', Zakharov, and Pontecorvo (1975),

(10.20)
$$e^+ + e^- \to B^0 + \bar{B}^0,$$
$$\bar{B}^0 \to l^-\bar{\nu}_l\pi^+$$
$$B^0 \to l^+\nu_l\pi^-$$

where the "$\Delta b = \Delta Q$" rule, which follows from the KM model (as does the $\Delta S = \Delta Q$ rule), has been invoked. This rule has the consequence that, as long as the B^0 and $\bar{B}^0$ retain their separate identities, the leptons have opposite signs of the electric charge, as in eq. (10.20).

However, after time intervals large enough for the B^0 to change into a $\bar{B}^0$ in accordance with eq. (10.17), and for the $\bar{B}^0$ to change into a B^0, there is a nonvanishing probability for both mesons to be in either the state $|B^0\rangle$ or the state $|\bar{B}^0\rangle$ so that both decay leptons have the same charge. The ratio of like- to unlike-charge lepton pairs depends on the probability of $B^0 \to \bar{B}^0$ conversion, and its time average is proportional to $(\Delta m_B/\Gamma_B)^2$, which is therefore the parameter determining whether like-charge leptons will be observed.

An order of magnitude estimate of $\Delta m_H/\Gamma_H$ can be made on the basis of any model that provides an estimate of the matrix elements M_{12}. This result can be established by noting that the eigenvalues of the H^0, $\bar{H}^0$ mass matrix $\mathbf{M}_{jk}$, $z_\alpha = m_\alpha - \frac{1}{2}i\Gamma_\alpha$ and $z_\beta = m_\beta - \frac{1}{2}i\Gamma_\beta$, are the solutions of the associated 2×2 secular problem. Therefore their sum is (by trace invariance)

(10.21a) $$z_\alpha + z_\beta = \mathbf{M}_{11} + \mathbf{M}_{22},$$

and their product is

(10.21b) $$z_\alpha z_\beta = \mathbf{M}_{11}\mathbf{M}_{22} - \mathbf{M}_{12}\mathbf{M}_{21}.$$

By combining the square of eq. (10.21a) with eq. (10.21b), we find

(10.22a) $$(z_\alpha - z_\beta)^2 = 4\mathbf{M}_{12}\mathbf{M}_{21}.$$

If there is no *CP* violation M_{12} and Γ_{12}, the "dispersive" and absorptive parts of $\mathbf{M}_{12}$ (see eqs. 9.6 and 9.7), are real, so that $\mathbf{M}_{12} = \mathbf{M}_{21}$ and the solution of eq. (10.22a) is

(10.22b) $$\Delta m_H = \pm 2M_{12}, \quad \Delta\Gamma_H = \pm 2\Gamma_{12}.$$

When *CP* violation is small, $M_{12} \approx M_{21}$, and eq. (10.22a) is easily solved to give $\Delta m_H = 2 \text{ Re } M_{12}$, which is commonly and appropriately used for the K^0, $\bar{K}^0$ case. However, if the *CP* violation is not small, the real and imaginary parts of eq. (10.22a) become

(10.23a) $$(\Delta m_H)^2 - \tfrac{1}{4}(\Delta\Gamma_H)^2 = 4|M_{12}|^2 - |\Gamma_{12}|^2$$

and

(10.23b) $$\Delta m_H \Delta\Gamma_H = 4 \text{ Re}(M_{12}\Gamma_{12}^*).$$

According to eq. (10.19) $\Delta\Gamma_H/\Gamma_H$ is small, and from eq. (10.23a) it follows that

(10.24a) $$|\Delta m_H| \approx 2|M_{12}|$$

can be used as a basis to determine whether $\Delta m_H/\Gamma_H$ could be large enough to lead to the mixing effects that are of interest. Of course, if $|\Gamma_{12}| \approx 2|M_{12}|$, eq. (10.24) overestimates Δm_H, but it will be found that, in the examples of mixing to be considered, $|\Gamma_{12}|/|M_{12}| \ll 1$ (see eq. 10.51), a result that is consistent with eqs. (10.23) and (10.19) combined with $\Delta m_H/\Gamma_H \gtrsim 1$, which collectively require that

(10.24b) $$|\text{Re}(\Gamma_{12}/M_{12})| \ll 1.$$

10.4 $\Delta m_H/\Gamma_H$ in the KM Model

The general form of the dependence of Δm_H on the parameters of the KM model can be obtained by examining the $H^0 \to \bar{H}^0$ diagram equivalent to the $K^0 \to \bar{K}^0$ diagrams of figure 10.2. Then M_{12} takes a form similar to eq. (10.12) with an appropriate change in the definition of the λ_f, which for the D mesons is

(10.25) $$\lambda_f^D = V_{cf}\, V_{uf}^*,$$

where $f = d$, s, or b rather than u, c, or t as in eq. (10.12):

(10.26) $$M_{21}^D = \sum_{ff'} D_{ff'}\, \lambda_f^D \lambda_{f'}^D.$$

By examination of V, eq. (10.11), we find that $\lambda_d \sim s_C$, $\lambda_s \sim s_C$, and $\lambda_b \sim s_C^5$, so that the dominant contributions arise from $f = d$, s and $f' = d$, s and we need to carry the summations only over these two values.

Although a detailed calculation is needed to determine the coefficients $D_{ff'}$, for $f, f' = d$ or s the dependence on f, f' will be found to be a function of $(m_f/m_D)^2$ and $(m_{f'}/m_D)^2$, both of which are small. Therefore $D_{ff'}$ can be expanded in powers of these small parameters, and the contribution of the leading term to eq. (10.26) is proportional to $(\lambda_d + \lambda_s)^2$. But according to the GIM condition eq. (10.7), $(\lambda_d + \lambda_s)^2 = \lambda_b^2 \sim s_C^{10}$, which is negligible. The next leading term proportional to $(m_s/m_D)^2$ therefore gives rise to the only significant contribution, so that

(10.27) $$\Delta m_D = 2\,|M_{12}^D| \sim (m_s/m_D)^2 s_C^2.$$

According to eqs. (10.18a) and (10.21a),

(10.28) $$\Gamma_H = \tfrac{1}{2}(\Gamma_{11} + \Gamma_{22}),$$

and the dependence of Γ_H on V may therefore be obtained by summing the squares of amplitudes such as those shown in figure 10.1. Thus

(10.29) $$\Gamma_D = \sum_{ff'} D'_{ff'}\, |V_{cf}|^2\, |V_{uf'}|^2.$$

However, only those families f, f' that are energetically accessible as final states can contribute to this sum. Since $m_b \gg m_c$ or m_u, the sum is limited to $f, f' = d$ and s. From eq. (10.11) we see that the dominant terms in $|V_{cf}|^2 |V_{uf'}|^2$ are order s_C^0. Therefore

(10.30) $$\Delta m_D/\Gamma_D \sim s_C^2 (m_s/m_D)^2,$$

and the $D^0 \to \bar{D}^0$ probability is suppressed by a factor of $(m_s/m_D)^4 s_C^4$. This "second-order Cabibbo suppression" combined with the GIM suppression factor makes it unlikely that time-dependent mixing effects, such as like-charge dilepton production, will be observed if the KM model is correct.

For the B mesons

(10.31a)
$$\lambda_f^B = V_{fb}^* V_{fd},$$

where $f = u, c, t$, and

(10.31b)
$$M_{21}^B = \sum_{ff'} B_{ff'} \lambda_f^B \lambda_{f'}^B.$$

Examination of eq. (10.11) indicates that the dominant contribution to λ_f^B arises from each of three intermediates u, c, and t and is of order s_C^3. Therefore one might conclude that M_{12}^B is of order s_C^6. However, the coefficients $B_{ff'}$ depend on the quark masses, just as in the case of the $K_{ff'}$, eq. (10.12). In fact, Hagelin (1979) has shown for the box diagram that there are terms proportional to m_f^2/m_H^2, where the flavor f runs over the intermediate states. Since $m_b/m_D \approx 2.5$ ($m_b \approx 5$ GeV, $m_c \approx 1.9$ GeV), this mass dependence was ignored in assessing the dominant factors determining the size of Δm_D. However, m_t is believed to be very large ($m_t \gtrsim 40$ GeV) compared with $m_B \approx 5.3$ GeV, the $B_{ff'}$ factors are important, and the term B_{tt} dominates so that

(10.32)
$$\Delta m_B = 2|M_{12}^B| \sim s_C^6(m_t/m_B)^2.$$

By analogy with eq. (10.29), Γ_B has the form

(10.33)
$$\Gamma_B = \sum_{ff'} B'_{ff'} |V_{fb}|^2 |V_{f'd}|^2,$$

where again the values of f, f' are limited to those flavors that are energetically accessible, namely, $f, f' = u$ or c. The dominant term is found from eq. (10.11) to be $|V_{cb} V_{ud}| \sim s_C^2$, so that

(10.34)
$$\Gamma_B \sim s_C^4,$$

and, from eq. (10.32),

(10.35a)
$$\Delta m_B/\Gamma_B \sim s_C^2(m_t/m_B)^2.$$

If one uses the box diagram estimate of Hagelin, the coefficient of proportionality is $(2\pi)^{-1}$ and, for $m_t \approx 40$ GeV,

(10.35b)
$$\Delta m_B/\Gamma_B \approx 0.5.$$

Therefore this model predicts an appreciable probability for the occurrence of like-charge dilepton pairs. Furthermore, interference between B_α and B_β should be observable. The important question is whether these effects can be used to detect the CP-violating terms in V. We shall come back to that question after estimating $\Delta m_T/\Gamma_T$ for the (hypothesized) T meson.

For the T meson,

(10.36a) $$\lambda_f^T = V_{tf} V_{uf}^*$$

and

(10.36b) $$M_{21}^T = \sum_{ff'} T_{ff'} \lambda_f^T \lambda_{f'}^T$$

where $f, f' = d, s, b$. Since $T_{ff'}$ is a function of m_T^2, m_f^2, and $m_{f'}^2$ we can take advantage of the fact that $m_T \gg m_f$ for all f and expand it in powers of m_f^2/m_T^2:

(10.37) $$T_{ff'} = C_0 + C_1(m_f/m_T)^2 + C_1'(m_{f'}/m_T)^2 + C_2(m_f/m_T)^4 + C_2'(m_f m_{f'}/m_T^2)^2 + C_2''(m_{f'}/m_T)^4 + \cdots.$$

Because V is a unitary matrix,

(10.38) $$\sum_f \lambda_f^T = 0.$$

Therefore the contributions to M_{12}^T from all terms in eq. (10.37) except the C_2' term and higher-order terms vanish:

(10.39) $$M_{21}^T = \sum_{f,f'} C_2'(m_f m_{f'}/m_T^2)^2 \lambda_f^T \lambda_{f'}^T + \cdots.$$

From eq. (10.11) it is evident that $\lambda_f^T \sim s_C^3$ for all f, so that the dominant term in Δm_T is

(10.40) $$\Delta m_T \sim (m_b/m_T)^4 s_C^6.$$

The form of Γ_T is obtained by analogy with eq. (10.29):

(10.41) $$\Gamma_T = \sum_{ff'} T'_{ff'} |V_{tf}|^2 |V_{uf'}|^2,$$

where $f, f' = d, s, b$, since $m_T > m_f, m_{f'}$. The unitarity of V yields

(10.42) $$\sum_f |V_{tf}|^2 = \sum_{f'} |V_{uf'}|^2 = 1,$$

so that we find

(10.43a) $$\Gamma_T \sim s_C^0$$

and

(10.43b) $$\Delta m_T/\Gamma_T \sim (m_b/m_T)^4 s_C^6.$$

Therefore the mixing effects are Cabibbo suppressed to a very high order *and* are suppressed further[14] by the very small factor $(m_b/m_T)^4$.

Repeating the analysis for the exotic neutral B_s mesons, eq. (10.14e), leads to

(10.44a)
$$\Delta m_{xB} \sim s_C^4(m_t/m_{xB})^2,$$

and

(10.44b)
$$\Gamma_{xb} \sim |V_{cb}|^2 |V_{cs}|^2 \sim s_C^4,$$

so that

(10.44c)
$$\Delta m_{xb}/\Gamma_{xB} \sim (m_t/m_{xB})^2.$$

In contrast to the other cases, $\Delta m_{xB}/\Gamma_{xB}$ may be large, since $m_t \gg m_{xB}$. However, in the absence of detailed information about the coefficient in eq. (10.44c), we cannot rule out the possibility $\Delta m_{xB}/\Gamma_{xB} \approx 1$.

Again for the exotic neutral T_c meson, the treatment is a repetition of that for the *T* and

(10.45)
$$\Delta m_{xT}/\Gamma_{xT} \sim (m_b/m_{xT})^4 s_C^4,$$

which precludes the possibility of mixing in this case too.

We must keep in mind that all these conclusions are based on the KM model, which, despite its attractiveness, is not firmly established. Confirmation of the qualitative behavior suggested by the model—for example, the appearance of like-charge lepton pairs in B^0, $\bar{B}^0$ production and their absence in D^0, $\bar{D}^0$ production—would be an essential step in ascertaining its viability.

10.5 Charge Asymmetries in the Decay of Heavy Mesons

The important question for our purposes is how *T* violation and *CP* violation may be measured experimentally by using the heavy mesons. If the results of section 10.4 are taken as a guide, the only opportunity to use time-dependent interference methods to evaluate *CP* violation appears to be offered by B^0, $\bar{B}^0$ or B_s^0, $\bar{B}_s^0$ experiments. Hagelin (1979) pointed out that a measurement of the average like-charge dilepton asymmetry in the production of B^0, $\bar{B}^0$ pairs (e.g., by $e^+e^- \rightarrow B^0\bar{B}^0$) is sensitive to the phase δ appearing in *V*. This asymmetry is defined by the ratio

(10.46)
$$\rho = \frac{N(l^+l^+) - N(l^-l^-)}{N(l^+l^+) + N(l^-l^-)},$$

[14] This argument that *T* meson interference effects are suppressed as a result of the unitarity of *V* *irrespective* of the magnitudes of the s_i is due to Hochberg and Sachs (1983, n. 29).

where $N(l^+l^+)$ is the total number of l^+l^+ events and $N(l^-l^-)$ is the total number of l^-l^- events in the process.

The time dependence of the probability of like-charge events can be obtained from eqs. (10.17) and (10.16) by use of the $\Delta b = \Delta Q$ rule, and the average asymmetry is found by integration over time to be

(10.47) $$\rho = 4\,\frac{\text{Re}\, M_{12}^B\, \text{Im}\, \Gamma_{12}^B - \text{Im}\, M_{12}^B\, \text{Re}\, \Gamma_{12}^B}{4|M_{12}^B|^2 + |\Gamma_{12}^B|^2}.$$

If V, given by eq. (10.11), is applied to Hagelin's (1979) expressions for M_{12}^B and Γ_{12}^B, the resulting estimate of $|\rho|$ is

(10.48) $$|\rho| \approx |\text{Im}(\Gamma_{12}/M_{12})| \approx m_c^2/m_t^2,$$

which is very small. It will be shown below that the small size of the effect depends only on the most general characteristics of the model. Therefore the largest asymmetry to be expected for this process is expected to be much less than 1 percent.

Although the use of B^0, $\bar{B}^0$ pair production offers the most immediate and graphic test of CP invariance, the use of "tagged" single B^0 or $\bar{B}^0$ production may be more promising for the future. Since leptons produced by decay of the B^0 and $\bar{B}^0$ will have opposite signs of charge, the time dependence of the sign of the charge on the decay lepton, which is governed by eq. (10.17), is different when the initial state is $|\bar{B}^0\rangle$ than when the initial state is $|B^0\rangle$. A comparison of the charge asymmetry of leptonic decay events for the two cases therefore should provide a possible measure of CP violation.

If the integrated (over time) number of leptonic decay events is denoted by $N(l^\pm)$ for the initial B^0 and by $\bar{N}(l^\pm)$ for the initial $\bar{B}^0$, a comparison of the ratios (asymmetries)

(10.49a) $$R_l = \frac{N(l^+) - N(l^-)}{N(l^+) + N(l^-)}$$

and

(10.49b) $$\bar{R}_l = \frac{\bar{N}(l^-) - \bar{N}(l^+)}{\bar{N}(l^-) + \bar{N}(l^+)}$$

will provide a measure of CP violation. Such a comparison can be made by means of the ratio $R_l/\bar{R}_l$, which, when calculated on the basis of eq. (10.17), turns out to be[15]

(10.50a)
$$R_l/\bar{R}_l = \frac{2\alpha_B + (1 - \alpha_B^2)\chi}{2\alpha_B - (1 - \alpha_B^2)\chi}.$$

Here α_B is a measure of the probability of interference,

(10.50b)
$$\alpha_B = \frac{1 - (\Delta\Gamma_B/2\Gamma_B)^2}{1 + (\Delta m_B/\Gamma_B)^2},$$

and χ is determined by *CP* violation in the mass matrix:

(10.50c)
$$\chi = \frac{\text{Im } M_{12}^B \text{ Re } \Gamma_{12}^B - \text{Re } M_{12}^B \text{ Im } \Gamma_{12}^B}{|\mathbf{M}_{12}^B \mathbf{M}_{21}^B|}.$$

When the conditions for mixing are met, that is, $|\alpha_B| \neq 1$, the size of the effect is determined by χ, which has essentially the same form as ρ, eq. (10.47). Therefore its magnitude is of the same order as that of ρ, and the effect is again very small.

This suppression of the asymmetry by the square of a very small mass factor has the same origin as the enhancement of the interference effects by the factor $(m_t/m_B)^2$, eq. (10.35a). The latter is measured by the ratio of the "dispersive" part M_{jk}^B of the mass matrix to the absorptive part Γ_{jk}^B and the former by the inverse of that ratio. Contributions to Γ_{jk}^B arise only from directly accessible intermediate states, that is, those that have the mass m_B. Since $m_t \gg m_B$, the t quark states are not accessible, and the m_t^2 term does not appear in Γ_{jk}^B. On the other hand, this term appears and dominates in M_{jk}^B because there is no such energy restriction on the virtual intermediate states. Therefore $|\Gamma_{12}|/|M_{12}|$, which governs the magnitudes of ρ and χ, is quite small:

(10.51)
$$|\Gamma_{12}| \ll |M_{12}|.$$

In the special case represented by eq. (10.48), ρ is even smaller than $|\Gamma_{12}/M_{12}|$, by a factor of $(m_c/m_B)^2$, because the $(m_B/m_t)^2$ term in Γ_{12}/M_{12} is real.

A more general statement of the result is that time-dependent mixing effects depending *only* on *CP* violation in the *mass matrix* are small for B mesons although conditions are such that there is time to develop interference. This discouraging result can be mitigated if there is a possibility of *direct* interference between B^0 and $\bar{B}^0$ decay amplitudes, a possibility that is

[15] This test was proposed by Hochberg and Sachs (1983). Note that there is an error in eq. (46) of that paper that should be replaced by my eq. (10.50a). The error does not have a substantial effect on the conclusions of the referenced paper.

excluded for the leptonic decays we have been considering because of the $\Delta b = \Delta Q$ rule.

Hadronic decay modes fortunately do offer an opportunity for direct interference experiments, as in the case of K_L, K_S interference in the 2π mode, figure 9.2. Possible interfering common modes in the cascade decay of B^0 and $\bar{B}^0$ were first identified by Carter and Sanda (1980). Detailed analyses on the basis of the KM model of charge asymmetry effects in such modes following B^0, $\bar{B}^0$ pair production have been presented by Carter and Sanda (1981) and Bigi and Sanda (1981, 1984). A generic treatment for the KM model based on eq. (10.11) of direct hadronic decays into CP eigenstates has been given by Wolfenstein (1984b).

The direct interference term arises as a result of the time dependence of the decay amplitude, $A(t)$, which, in terms of the amplitudes A_α and A_β of the mass eigenstates $|\alpha\rangle$ and $|\beta\rangle$, has the form, as in eq. (9.32),

$$A(t) = C_\alpha A_\alpha e^{-im_\alpha t} e^{-\Gamma_\alpha t/2} + C_\beta A_\beta e^{-im_\beta t} e^{-\Gamma_\beta t/2}, \tag{10.52}$$

where C_α, C_β are determined by the initial conditions. The interference effect is then determined, as in eq. (9.33), by the ratio

$$\eta_H = A_\beta / A_\alpha. \tag{10.53}$$

Since eqs. (10.16) yield

$$A_\alpha = pA + q\bar{A}, \tag{10.54a}$$

$$A_\beta = pA - q\bar{A}, \tag{10.54b}$$

we have

$$\eta_H = (pA - q\bar{A})/(pA + q\bar{A}), \tag{10.55}$$

where, in contrast to the leptonic decay amplitudes, both A and $\bar{A}$ refer to one common decay channel.

The ratio η_H is independent of the choice of phase conventions, as it must be since it is a measurable complex number. This statement can be confirmed by noting that any relative phase change of the amplitudes A and $\bar{A}$ corresponds to an equal change in relative phase of $|H^0\rangle$ and $|\bar{H}^0\rangle$, since A and $\bar{A}$ refer to a common final state, and the change in phase of q/p is then the same according to eq. (10.16c). Thus there is no net change of the phase of η_H, eq. (10.55). In the context of the KM model such phase changes relate to the arbitrary choices of the phases of quark states of different flavors; therefore they correspond to changes in the way phases appear in the KM matrix, eq. (10.9).

When the condition eq. (10.51) is applicable, Γ_{12} may be neglected in eq. (10.16c), determining p and q, and

(10.56) $$p = q^* = [M_{12}/2|M_{12}|]^{1/2}$$

because the matrix M is Hermitian.

Although in the general case of hadronic decay the value of η_H will depend on the interactions among the various possible final states, the origin of hadronic charge asymmetry and an estimate of the order of magnitude of the effect to be expected may be understood by considering a highly oversimplified model, one in which there are only two hadronic decay channels, a hadronic mode f and its CP conjugate $\bar{f}$, for each of which there are two nonzero decay amplitudes $A(f)$, $\bar{A}(f)$ and $A(\bar{f})$, $\bar{A}(\bar{f})$.[16] Then, under the assumption of CPT invariance eq. (9.60), these amplitudes are related:

(10.57a) $$|A(f)|^2 = |\bar{A}(\bar{f})|^2$$

and

(10.57b) $$\bar{A}(f)/A(f) = [A(\bar{f})/\bar{A}(\bar{f})]^*.$$

From eqs. (10.56) and (10.57b) it follows that

(10.58a) $$\eta_H(\bar{f}) = -\eta_H^*(f),$$

and the A_α, A_β given by eqs. (10.54) satisfy

(10.58b) $$|A_{\alpha,\beta}(\bar{f})|^2 = |A_{\alpha,\beta}(f)|^2.$$

We now have the possibility of using hadronic charge asymmetry as a measure of η_H. We compare two experiments, one in which the initial state is $|H^0\rangle$ and the other in which the initial state is $|\bar{H}^0\rangle$. Then, in eq. (10.52)

(10.59a) $$C_\alpha = C_\beta = 1/2p$$

in the former case and

(10.59b) $$C_\alpha = -C_\beta = 1/2q$$

in the latter. If we call the time-dependent amplitude $A_f(t)$ for production of the final state f in the first case and call it $\bar{A}_f(t)$ for production of the state f

[16] The important case concerns the B mesons, and this model is applicable to a determination of the charge asymmetry in the exclusive channels $D^-\pi^+$ and $D^+\pi^-$ if only the diagrams of figure 10.4 contribute. However, there are other contributions (diagrams in which the d or $\bar{d}$ quarks are "spectators") for which the final state interactions with $\bar{D}^0\pi^0$ or $D^0\pi^0$ channels must be taken into account to obtain quantitative results.

in the second case, we find from eqs. (10.52) and (10.59):

(10.60a) $$|A_f(t)|^2 = \frac{|A_\alpha(f)|^2}{4|p|^2} e^{-\Gamma_H t}\{1 + 2\mathrm{Re}[\eta_H(f)e^{-i\Delta m_H t}] + |\eta_H(f)|^2\}$$

and

(10.60b) $$|\bar{A}_f(t)|^2 = \frac{|A_\alpha(f)|^2}{4|q|^2} e^{-\Gamma_H t}\{1 - 2\mathrm{Re}[\eta_H(f)e^{-i\Delta m_H t}] + |\eta_H(f)|^2\}.$$

The corresponding rates for production of the state $\bar{f}$ are obtained by replacing f by $\bar{f}$. It is assumed in accordance with eq. (10.51) that $\Delta\Gamma \ll \Delta m$.

The overall asymmetries are defined in terms of the integrated rates

(10.61a) $$N(f) = \int_0^\infty dt\,|A_f(t)|^2,$$

(10.61b) $$\bar{N}(f) = \int_0^\infty dt\,|\bar{A}_f(t)|^2$$

in analogy with eqs. (10.49) as

(10.62a) $$R_{f\bar{f}} = \frac{N(f) - N(\bar{f})}{N(f) + N(\bar{f})}$$

and

(10.62b) $$\bar{R}_{\bar{f}f} = \frac{\bar{N}(\bar{f}) - \bar{N}(f)}{\bar{N}(\bar{f}) + \bar{N}(f)},$$

the first corresponding to a tagged H^0 beam and the second to a tagged $\bar{H}^0$ beam. From eqs. (10.56), (10.58), and (10.60) it follows that

(10.63a) $$R_{f\bar{f}} = 2\Gamma_H^2 \,\mathrm{Re}\,\eta_H(f)[\{\Gamma_H^2 + (\Delta m_H)^2\}(1 + |\eta_H(f)|^2) + 2\Delta m_H \Gamma_H \,\mathrm{Im}\,\eta_H(f)]^{-1}$$

and

(10.63b) $$\bar{R}_{\bar{f}f} = 2\Gamma_H^2 \,\mathrm{Re}\,\eta_H(f)[\{\Gamma_H^2 + (\Delta m_H)^2\}(1 + |\eta_H(f)|^2) - 2\Delta m_H \Gamma_H \,\mathrm{Im}\,\eta_H(f)]^{-1}.$$

CP invariance would imply that there should be no distinction between the two cases (initial H^0 and initial $\bar{H}^0$), so that comparing $\bar{R}_{\bar{f}f}$ with $R_{f\bar{f}}$ provides a direct measure of CP violation, which therefore may be taken to be $\mathrm{Im}\,\eta_H(f)$.

These general results may now be applied to the hadronic B^0, $\bar{B}^0$ decay. For this case eq. (10.31b) may be used to determine M_{21}^B and, as we found in calculating Δm_B, eq. (10.32), the sum is dominated by the term $(\lambda_t^B)^2 =$

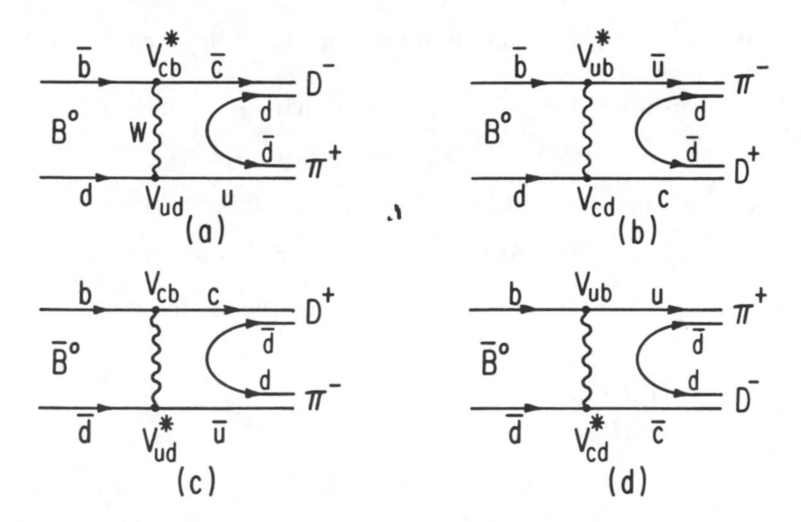

FIG. 10.4 Diagrams for B^0, $\bar{B}^0 \to D^{\pm}\pi^{\mp}$ that are used as the basis of a simple model to illustrate hadronic interference effects. Corresponding diagrams for B_s^0, $\bar{B}_s^0$ are obtained by replacing the incoming d and $\bar{d}$ quarks by s and $\bar{s}$, respectively.

$(V_{tb}^* V_{td})^2$. From eqs. (10.11) and (10.56) it is then found that

(10.64)
$$q/p = \frac{1 - \mu - i\xi}{1 - \mu + i\xi}.$$

Now, as an illustration, consider a particularly simple model of the exclusive hadronic modes of decay

(10.65a)
$$B^0 \to D^{\mp} + \pi^{\pm}$$

and

(10.65b)
$$\bar{B}^0 \to D^{\pm} + \pi^{\mp}.$$

Figure 10.4 illustrates the simplified hypothetical model that will be used here for these modes.[17] In particular, it can be seen that both B^0 and $\bar{B}^0$ can decay into either mode. Furthermore, the ratio of amplitudes $\eta_B(D^{\pm})$ depends only on the KM matrix elements,[18] so that the charge asymmetry

[17] See note 16.

[18] This statement is based on the usual assumption that, because of the high momentum transfer, QCD corrections to the weak quark vertices are negligible. Then in the special case of the diagrams of figure 10.4 the matrix elements of the quark current operators between hadron states may be factored and shown to be equal in magnitude for the two cases: $B^0 \to f$ and $\bar{B}^0 \to f$, where $|f\rangle$ is the common hadron final state. Furthermore, for this model the ratio of matrix elements turns out to be independent of the final state isotopic spin so that both the charged $D^{\pm}\pi^{\mp}$ and neutral $D^0\pi^0$ or $\bar{D}^0\pi^0$ channels satisfy the conditions eqs. (10.57) and (10.58). For that reason, the results obtained with this model are more realistic than might appear to be the case. It is for the same reason that the results are independent of final state interactions, which is not realistic but suggests that final state interaction effects may not strongly affect the order of magnitude of real phenomena.

takes on an especially simple form for this model of the exclusive channels described by eqs. (10.65).

From figure 10.4 and eq. (10.11) it is found that

(10.66) $$\bar{A}(D^-\pi^+)/A(D^-\pi^+) = -s_C^2(\mu - i\xi),$$

and eqs. (10.55) and (10.64) yield, to first order in s_C^2,

(10.67) $$\eta_B(D^-\pi^+) = 1 + 2s_C^2(\mu - i\xi)(1 - \mu - i\xi)(1 - \mu + i\xi)^{-1}.$$

Since Im $\eta_B(D^-)$ is small, of order s_C^2, we can write for the measure of *CP* violation

(10.68a) $$\frac{\bar{R}_B(D^+D^-)}{R_B(D^-D^+)} - 1 = \frac{-2\Delta m_B \Gamma_B}{\Gamma_B^2 + (\Delta m_B)^2} \text{ Im } \eta_B(D^-\pi^+),$$

with

(10.68b) $$\text{Im } \eta_B(D^-\pi^+) = -2s_C^2(\mu \sin 2\phi + \xi \cos 2\phi),$$

where

(10.68c) $$\phi = \arctan[\xi/(1 - \mu)].$$

Since $s_C^2 \approx 0.05$, the upper limit on the *CP*-violating effect measured by the ratio of the asymmetries would be expected to be of the order of 10 percent on the basis of this oversimplified model if $\Delta m_B/\Gamma_B \approx 1$. However, $\Delta m_B/\Gamma_B$ may turn out to be much smaller than this, in which case the ratio would be correspondingly decreased.

In section 10.4 we found that $\Delta m/\Gamma$ may be much larger for the exotic B_s^0, $\bar{B}_s^0$ mesons. Since the diagrams of figure 10.4 may be applied equally well to the B_s case by simply replacing the incident d quark with an s quark, comparison of the $D^\pm$ charge asymmetries in the exclusive channels of eqs. (10.65) may be a more opportune example of a way to measure *CP* violation.

The calculation of these asymmetries $R_{f\bar{f}}^x$ and $\bar{R}_{\bar{f}f}^x$ proceeds as before. On the basis of eq. (10.11), we find that M_{12} is real to lowest order in s_C, so that

(10.69) $$q_x = p_x$$

to that order. Furthermore, for the simplified model based on figure 10.4 (with d replaced by s),

(10.70) $$\bar{A}_x(D^-\pi^+)/A_x(D^-\pi^+) = \mu - i\xi;$$

there is no Cabibbo suppression of this ratio, so that large asymmetries are to be expected even in the absence of *CP* violation ($\xi = 0$). However, the asymmetries should be the *same* (equal and opposite) for a tagged $\bar{B}_s^0$ beam as for a tagged B_s^0 beam in that case. From eqs. (10.69) and (10.70) we find

$$\eta_{xB} = \frac{1 - \mu + i\xi}{1 + \mu - i\xi}. \tag{10.71}$$

Therefore the measure of *CP* violation is, according to eqs. (10.63),

$$\frac{\bar{R}_B^x(D^+D^-)}{R_B^x(D^-D^+)} - 1 = -\frac{4\Delta m_{xB}\Gamma_{xB}\xi}{[\Gamma_{xB}^2 + (\Delta m_{xB})^2][1 + \mu^2 + \xi^2] + 2\Delta m_{xB}\Gamma_{xB}\xi}, \tag{10.72}$$

which can clearly be quite a large effect unless it should turn out that $\Delta m_{xB}/\Gamma_{xB} \gg 1$, in which case the wavelength for oscillation would be short compared with the decay length and B_s^0, $\bar{B}_s^0$ mixing would be large. Therefore the average of the difference of interference effects between B_s^0 and $\bar{B}_s^0$ beams over an appreciable time interval would be suppressed by a factor $\Gamma_{xB}/\Delta m_{xB}$, as shown by eq. (10.72). Note, further, that if $\mu^2 + \xi^2 = 1$, $\mathrm{Re}\,\eta_{xB} = 0$ and the charge asymmetries R_{xB} and $\bar{R}_{xB}$ vanish.

These particularly simple examples illustrate how hadronic charge asymmetries arise in neutral heavy meson decay according to the KM model and how they might be used to measure the parameter (phase) governing *CP* violation, but the required measurements for these specific exclusive channels will be very difficult indeed. The results do suggest that even in the experimentally more manageable case of *inclusive* $D^{\pm}$ decay a measurement of the ratio of charge asymmetries could provide a significant test of the model.

In addition to the time-dependent charge asymmetry effects that have been treated here, the KM model leads to asymmetries in the partial hadronic decay rates of charged *B* mesons (Carter and Sanda 1981; Chau and Cheng 1984). These asymmetries depend on the occurrence of more than one final hadronic state and on the difference in final state interaction phases between them. Their measurement would be a clear indication of *CP* violation, but the interpretation in terms of the KM matrix is likely to be obscured by a lack of detailed information concerning the eigenphases of the *S* matrix in these processes.

10.6 Concluding Remarks

It seems likely that charge asymmetry experiments on the decay products of the heavy mesons will be carried out in the coming years, and they should resolve some questions concerning the origins of *CP* violation—in particular, whether the KM model provides a completely satisfactory phenomenological description of the weak interactions. My emphasis has been on the KM model because of its basic simplicity in the context of the Weinberg-

Salam-Glashow standard model of the weak interactions and its richness of content. We should, of course, be mindful that the weak currents described by the KM model may themselves be *effective* currents, describing the effects of an underlying and more fundamental theory.

Some alternatives to the KM model have been suggested and subjected to analysis, including in particular Lee's (1973) and Weinberg's (1976) suggestion before the discovery of the fifth quark that the origin of *CP* violation in the four-quark model could be associated with a proliferation of out-of-phase Higgs doublets. The effects of exchange of Higgs bosons and variations on this model have been included in some of the analyses quoted here.[19]

When the day comes that we can be certain of the phenomenological model, we will still not have resolved the basic question, which is the origin of the *CP*-violating phase. For example, what is the source of the interactions that generate the mixing of quark families expressed in the KM matrix? This mixing is usually ascribed to the coupling of the quarks to the Higgs doublets, but at this point each such coupling introduces a new ad hoc parameter, and the mysterious phase is included among the parameters introduced in this way. These parameters must have a deeper origin.

Although the introduction of a *CP*-violating phase also implies *T* violation in the *CPT*-invariant model, all our attention has been given to experiments on *CP*. To confirm the *T* violation and the *CPT* invariance, it may be necessary, as in the case of the *K* mesons, to supplement the *CP* experiments with enough additional information on *B* meson decays to verify a unitarity condition. That would certainly be a formidable task, and it would be much more satisfactory to have direct measurements of *T* violation, that is, measurements of the violation of motion reversal.

As we learned in section 5.3, the difficulty with motion reversal experiments is the *T*-violation mimicry associated with the final state interactions. The determination of *S* matrix eigenphases among the hadrons in high-energy experiments that is needed to untangle these effects appears to be beyond the state of the art for some time to come. The one hope would be the measurement of motion reversal terms in leptonic decays, because the final state interactions of the leptons are electromagnetic and therefore quite small. An example would be the determination of the component of lepton polarization normal to the decay plane. A nonvanishing transverse polarization would imply *T* violation, as for the neutron (section 5.1).

Unfortunately, within the context of the KM model, that specific effect is not expected to occur, because the only *T*-violating phase is a common

[19] Wolfenstein (1986b) summarizes the present situation with regard to these and other gauge models in this review of the present status of *CP* violation.

phase of the total amplitude. Transverse polarization requires a phase difference between interfering amplitudes, as in eq. (5.64).

The alternative of determining terms that are odd under motion reversal in hadronic decays, such as a momentum correlation term $(\vec{p}_1 \cdot \vec{p}_2 \times \vec{p}_3)$ among three hadrons or a baryon spin correlation term, is subject to the limitations imposed by (unknown) strong final state interactions. The predictions of the KM model for such correlations would therefore be expected to be strongly dependent upon the approximations used to treat QCD effects, which could not easily be separated from effects due to the KM phase δ.

That the KM model does not lead to lepton transverse polarization should not be allowed to discourage attempts to make the measurement. Nor should the indications that leptonic charge asymmetries are very small stand in the way of trying to measure them. These negative predictions are special features of the model, which is not so well founded that it is inviolate. It may be that relatively small modifications of the model would change the results significantly, and the need for such modifications or the fact that the model is inadequate can be established only by such experiments.

It is only by exploiting every opportunity to detect additional effects of both *CP* violation and *T* violation that we can hope to arrive at an unambiguous resolution of the larger questions concerning the origins of these aberrations of the theory.

11 | Perspectives, Problems, and Prospects

We have dealt here with the physics of time reversal and the related improper transformation CP in many of their different aspects, but in doing so we have had little opportunity to take a larger perspective and reflect on the way our considerations relate to the grand scheme of physics past, present, and future. I shall now consider some of the general issues whose existence may easily have been missed in the welter of discussion in the foregoing chapters.

Two aspects of time reversal have been the subjects of those chapters: the consequence of T invariance for physical phenomena and the violation of T invariance in certain very special weak interaction phenomena. The questions to be raised now bear on both these issues. The first of the questions to be dealt with is the relationship between T invariance in classical physics and the so called Arrow of Time. I reiterate the conclusions of section 2.4 and add some related remarks on classical cosmology. In the remainder of the chapter I raise questions that must be answered if recent and future developments in theoretical physics are to encompass the physics of time reversal and the CPT theorem.

These new questions arise because the theoretical basis for our discussion has been limited to the reasonably well established methods of quantum mechanics and "conventional" quantum field theory. Such methods appeared to be quite fundamental until recently, but now developments in the theory of particles and fields suggest they may represent only an approximation that is related to the underlying fundamental theory by some sort of correspondence principle, a principle requiring that a correct theory must be well approximated by the conventional theory at "large" distances. In that sense, conventional relativistic quantum field theory might be viewed as a "phenomenological theory" that encompasses the known (long-distance) phenomena but leaves open the possibility of the discovery of some hitherto unknown small-distance phenomena that could not be described by such a theory.

The possible impact of the short-distance viewpoint on the definitions of *CP* and *T* has not been explored, and in the last two sections of this brief final chapter I shall indulge in some speculation about the directions in which such exploration might lead.

11.1 *T* Invariance in Classical Physics

The only general question concerning *T* invariance that seems to have arisen in the context of classical physics concerns the paradox of the Arrow of Time, a concept conveying a sense of the direction of the flow of time, which is usually referred to in terms of the inevitable increase of entropy. As noted in section 2.4, it was Loschmidt (1876) who first brought out the paradox, that is, the inconsistency between Boltzmann's (1872) microscopic mechanical model (as in the kinetic theory of gases) of the entropy increase and the *T* invariance of the microscopic dynamics.

In section 2.4 I summarized briefly the resolution of the paradox that is implicit in Gibbs's introduction in 1902 of the concept of coarse-grained average distributions in place of the detailed microscopic distributions of the states of individual particles (Gibbs 1931). For the increase in entropy the pertinent inequality is that between the probability associated with the coarse-grained equilibrium distribution and a nonequilibrium coarse-grained distribution. In the way Gibbs demonstrated it, the inequality relates to *any* change in the distribution from equilibrium without reference to a time sequence of events, so that *T* invariance is irrelevant. In fact Gibbs's inequality holds *whether or not* the microscopic dynamics are *T* invariant. There is no direct connection between the Arrow of Time and the kinematic time reversal transformation *T*.

Of course the Arrow of Time also plays an important role in classical cosmology, for which the underlying mechanics is Einstein's general theory of relativity. If classical relativity is taken to be a fundamental theory of matter (i.e., if questions concerning quantum relativity are ignored) the question whether it is a *T*-invariant theory should be addressed. However, even the way that question is to be formulated is obscure because of the purely geometric nature of the general theory.

Although classical general relativity is a classical field theory with the metric tensor serving as the field variable, the space-time point at which the field is evaluated is defined in a geometry governed by the solutions to the field equations. Since the source term in the field equations is the stress-energy-momentum tensor that includes all interactions, the geometric, kinematic, and dynamic properties of general relativity are thoroughly entangled, and it is not evident that a global kinematically admissible transformation T can be defined.

It is possible to define a *local* transformation T in terms of the timelike tangent vector,[1] just as one can define a Lorentz transformation locally in those terms, but the relationship of this transformation at one point on the space-time manifold to that at another such point depends on the dynamics except in the case of flat space, that is, when the curvature of the space-time manifold vanishes globally.

From the viewpoint of general relativity, we have used throughout this book the linear approximation[2] in which the fields are embedded in a flat space-time geometry. For most of the physics this is an excellent approximation because of the small size of the gravitational constant G. However, when it comes to cosmological questions, the big bang and all that, extremely small time and distance scales (Planck scale) enter, and the curvature of space-time plays an essential role. In treating that era, since it is not clear that time reversal can be defined, it certainly is not clear that T invariance is relevant.

The way the cosmological question can be dealt with in classical physics is by having patience, namely by starting the dynamic treatment at so late a time, t_0, that the linear approximation is adequate. Then the dynamic evolution of the early universe for $t > t_0$ is determined by the initial conditions at t_0, and the subsequent cosmological Arrow of Time is set by those initial conditions. Again the issue of T invariance of the dynamics for $t > t_0$ is irrelevant to the issue of the sense of time because the direction of motion has been set by the initial conditions.

Although that procedure may circumvent the cosmological issue, there still remains the question of the role of T invariance in other gravitational phenomena for which the curvature of space-time cannot be neglected, as, for example, in those associated with the dynamics of black holes. Such questions appear to be concerned with quantum relativity[3] rather than classical relativity and remain to be resolved in the future.

11.2 *T* Invariance in Quantum Physics: The *CPT* Theorem

Three important assumptions that have been made about quantum field theory in our treatment of improper transformations require reexamination in the light of recent developments. The first is the assumption that, as long

[1] See, e.g., Wald (1984, p. 14).
[2] Wald (1984, p. 74).
[3] Compare Wald (1980).

as the theory is formally renormalizable, an iteration procedure will converge and lead to unique solutions of the equations of motion, no matter how strong the interactions may be. The second, which will be found to be relevant to the first, is the assumption of a unique vacuum state. And the third is the assumption that the fields are local.

Since we have been considering the implications of a possible violation of P, T, and C at all levels of interaction, we have found it necessary to introduce the concept of the kinematically admissible transformation, which is defined independent of the properties of the interaction. Before the discovery of parity violation in 1956, the definitions of the transformations P, T, and C appeared much more straightforward because the invariance of all interactions was taken as a basic assumption. P violation by the weak interactions showed that this assumption is untenable, and it was replaced by the assumption that the only interactions for which there is any question about invariance are weak enough to be treated as small perturbations. Then the definition of the transformations could be established in the context of perturbation theory by defining them as transformations on the zero-order states, which are associated with P-, T-, and C-invariant (strong) interactions.

However, this leaves unanswered the question of how to define the transformations in the event there is a violation of one or more of these symmetries by strong interactions. In such a situation there is a question whether it is meaningful to assert the *existence* of a *kinematic* (i.e., independent of the interaction) transformation associated with the violated symmetry. We have found that the existence of such transformations can be established within the context of conventional quantum field theories by constructing them explicitly in terms of what we have called the kinematic variables, and that when so structured they are independent of questions concerning the invariance of the dynamics.

The point is of special importance with regard to the proof of the CPT theorem. If the existence of each of the three transformations required an assumption of invariance of the interaction under each, the CPT theorem would be the trivial statement that it is then necessarily invariant under all. It would mean the theorem is a tautology: we are assuming the result and then using it to prove itself. Again, our use of the kinematically admissible transformations circumvents that tautological situation by providing a means to establish the existence of the transformation that is independent of assumptions about invariance of the interactions.

The success of this procedure has been demonstrated only in the context of conventional quantum field theory. Recent developments in high-energy and particle physics have led to new perspectives on, and new formulations of, field theory for which there may be reason to question the existence of kinematically admissible transformations.

For example, quantum chromodynamics (QCD), the currently accepted basis of the theory of strong interactions,[4] is presumed to lead to quark confinement, which would not permit the existence of free quarks, so that a question arises concerning the existence of free fields $\psi_0(x)$ that can serve as kinematic variables.

The property[5] of the theory called "asymptotic freedom," which results because the renormalized coupling constant between quarks and gluons goes to zero in the limit of large momentum transfer (small distance scale), suggests that there is a domain (the "femtouniverse")[6] in which the solutions do behave as perturbed free waves. Although the free (quark) fields $\psi_\tau(x)$ that correspond to physical solutions of the equations of motion at a given instant of time are not plane waves, they can be described as wave packets within the region of confinement that can be Fourier analyzed. Therefore it seems reasonable to assume that it is possible to define a complete set of confined, noninteracting wave packets.[7] Then the equations of motion of the quarks may be expressed in the form eq. (6.33a), where $\psi_\tau(x)$ is made up of a linear combination of annihilation and creation operators for these wave packets and represents the confined state of a quark.

In the femtouniverse it is certainly possible to use our methods for defining improper transformations and determining the transformation properties of very small (free) wave packets. These definitions can then be formally extended to the complete set of free (noninteracting) packets and, *if the iteration procedure converges*, the behavior under C, P, and T of physically meaningful wave packets may be determined by means of eq. (6.33a).

Confinement of the quarks implies that the observables depend on composites of the quark field (and gluon) wave packets that may be treated as nonlocal hadron field operators. The transformation properties of the hadron fields are determined by the transformation properties of the primi-

[4] Quigg (1983) provides a good summary and references.

[5] See note 4.

[6] This term was introduced by Bjorken (1980) to encompass the universe within distances of order 10^{-15} cm.

[7] This would be in the spirit of the bag model (Chodos et al. 1974) and is analogous to the construction of complete sets of confined nucleon states (including, however, the interactions) by Wigner and Eisenbud (1947) to describe nuclear properties (see chap. 4, note 16). Sachs (1954) attempted to construct and quantize a complete set of confined pion wave packets to describe nucleon properties in an analogous manner.

tive quark and gluon packets of which they are composed. As long as the operators representing observables can be expressed in terms of the irreducible tensors or spinors, their behavior under improper transformation will follow from that of the constituent fields. However, if the free *hadron* fields are taken to be the phenomenological kinematic variables, their transformations under C, P, and T will be complicated by the fact that they are nonlocal combinations of the constituent (quark) fields. This will affect the behavior of the observable and, in particular, may cast doubt on the proof of the CPT theorem, since existing proofs depend so crucially on the assumption of locality.

Another important assumption, that the vacuum is a nondegenerate state of the system, is obscured in our treatment of the CPT theorem[8] but is explicit in Jost's (1957) general proof. For QCD it turns out that there is a degeneracy of the vacuum that, among other things, leads to "strong CP violation."[9]

This result appears to lead to a paradox, because it occurs in spite of the fact that the original interactions are structured so as to appear to be CP invariant.

We recall again that our method of establishing either the CP invariance or the T invariance of the interactions makes use of an iteration of the equations of motion starting from an approximation in which the solutions (the interacting fields) are replaced in the interaction by the kinematic variables (the free field operators). This procedure depends on the underlying assumption that, even for strong interactions, an infinite sequence of such iterations will converge to a well-defined solution.

If the (zero-order) interaction expressed in terms of the free fields is invariant under CP or T, as it is in this case, the interacting fields obtained by iteration should transform in exactly the same way as the free fields. Then it would follow that the full interaction (expressed in terms of the interacting fields) is also invariant, thereby establishing the consistency of the assumption of CP and T invariance.

The appearance of strong CP violation in QCD is contrary to this

[8] Although it is not included in the list of assumptions at the end of section 6.1, it is built into our use of the creation and annihilation operators to construct the transformation $\mathbf{\Theta}$. The completeness of the set of states generated out of the nondegenerate vacuum by the creation operators, as in eq. (7.56), is an important underlying assumption. Another assumption that is not made explicit is that the field operators must belong to a finite representation of the Lorentz group. Oksak and Todorov (1968) and Stoyanov and Todorov (1968) have presented an example of a free field theory with infinite component fields that explicitly violates the CPT theorem.

[9] See Quigg (1983, pp. 253–54).

conclusion.[10] It implies that the solutions of the equations of motion (interacting fields) do *not* transform in the same way as the free field operators, so that the full interaction is *not* invariant. The resulting inconsistency with the original argument suggests that the iteration procedure fails to converge. Failure of the iteration procedure is to be expected when there is a vacuum degeneracy unless the zero-order vacuum states are based on an appropriate linear combination of the degenerate states.

The appropriate linear combinations describe vacuum states that are not invariant under T as we have defined the transformation, because they are linear combinations of states characterized by a phase difference θ between terms ("theta vacuum"). The conclusion that CP invariance must be violated follows from the assumption of the validity of the CPT theorem. However, the proof of the CPT theorem starting from a kinematically admissible CPT transformation suffers not only from the doubts raised earlier concerning the locality question, but also from doubts about the effect of the theta vacuum, which is again transformed by the antiunitary CPT transformation.

These effects of the choice of a vacuum state on the iterative solution of the equations of motion offer further examples of the importance of making a distinction between the invariance of the equations of motion and the invariance of the "actual motion," a distinction discussed at some length in chapter 2. In the present case, the "actual motion" is to be interpreted as the solution of the field equations. What we are seeing here is noninvariance of the iterative solution resulting from the fact that the boundary conditions are not invariant, although the equations of motion are.

The boundary conditions, which are governed by the choice of a vacuum state from among the set of degenerate states, play the same role here as do the initial conditions in the classical case. In QCD, the question is which value of θ is to be used to identify the *physical* vacuum state

$$|\theta\rangle = \sum_n e^{in\theta} |n\rangle. \tag{11.1}$$

Here the degenerate vacuum states $|n\rangle$ correspond to the different choices of the winding number $n = 0, \pm 1, \pm 2, \ldots$, characterizing the topological structure of the *classical* non-Abelian gauge fields of QCD.

The classical fields define the field manifold underlying the quantized functional space of the theory, and the winding number n is a property of that manifold. Therefore the real numbers n are invariant under the unitary

[10] Quigg (1983, pp. 250 ff.). I am raising a question here concerning the way the *definition* of CP is propagated from a fundamental level to the phenomenological level of effective interactions, *not* a question on the propagation of CP-violating interactions.

and antiunitary transformations **CP** and **T**. By consequence $|\theta\rangle$ is invariant under **CP** and is transformed into $|-\theta\rangle$ by **T**. Since **CPT** is also antiunitary, it will also transform $|\theta\rangle$ into $|-\theta\rangle$. Therefore, although the basic equations of motion of QCD may be invariant under T and CP, the effective (phenomenological) strong interactions among the hadrons appear to violate both T and CPT invariance if θ is not equal to 0 or π. This T violation is what is usually referred to as "strong CP violation" because of an incorrect presumption of the inviolability of the CPT theorem at the phenomenological level. The CPT theorem applies at this level only if the physically defined boundary conditions are CPT invariant.

Since no evidence exists for either strong T violation or strong CPT violation (or for that matter for strong CP violation), we can conclude that the physical vacuum of QCD must correspond to $\theta = 0$ or $\theta = \pi$ or, at least, that the difference $\Delta\theta$ from either of those values must be very small indeed. Just how small it must be may be judged by considering the experimental limit, eq. (4.29), on the size of the neutron static electric dipole moment, which, as shown in chapter 4, is an unambiguous measure of T violation. Estimates of the limit on $\Delta\theta$ on that basis are given as (Baluni 1979; Crewther et al. 1979)

$$\Delta\theta < 10^{-8} \text{ to } 10^{-9}. \tag{11.2}$$

The value of θ is not determined by the theory in its simplest form. However, the theory may be modified so as to remove the degeneracy and establish the $\theta = 0$ state as the unique vacuum state, thereby restoring the T invariance. Such modifications lead to other verifiable physical consequences, such as the existence of the particles called "axions," but at this time there is no generally accepted evidence that they do exist.

Clearly there are also implications here that call for careful attention to experiments on CPT invariance and for reexamination of the proof of the theorem because existing proofs depend on the assumption of both a nondegenerate vacuum and locality of the physical field operators.

At this time the theory of string[11] appears to hold much promise as a basis for the Grand Design of physics, including the unification of electroweak dynamics, QCD, and gravitation with a minimum of parameters. However, any attempt to define the improper transformations in the context of string theory must wait until the subject has matured to the point that its axiomatic structure and conceptual structure have been clarified. In particular, in order to follow our approach to these definitions we need a

[11] Readings on string theory may be found in Schwarz (1985), Bardeen and White (1985), and Green and Gross (1985).

clear distinction between the kinematic and dynamic aspects of the theory. If string theory is indeed a path to a quantized version of the general theory of relativity, the associated jumbling of geometry, kinematics, and dynamics will make the definition of T or CP even more obscure, and the question raised in section 11.1 about the meaning of global time reversal when the direction of time is defined only locally also must be confronted.

An inability to define the transformations associated with T and CP renders meaningless questions concerning T invariance and CP invariance at this fundamental level. It also appears to undermine a conventional approach to a proof of the CPT theorem, because a clear definition of the CPT transformation is required for this purpose. Therefore determining the improper transformations in the theory of string is an interesting challenge for the future.

11.3 The Origins of *CP* Violation and *T* Violation

Our analyses of the observed CP violation and the associated T violation in chapters 9 and 10 have made use of the notion of an effective or phenomenological weak interaction Hamiltonian involving a T-violating phase. For example, this phase was introduced in the KM matrix describing the coupling between quarks and the intermediate vector bosons $W^{\pm}$. It was emphasized that this is an ad hoc procedure; the phase that is introduced in this way is not determined by the theory but simply adjusted to fit experimental data. The same can be said of the Lee (1973) and Weinberg (1976) approach that makes use of arbitrary phase differences between different Higgs fields.

A fundamental theory explaining the origin of these phases would be expected to provide a way to determine its value in terms of the fundamental constants of the theory or in terms of numerical ratios characteristic of the mathematical structure of the theory. Since from the viewpoint of quantum field theory the fundamental character of interactions of the fields is to be obtained by examining their behavior in the limit of very small distance or very large momentum transfer, it is possible that the origin of CP violation is to be found in a "naturally occurring" phase of the primitive interactions at the high-momentum transfer limit. However, this is not the only possibility. At small distances the interactions may turn out to involve no such phase directly, but the observed phase could have a cosmological origin. We shall now look at some of the prospects and problems associated with these two approaches to the issue.

Current ideas about the behavior of interactions at short distances suggest a convergence of all interactions of physics to some common origin at a limiting distance that, if the theory is to include gravitation, is likely to be of the order of the Planck length, about 10^{-33} cm. Such a theory would correspond to the aforementioned Grand Design in which weak, electromagnetic, strong, and gravitational interactions are unified. The hope is that at this fundamental level the theory can be formulated in terms of massless spinor, scalar, and gauge fields, all of them related by a symmetry group of high enough order to encompass all known particles and fields of physics.

The distinct behavior of the different particles and fields, including the occurrence of finite masses, on the distance scale at which they are observed is presumed to be due to a process of "spontaneous symmetry breaking."[12] This process occurs because the self-interaction of certain scalar fields (e.g., the Higgs fields) leads to a symmetrical set of degenerate vacuum states but only one of the vacuum states occurs in nature, so that the underlying symmetry is obscured, just as the specification of initial conditions leads to physical behavior (motion) in which the underlying symmetry of the interaction is obscured. The self-interaction of these symmetry-breaking scalar fields generates mass for the particles associated with the gauge fields, and the masses of the spinor particles are generated by interactions between the spinor and scalar fields.

The distance scale at which these ideas can be tested experimentally is limited by the energy range over which experiments can be performed. The reason for some optimism about the general theoretical approach is that at the highest available energies the observed production of the $W^{\pm}$ and Z^0 particles (UA1-Collaboration 1983; UA2-Collaboration 1983) confirmed the Glashow (1980), Salam (1980), and Weinberg (1980) unification by such a mechanism of electromagnetic and weak interactions. The corresponding distance scale is 10^{-16} cm, corresponding to $W^{\pm}$ and Z^0 masses of about 100 GeV. These masses are presumed to be generated by the Higgs field, which breaks the electroweak symmetry at lower energies.[13] The associated Higgs particle has not been observed.

The Higgs field also interacts with the quark fields and contributes to the quark masses. It is possible to use this mechanism to generate a quark mass matrix (labeled by quark flavors) that leads to a Cabibbo-Kobayashi-Maskawa mixing of quark flavors of the form eq. (10.6), thereby accounting for the KM model of *CP* violation. However, the Higgs-quark coupling matrix involves the same number of undetermined parameters as the KM matrix, including the *CP*-violating phase. We must therefore look more

[12] See, e.g., Quigg (1983, chaps. 5 and 6).

[13] See note 12.

deeply for the source of this elusive phase, presumably at smaller distance scales.

The same can be said, of course, about the relative phases of the Higgs fields suggested by the Lee-Weinberg models of the *CP* violation.

We should keep in mind that at 10^{-16} cm we are well within the femtouniverse,[14] so that the standard procedure for defining *CP* and *T* appears to be justified in arriving at a form for *CP*- and *T*-violating interactions.

Evidently the search for the "original" *CP*-violating interactions must be pursued to much smaller distances, well beyond the level of electroweak unification. There are at least two possible paths. One would treat the quarks as composite systems and ascribe the *CP* properties to the subunits. The other would continue to treat the quark fields as elementary and assume that the *CP* violation arises in some way out of the unification of the electroweak and strong interactions, known as "grand unification."[15]

The former path would simply shift the burden of the theory from the quarks to their subunits. Until there is a specific theory to suit the purpose, that path does not offer any new perspective on how the *CP*-violating phase originates.

The ultimate form of the Grand Unified Theories (GUTS) would introduce a high enough symmetry to encompass all the known spinor fields (quarks and leptons), and the interactions of these zero-mass fields with the associated gauge fields would be characterized by a single coupling constant. This possibility is suggested by the asymptotic freedom of QCD, the decrease of the QCD coupling at decreasing distance. An extrapolation of this behavior of the coupling constant indicates that it becomes equal to the electroweak coupling at about 10^{-29} cm.[16]

Unfortunately, breaking the GUTS symmetry in order to account for the distinct behavior of the observed particles and to generate their masses requires introducing symmetry-breaking scalar fields with myriad arbitrary coupling constants. Therefore, as a source of *CP* violation this symmetry breaking suffers at least the same degree of arbitrariness as the breaking of the electroweak symmetry. If the *CP*-violating phase is somehow introduced into the original interaction with the gauge field before breaking the GUTS symmetry, it would be expected to lead to *CP* violation in strong as well as weak interactions.

[14] See note 6.

[15] An excellent descriptive presentation of one version of the Grand Unified Theories (GUTS) is to be found in the article by Georgi and Glashow (1980) in *Physics Today*. See also Quigg (1983, chap. 9).

[16] See note 15.

My conclusion is that the origin of the *CP* phenomenon is not likely to be found in the form of a primitive interaction at extremely small distances.

The possibility that the observed *CP* violation has a cosmological origin has emerged very recently in connection with speculations concerning the relation between the early universe and the complete unification of theoretical physics in a Grand Design. These ideas are in such a state of flux that we can address the questions of interest here only in the vaguest terms. We should recognize that a move in this direction entails a substantial departure from a popular view that the intrinsic *CP* violation of particle dynamics is *the cause of* the very small imbalance between the matter and antimatter content of the universe.[17] In contrast to this notion we must also consider the possibility that spontaneous symmetry-breaking processes, of which the matter-antimatter imbalance may be a symptom, are the cause of the observed *CP* violation.

We start from the premise that the Grand Design is a theory based on a symmetry encompassing all existing fields. This symmetry would be evident only at extremely small distances (high momentum transfer), and on that distance scale the fields are subject to a common coupling constant. The bare masses of all particles would vanish. In addition to the common dimensionless coupling constant (measured in units of $\hbar c$) and the distance scale, presumably of the order of the "extremely small" Planck scale, 10^{-33} cm, the only additional parameters would be the dimensionalities of the representations of the orders of the underlying symmetry groups and its subgroups.

From the vantage point of Big Bang cosmology the conditions in the universe within a Planck time ($\sim 10^{-44}$ sec) of the beginning correspond to the scale of both length and energy when the effects of symmetry breaking and finite particle masses would not have been appreciable. As the universe expanded and cooled, effects of renormalization of coupling constants and of spontaneous symmetry breaking would have determined the form of the phenomenological theory, that is, the theory as we know it.

A possible connection between the spontaneous symmetry breaking and the observed *CP* violation can be imagined if, at some stage, the symmetry breaking leads to degenerate vacuum states having a characteristic symmetry that is parameterized by a physically meaningful[18] phase η, as in the case of the theta vacuum of section 11.2, where $\eta \equiv \theta$. On the assumption

[17] This possibility was mentioned specifically by the Nobel Committee in its announcement of the 1980 award to Cronin and Fitch for the discovery of *CP* violation.

[18] That is, a relative phase between states that cannot be removed by a change in phase conventions or by a gauge transformation.

that the dynamics lead to this symmetry of the vacuum states, there is no dynamic basis for selecting one value of θ over another. However, the evolution of the universe is determined not only by the dynamics, but also by the initial conditions. Therefore we could ascribe the choice of η, and therefore the occurrence of *CP* violation, to an initial condition in the very early universe. We may go further and account for that initial condition as the result of the phase transition that causes the freezing in of a statistical fluctuation in the symmetry-breaking processes. Such a statistical fluctuation would be analogous to the one that determines the orientation of the magnetization of a ferromagnet as it cools through the Curie temperature in the absence of any external fields.

This mechanism would show no favoritism for one value of η over another if the vacuum symmetry in η was exact. Therefore the value of η that is implied by the observation on *CP* violation and *T* violation is just as "good" as any other value. Nevertheless, this explanation of *CP* violation is not very satisfactory, since it leaves the choice of η to an unverifiable chance happening in the early universe.

Another possibility is that the symmetry breaking in the early universe leads to a discrete set of degenerate vacuum states and that the choice of a phase results from the freezing in of a fluctuation to one of these discrete states. Consider, for example, the possibility that, owing to a symmetry property of the symmetry-breaking scalar fields, this degeneracy is associated with the vacuum states of two distinct scalar fields (Higgs fields?) and that the symmetry is characterized by the phase difference η between the expectation values of the two fields in such a way that the energy minima (vacuum states) occur at values of η given by

(11.3) $$\eta_j = 2\pi j/n,$$

where j is a positive integer or zero and n is the order of the symmetry.

Since η_j is a relative phase, it cannot be eliminated by means of a global gauge transformation. Therefore if j is not equal to either 0 or $\frac{1}{2}n$ for the physical vacuum, the interactions involving these scalar fields will in general carry the *CP*- and *T*-violating phase η_j. The question then is: Which value of j is associated with the physical vacuum? An answer to that question could be, again, that j is determined either by an initial condition or by a frozen-in fluctuation.

An explanation of *CP* violation along these lines suggests an "ultimate" *CP*-violating phase that is a rational fraction of π, and the denominator of the rational fraction is directly related to the degree of symmetry of the vacuum. For example, if electroweak symmetry breaking is the relevant level, η_j would be the relative phase of the Higgs fields in the Lee (1973) and Weinberg (1976) models. A determination that that phase is a rational fraction of π could be used to determine the symmetry of the Higgs vacuum, which would provide a new order of symmetry to be included in the constellation of symmetries to be encompassed in the Grand Design. Of course, the relevant symmetry breaking may occur at much smaller distances, making experimental access to the fundamental phase η_j much more difficult.

This, or alternative mechanisms based on the resolution of a degeneracy of the fundamental vacuum state, would place the burden for *CP* violation on what is in effect a cosmological initial condition. Therefore further information about *CP* violation at high-momentum transfer can make a substantial contribution to our understanding of physics in the early universe. Although this theoretical approach to *CP* violation treats it as a trickle-down effect from extremely high energies, experiments on high-energy *CP*-violating or *T*-violating phenomena are needed to provide guidance to the theory from the underside—that is, from the low-energy side. For example, if high-energy phenomena of the kind discussed in chapter 10 could be carried out with enough precision to obtain a value of the phase between Higgs fields, it would indicate the degree of symmetry, if there is any, required to account for a vacuum degeneracy that is the origin of *CP* violation at this level. At this time the experiments are approaching that level. Whatever the results may be, we need much more experimental information concerning *CP* violation and *T* violation in order to arrive at an understanding of their origin.

References

Abashian, A., and Hafner, E. M. 1958. *Phys. Rev. Lett.* 1:255–56.

Abegg, R.; Cameron, J. M.; Hutcheon, D. A.; Kitching, P.; McDonald, W. J.; Miller, C. A.; Pasos, J. W.; Soukup, J.; Thekkumthala, J.; Wilson, H. S.; Stetz, A. W.; and van Heerden, I. J. 1982. *Phys. Lett.* 118B:55–58.

Abragam, A., and Bleaney, B. 1970. *Electron paramagnetic resonance of transition ions.* Oxford: Clarendon Press.

Achiman, Y. 1969. *Lett. Nuovo Cim.* 2:301–6.

Adler, S. L., and Dashen, R. 1968. *Current algebras.* New York: W. A. Benjamin.

Aharony, A. 1970. *Lett. Nuovo Cim.* 3:791–92.

Alterev, I. S.; Borisov, Yu. V.; Borovikova, N. V.; Brandin, A. B.; Egorov, A. I.; Ezhov, V. F.; Ivanov, S. N.; Lobashev, V. M.; Nazarenko, V. A.; Ryabov, V. L.; Sevelerov, A. P.; and Taldaev, R. R. 1981. *Phys. Lett.* 102B:13.

Ames, Joseph Sweetman, and Murnaghan, Francis D. 1929. *Theoretical mechanics.* Boston: Ginn.

Anandan, J. 1982. *Phys. Rev. Lett.* 48:1660–63.

Anderson, R. L.; Prepost, R.; and Wiik, B. H. 1969. *Phys. Rev. Lett.* 22:651–54.

Aprile, E.; Eisenegger, C.; Hausammann, R.; Heer, E.; Hess, R.; Lechanoine-Leluc, C.; Leo, W. R.; Morenzoni, S.; Onel, Y.; and Rapin, D. 1981. *Phys. Rev. Lett.* 47:1360–65.

Arash, F.; Moravcsik, M. J.; and Goldstein, G. R. 1985. *Phys. Rev. Lett.* 54:2649–52.

Aronson, S. H.; Ehrlich, R. D.; Hofer, H.; Jensen, D. A.; Swanson, R. A.; Telegdi, V. L.; Goldberg, H.; Solomon, J.; and Fryberger, D. 1970. *Phys. Rev. Lett.* 25:1057–61.

Ashkin, J., and Kabir, P. K. 1970. *Phys. Rev.* D1:868–73.

Atac, M.; Chrisman, B.; Debrunner, P.; and Frauenfelder, H. 1968. *Phys. Rev. Lett.* 20:691–94.

Austern, N., and Sachs, R. G. 1951. *Phys. Rev.* 81:710–16.

Azimov, Ya. I., and Iogansen, A. A. 1981. *Yad. Fiz.* 33:388–96. (English translation *Sov. J. Nucl. Phys.* 33:205–10.)

Baltrusaitis, R. M., and Calaprice, F. P. 1977. *Phys. Rev. Lett.* 38:464–68.

Baluni, V. 1979. *Phys. Rev.* D19:2227–30.

Bardeen, W. A., and White, A. R., eds. 1985. *Symposium on anomalies, geometry, topology.* Singapore: World Scientific Publications.

Barshay, S. 1966. *Phys. Rev. Lett.* 17:49–53.

Bartlett, D. F.; Friedberg, C. E.; Goulianos, K.; Hammerman, I. S.; and Hutchinson, D. P. 1969. *Phys. Rev. Lett.* 23:893–97, 1205(E).

Baz', A. I. 1957. *JETP* 32:628–30. (Translation 1958, *Soviet Phys. JETP* 5:521–23.)

Bell, J., and Mandl, F. 1958. *Proc. Phys. Soc.* (London) 71:272–74.

Bell, J. S., and Steinberger, J. 1966. In *Proc. 1965 Int. Conf. Elem. Part., Oxford*, pp. 195–222. Chilton, Didcot, U.K.: Rutherford High Energy Laboratory.

Bernstein, J. 1968. *Elementary particles and their currents.* San Francisco: W. H. Freeman.

Bernstein, R. H.; Bock, G. J.; Carlsmith, D.; Coupal, D.; Cronin, J. W.; Gollin, G. D.; Keling, W.; Nishikawa, K.; Norton, H. W. M.; and Winstein, B. 1985. *Phys. Rev. Lett.* 54:1631–34.

Bhatia, T. S.; Glass, G.; Hilbert, J. C.; Northcliffe, L. C.; Tippens, W. B.; Bonner, B. E.; Simmons, J. E.; Hollas, C. L.; Newson, C. R.; Riley, P. J.; and Ransome, R. D. 1982. *Phys. Rev. Lett.* 48:227–30.

Biedenharn, L. C. 1959. *Nucl. Phys.* 10:620–25.

Bigi, I. I., and Sanda, A. I. 1981. *Nucl. Phys.* B193:85–108.

———. 1984. *Phys. Rev.* D29:1393–98.

———. 1985. *Comments Nucl. Part. Phys.* 14:149–84.

Bincer, A. M. 1960. *Phys. Rev.* 118:855–63.

Binstock, J.; Bryan, R.; and Gersten, A. 1981. *Ann. Phys.* 133:355–89.

Bjorken, J. D. 1980. In *Proceedings of Summer Institute on Particle Physics: Quantum chromodynamics*, ed. A. Mosher, pp. 219–89. Stanford, Calif.: Stanford Linear Accelerator Center-224.

Bjorken, J. D., and Drell, S. D. 1964. *Relativistic quantum mechanics.* New York: McGraw-Hill.

———. 1965. *Relativistic quantum fields.* New York: McGraw-Hill.

Blatt, J. M., and Weisskopf, V. F. 1952. *Theoretical nuclear physics.* New York: John Wiley.

Blin-Stoyle, R. J. 1952. *Proc. Phys. Soc.* (London) A65:452–53.

Blume, M., and Kistner, O. C. 1968. *Phys. Rev.* 171:417–25.

Bodansky, D.; Eccles, S. F.; Farwell, G. W.; Rickey, M. E.; and Robison, P. C. 1959. *Phys. Rev. Lett.* 2:101–4.

Boehm, F. H. 1968. *Hyperfine structure and nuclear radiations.* Amsterdam: North-Holland.

Boltzmann, L. 1872. *Wien Ber.* 66:275.

Brodine, J. C. 1970. *Phys. Rev.* D1:100–106; 2:627; 2:2090–93.

Buras, A. J.; Slominski, W.; and Steger, H. 1984. *Nucl. Phys.* B238:529–60.

Burgy, M. T.; Krohn, V. E.; Novey, T. B.; Ringo, G. R.; and Telegdi, V. L. 1958. *Phys. Rev. Lett.* 1:324–26.

Cabibbo, N. 1963. *Phys. Rev. Lett.* 10:531–33.

Calaprice, F. P.; Commins, E. D.; Gibbs, H. M.; Wick, G. L.; and Dobson, D. A. 1969. *Phys. Rev.* 184:1117–29.

Calaprice, F. P.; Commins, E. D.; and Girvin, D. C. 1974. *Phys. Rev.* D9:519–29.

Callan, C. C., Jr., and Treiman, S. B. 1967. *Phys. Rev.* 162:1494–96.

Carnegie, R. K.; Cestor, R.; Fitch, V. L.; Strovink, M.; and Sulak, L. R. 1971. *Phys. Rev.* D4:1–6.

Carter, A. B., and Sanda, A. I. 1980. *Phys. Rev. Lett.* 45:952–54.

———. 1981. *Phys. Rev.* D23:1567–79.

Casella, R. C. 1968. *Phys. Rev. Lett.* 21:1128–31.

———. 1969. *Phys. Rev. Lett.* 22:554–56.

Chau, L.-L. 1983. *Phys. Rep.* 95:1–94.

Chau, L.-L., and Cheng, H.-Y. 1984. *Phys. Rev. Lett.* 53:1037–40.

Chau, L.-L.; Cheng, H.-Y.; and Keung, W.-Y. 1985. *Phys. Rev.* D32:1837–40.

Chau, L.-L., and Keung, W.-Y. 1984. *Phys. Rev.* D29:592–95.

Chau, L.-L.; Keung, W.-Y.; and Tran, M. D. 1983. *Phys. Rev.* D27:2145–66.

Cheung, N. K.; Henrikson, H. E.; and Boehm, F. 1977. *Phys. Rev.* C16:2381–93.

Cheung, N. K.; Henrikson, H. E.; Cohen, E. J.; Becker, A. J.; and Boehm, F. 1976. *Phys. Rev. Lett.* 37:588–91.

Chew, G. F., and Goldberger, M. L. 1952. *Phys. Rev.* 87:778–82.

Chodos, A.; Jaffe, R. L.; Johnson, K.; Thorne, C. B.; and Weisskopf, V. F. 1974. *Phys. Rev.* D9:3471–95.

Christenson, J. H.; Cronin, J. W.; Fitch, V. L.; and Turlay, R. 1964. *Phys. Rev. Lett.* 13:138.

———. 1965. *Phys. Rev.* B140:74.

Clark, M. A., and Robson, J. M. 1960. *Can. J. Phys.* 38:693–95.

———. 1961. *Can. J. Phys.* 39:13–21.

Cleland, W. E.; Conforto, G.; Eaton, G. H.; Gerber, H. J.; Reinharz, M.; Gautschi, A.; Heer, E.; Revillard, C.; and von Dardel, G. 1972. *Nucl. Phys.* B40:221–54.

Close, F. E. 1979. *An introduction to quarks and partons.* New York: Academic Press.

Coester, F. 1951. *Phys. Rev.* 84:1259.

———. 1953. *Phys. Rev.* 89:619–20.

Condon, E. U., and Shortley, G. H. 1951. *The theory of atomic spectra.* Cambridge University Press.

Crewther, R.; diVecchia, P.; Veneziano, G.; and Witten, E. 1979. *Phys. Lett.* 88B:123–27.

Cronin, J. W. 1981. *Rev. Mod. Phys.* 53:373–83.

Cronin, J. W., and Overseth, O. E. 1963. *Phys. Rev.* 129:1795–1807.

Cullen, M.; Darriulat, P.; Deutsch, J.; Foeth, H.; Groves, G.; Holder, M.; Kleinknecht, K.; Rademacher, E.; Rubbia, C.; Shambroon, D.; Saire, M.; Stande, A.; and Tittel, K. 1970. *Phys. Lett.* 32B:523–27.

Dalitz, R. H. 1952. *Proc. Phys. Soc.* A65:175–78.

Dass, G. V., and Kabir, P. K. 1972. *Proc. Roy. Soc.* (London) A330:331–47.

Davis, B. R.; Koonin, S. E.; and Vogel, P. 1980. *Phys. Rev.* C22:1233–44.

Day, T. B. 1961. *Phys. Rev.* 121:1204–6.

Dehn, J. T.; Marzolf, J. G.; and Salmon, J. F. 1964. *Phys. Rev.* 5:B1307–10.

Dress, W. B.; Miller, P. D.; Pendelbury, J. M.; Perrin, P.; and Ramsey, N. F. 1977. *Phys. Rev.* D15:9–21.

Driller, H.; Blanke, E.; Geny, H.; Richter, A.; Schrieder, G.; and Pearson, J. M. 1979. *Nucl. Phys.* A317:300–312.

Dunietz, I.; Greenberg, O. W.; and Wu, D.-d. 1985. *Phys. Rev. Lett.* 55:2935–37.

Ebel, M. E., and Feldman, G. 1957. *Nucl. Phys.* 4:213–14.

Ehrenfest, Paul, and Ehrenfest, Tatiana. 1912. *Encyklopädie der mathematischen Wissenschaften IV 2 II*, no. 6. Leipzig; B. G. Teubner, (English translation, *The conceptual foundations of the statistical approach in mechanics.* Ithaca, N.Y.: Cornell University Press, 1959.)

Eichler, J. 1968. *Nucl. Phys.* A120:535–39.

Ellis, J.; Gaillard, M. K.; and Nanopoulos, D. V. 1976. *Nucl. Phys.* B109:213–43.

Enz, C. P., and Lewis, R. R. 1965. *Helv. Phys. Acta* 38:860–76.

Ericson, T. E. O. 1966. *Phys. Lett.* 23:97–99.

Ernst, F. J.; Sachs, R. G.; and Wali, K. C. 1960. *Phys. Rev.* 119:1105–14.

Erozolimisky, B. G.; Bondarenko, L. N.; Mostovoy, Yu. A.; Obinyakov, B. A.; Zacharova, V. P.; and Titov, V. V. 1968. *Phys. Lett.* 27B:557–59.

———. 1970. *Yad. Fiz.* 11:1049–57. (Translation, *Soviet J. Nucl. Phys.* 11:583–87.)

Fano, U. 1957. *Rev. Mod. Phys.* 29:74–93.

Feynman, R. P., and Gell-Mann, M. 1958. *Phys. Rev.* 109:193–98.

Fitch, V. L.; Roth, R. F.; Russ, J. S.; and Vernon, W. 1965. *Phys. Rev. Lett.* 15:73–76.

Foldy, L. L. 1952. *Phys. Rev.* 87:688–96.

Ford, W. T.; Lemonick, A.; Nauenberg, U.; and Piroué, P. A. 1967. *Phys. Rev. Lett.* 18:1214–18.

Ford, W. T.; Piroué, P. A.; Remmel, R. S.; Smith, A. J. S.; and Souder, P. A. 1970. *Phys. Rev. Lett.* 25:1370–73.

Fry, W. F.; Schneps, J.; and Swami, M. S. 1956. *Phys. Rev.* 103:1904–5.

Garrell, M.; Frauenfelder, H.; Ganek, D.; and Sutton, D. C. 1969. *Phys. Rev.* 187:1410–13.

Gasiorowicz, S. 1966. *Elementary particle physics.* New York: John Wiley.

Gell-Mann, M., and Pais, A. 1955. *Phys. Rev.* 97:1387–89.

Georgi, H., and Glashow, S. L. 1980. *Physics Today* 33, no. 9:30–39.

Gibbs, J. Willard. 1931. *Elementary principles of statistical mechanics.* In *The collected works of J. Willard Gibbs.* New York: Longmans, Green. (Originally published 1902.)

Gilman, F. J., and Hagelin, J. S. 1983. *Phys. Lett.* 126B:111–16; 133B:443–48.

Gilman, F. J., and Wise, M. B. 1979. *Phys. Lett.* 83B:83–86.

Gimlett, J. L.; Henrikson, H. E.; and Boehm, F. 1982. *Phys. Rev.* C25:1567–74.

Gimlett, J. L.; Henrikson, H. E.; Cheung, N. K.; and Boehm, F. 1981. *Phys. Rev.* C24:620–30.

Glashow, S. L. 1980. *Rev. Mod. Phys.* 52:539–43.

Glashow, S. L.; Iliopoulos, J.; and Maiani, L. 1970. *Phys. Rev.* D2:1285–92.

Goddard, P., and Olive, D. 1978. *Rep. Prog. Phys.* 41:1357–1437.

Goldberger, M. L., and Treiman, S. B. 1958. *Phys. Rev.* 111:354–61.

Goldberger, M. L., and Watson, K. M. 1964. *Collision theory.* New York: John Wiley.

Goldstein, H. 1980. *Classical mechanics.* Reading, Mass.: Addison-Wesley.

Goldwire, H. C., and Hannon, J. P. 1977. *Phys. Rev.* B16:1875–88.

Good, M. L. 1957. *Phys. Rev.* 106:591–95.

———. 1958. *Phys. Rev.* 110:550–51.

Good, R. H.; Matsen, R. P.; Muller, F.; Piccioni, O.; Powell, W. M.; White, H. S.; Fowler, W. B.; and Birge, R. W. 1961. *Phys. Rev.* 124:1223–39.

Gourdin, M. 1967. *Nucl. Phys.* B3:207–33.

Green, M. B., and Gross, D. J. 1985. *Unified string thories.* Singapore: World Scientific Publications.

Griffith, J. S. 1961. *The theory of transition metal ions.* Cambridge: Cambridge University Press.

Gronau, M., and Schechter, J. 1985. *Phys. Rev. Lett.* 54:385–88, 1209(E).

Gross, E. E.; Malanify, J. J.; van der Wurde, A.; and Zucker, A. 1968. *Phys. Rev. Lett.* 21:1476–79.

Hagelin, J. S. 1979. *Phys. Rev.* D20:2893–98.

Handler, R.; Wright, S. C.; Pondrom, L.; Limon, P.; Olsen, S.; and Kloeppel, P. 1967. *Phys. Rev. Lett.* 19:933–36.

Hannon, J. P., and Trammell, G. T. 1968. *Phys. Rev. Lett.* 21:726–32.

Hardekopf, R. A.; Keaton, P. W.; Lisowski, P. W.; and Vesser, L. R. 1982. *Phys. Rev.* C25:1090–93.

Harrison, G. E.; Sandars, P. G. H.; and Wright, S. J. 1969. *Phys. Rev. Lett.* 22:1263–65.

Harte, J., and Sachs, R. G. 1964. *Phys. Rev.* 135:B459–66.

Henley, E. M. 1969. *Ann. Rev. Nucl. Sci.* 19:367–432.

Henley, E. M., and Huffman, A. H. 1968. *Phys. Rev. Lett.* 20:1191–94.

Henley, E. M., and Jacobsohn, B. A. 1957. *Phys. Rev.* 108:502–3.

———. 1959. *Phys. Rev.* 113:225–33.

Herzberg, Gerhard. 1950. *Spectra of diatomic molecules.* New York: D. Van Nostrand.

Herzo, D.; Banner, D.; Beier, E.; Bertram, W. K.; Edwards, R. T.; Frauenfelder, H.; Koester, L. J.; Rosenberg, E.; Russell, A.; Segler, S.; and Wattenberg, A. 1969. *Phys. Rev.* 186:1403–14.

Hillman, P.; Johansson, A.; and Tibell, G. 1958. *Phys. Rev.* 110:1218–19.

Hinds, E. A., and Sandars, P. G. H. 1980. *Phys. Rev.* A21:480–87.

Hochberg, D., and Sachs, R. G. 1983. *Phys. Rev.* D27:606–15.

Holstein, B. R. 1972. *Phys. Rev.* C5:1529–34.

Hsueh, S. Y.; Muller, D.; Tang, J.; Winston, R.; Zapalac, G.; Swallow, E. C.; Berge, J. P.; Brenner, A. E.; Grafstrom, P.; Jastrzembski, E.; Lach, J.; Marriner, J.; Raja, R.; Smith, V. J.; McCliment, E.; Newson, C.; Anderson, E. W.; Denisov, A. S.; Grachev, V. T.; Kulikov, A. V..; Schegelsky, V. A.; Seliverstov, D. M.; Smirnov, N. N.; Terentyev, N. K.; Tkatch, I. I.; Vorobyov, A. A.; Cooper, P. S.; Razis, P.; and Teig, L. J. 1985. *Phys. Rev. Lett.* 54:2399–2402.

Hwang, C. F.; Ophel, T. R.; Thorndike, E. H.; and Wilson, R. 1960. *Phys. Rev.* 119:352–61.

Jackson, J. D.; Treiman, S. B.; and Wyld, H. W., Jr. 1957a. *Phys. Rev.* 106:517–21.

———. 1957b. *Nucl. Phys.* 4:206–12.

Jacob, R., and Sachs, R. G. 1961. *Phys. Rev.* 121:350–56.

Jacobsohn, B. A., and Henley, E. M. 1959. *Phys. Rev.* 113:234–38.

Jarlskog, C. 1985a. *Phys. Rev. Lett.* 55:1039–42.

———. 1985b. *Z. Phys. C.* 29:491–97.

Jensen, D. A.; Aronson, S. H.; Ehrlich, R. D.; Fryberger, D.; Nissin-Sabat, C.; Telegdi, V. L.; Goldberg, H.; and Solomon, J. 1969. *Phys. Rev. Lett.* 23:615–19.

Jost, R. 1957. *Helv. Phys. Acta* 30:409–16.

Kabir, P. K. 1968a. *The CP puzzle.* New York: Academic Press.

———. 1968b. *Nature* (London) 220:1310–13.

———. 1970. *Phys. Rev.* D2:540–42.

———. 1982. *Phys. Rev.* D25:2013–14.

Kajfosz, J.; Kopecky, J.; and Honzatko, J. 1965. *Phys. Lett.* 20:284–86.

———. 1968. *Nucl. Phys.* A120:225–33.

Karpman, G.; Leonardi, R.; and Strocchi, F. 1968. *Phys. Rev.* 174:1957–68.

Kenny, B. G. 1967. *Ann. Phys.* 45:25–71.

Kenny, B. G., and Sachs, R. G. 1965. *Phys. Rev.* 138:B943–46.

———. 1973. *Phys. Rev.* D8:1605–7.

Kistner, O. C. 1967. *Phys. Rev. Lett.* 19:872–77.

Kleinknecht, K. 1976. *Ann. Rev. Nucl. Sci.* 26:1–50.
Kobayashi, M., and Maskawa, T. 1973. *Prog. Theo. Phys.* (Japan) 49:652–57.
Kramers, H. A. 1930. *Koninkl. Ned. Akad. Wetenschap., Proc.* 33:959.
Krane, K. S.; Murdoch, B. T.; and Steyert, W. A. 1974. *Phys. Rev.* C10:840–52.
Landau, L. 1957. *Zhur. Eksptl. Teort. Fiz.* 32:405–6. (Translation *Soviet Phys. JETP* 5:336–37); *Nucl. Phys.* 3:127–31.
Lande, K.; Booth, E. T.; Impeduglia, J.; Lederman, L. M.; and Chinowsky, W. 1956. *Phys. Rev.* 103:1901–4.
Lee, B. W. 1972. *Chiral dynamics.* New York: Gordon and Breach.
Lee, T. D. 1973. *Phys. Rev.* D8:1226–39.
Lee, T. D.; Oehme, R.; and Yang, C. N. 1957. *Phys. Rev.* 106:340–45.
Lee, T. D., and Wu, C. S. 1966. *Ann. Rev. Nucl. Sci.* 16:511–90.
Lee, T. D., and Yang, C. N. 1956. *Phys. Rev.* 104:254–58.
———. 1957. *Phys. Rev.* 108:1645–47.
Lehmann, H. 1954. *Nuovo Cim.* 11:342–57.
Lehmann, H.; Symanzik, K.; and Zimmermann, W. 1955. *Nuovo Cim.* 1:205–25.
———. 1957. *Nuovo Cim.* 6:319–33.
Lippmann, B. A., and Schwinger, J. 1950. *Phys. Rev.* 79:469–80.
Lipschutz, N. R. 1966, 1967. *Phys. Rev.* 144:1300–1304; 158:1491–97.
Lloyd, S. P. 1951. *Phys. Rev.* 81:161–62.
Lobashev, V. M.; Nazarenko, V. Z.; Saenko, L. F.; Smotritsky, L. M.; and Kharkevitch, G. I. 1967. *Zh. Eksperim. Teor. Fiz—Pis'ma Redakt.* 5:73–75. (Translation *JETP Lett.* 5:59–61.)
Lobkowicz, F.; Melissinos, A. C.; Nagashima, Y.; Tewksbury, S.; and von Briesin, H., Jr. 1969. *Phys. Rev.* 185:1676–87.
Loschmidt, J. 1876. *Wien Ber.* 73:139.
———. 1877. *Wien Ber.* 75:67.
Lyuboshitz, L. 1965. *Yadern. Fiz.* 1:497–506. (Translation *Soviet J. Nucl. Phys.* 1:354–60.)
Mahaux, C., and Weidenmüller, H. A. 1966. *Phys. Lett.* 23:100–103.
Mandl, F. 1958. *Proc. Phys. Soc.* (London) 71:177–93.
March, A. 1951. *Quantum mechanics of particles and fields.* New York: John Wiley.
Marshak, R. E.; Riazuddin; and Ryan, C. P. 1969. *Theory of weak interactions in particle physics.* New York: Wiley-Interscience.
Messiah, A. 1961. *Quantum mechanics.* Vol. 1. Amsterdam: North-Holland.
———. 1962. *Quantum mechanics.* Vol. 2. Amsterdam: North-Holland.
Moldauer, P. A. 1968a. *Phys. Rev.* 165:1136–46.
———. 1968b. *Phys. Lett.* 26B:713–15.
Møller, C. 1945. *I. D. Kgl. Danske Vidensk. Selskab, Mat-fys. Medd.* 23, no. 1.
———. 1946. *I. D. Kgl. Danske Vidensk. Selskab, Mat-fys. Medd.* 22, no. 19.
Moravcsik, M. J. 1982. *Phys. Rev. Lett.* 48:718–21.
Mott, N. F., and Massey, H. S. W. 1933. *The theory of atomic collisions.* Oxford: Clarendon Press. (2d ed. 1949, 3d ed. 1965.)
Murdoch, B. T.; Olsen, C. E.; Rosenblum, S. S.; and Steyert, W. A. 1974. *Phys. Lett.* 52B:325–28.
Murdoch, B. T.; Olsen, C. E.; and Steyert, W. A. 1974. *Phys. Rev.* C10:1475–83.
Niebergall, F.; Regler, M.; Stier, H. H.; Winter, K.; Aubert, J. J.; De Bonard, X.; Lepeltier, V.; Massonet, L.; Pessard, H.; Vivargent, M.; Willitts, T. R.; Yvert, M.; Bartl, W.; Neuhofer, G.; and Steuer, M. 1974. *Phys. Lett.* 49B:103–8.
Oksak, A. I., and Todorov, I. T. 1968. *Commun. Math. Phys.* 11:125–30.

Okun', L. 1966. *Usp. Fiz. Nauk.* 89:603–46. (Translation 1967, *Soviet Physics Uspekhi* 9:574–601.)
———. 1968. *Uspekhi Fiz. Nauk.* 95:402–16. (Translation 1969, *Soviet Phys. Uspekhi* 11:462–69.)
Okun', L. B.; Zakharov, V. I.; and Pontecorvo, B. M. 1975. *Lett. Nuovo Cim.* 13:218–20.
Overseth, O. E., and Roth, R. F. 1967. *Phys. Rev. Lett.* 19:391–93.
Painlevé, Paul. 1904. *Comptes Rendus* 139:1170–74.
Pais, A., and Piccioni, O. 1955. *Phys. Rev.* 100:1487–89.
Pais, A., and Treiman, S. B. 1975. *Phys. Rev.* D12:2744–50.
Pauli, W. 1928. In *Probleme der modernen Physik: Arnold Sommerfeld zum 60.* Leipzig: S. Hirzel. (Reprinted in *Collected scientific papers of W. Pauli*, pp. 30–45. New York: Interscience, 1964.)
———. 1955. In *Niels Bohr and the development of physics*, ed. W. Pauli, pp. 30–51. New York: Pergamon Press.
———. 1958. Die Allegemeinen Principien der Wellenmechanik. In *Encyclopedia of Physics*, 5:1–168. Berlin: Springer.
Pauli, W., and Weisskopf, V. 1934. *Helv. Phys. Acta* 7:709–31.
PDG (Particle Data Group). 1986. *Phys. Lett.* 170B:1–350.
Peierls, R. E. 1955. In *Proceedings of the 1954 Glasgow Conference on Nuclear and Meson Physics*, ed. E. H. Bellamy and R. G. Moorhouse, pp. 296–99. New York: Pergamon Press.
Perkins, R. B., and Ritter, E. T. 1968. *Phys. Rev.* 174:1426–28.
Phillips, R. J. N. 1958. *Nuovo Cim.* 8:265–70.
Player, M. A., and Sandars, P. G. H. 1970. *J. Phys.* B3:1620.
Purcell, E. M., and Ramsey, N. F. 1950. *Phys. Rev.* 78:807.
Quigg, C. 1983. *Gauge theories of the strong, weak, and electromagnetic interactions.* Reading, Mass.: Benjamin/Cummings.
Ramsey, N. F. 1957. *Phys. Rev.* 109:225.
———. 1978. *Phys. Rep.* 43:410.
———. 1981. *Comments Nucl. Part. Phys.* 10:227–41.
Renner, B. 1968. *Current algebras and their applications.* London: Pergamon Press.
Robson, D. 1968. *Phys. Lett.* 26B:117–19.
Rosen, L., and Brolley, J. E., Jr. 1959. *Phys. Rev. Lett.* 2:98–101.
Sachs, R. G. 1953. *Nuclear theory.* Cambridge, Mass.: Addison-Wesley.
———. 1954. *Phys. Rev.* 95:1065–78.
———. 1963a. *Phys. Rev.* 129:2280–85.
———. 1963b. *Ann. Phys.* 22:239–62.
———. 1963c. *Science* 140:1284–90.
———. 1964. *Phys. Rev. Lett.* 13:286–88.
———. 1975. *Prog. Theo. Phys.* (Japan) 54:809–22.
———. 1976. *Phys. Rev. Lett.* 36:1014–17.
———. 1986. *Phys. Rev.* D33:3283–92.
Sachs, R. G., and Austern, N. 1951. *Phys. Rev.* 81:705–9.
Sachs, R. G., and Treiman, S. B. 1962. *Phys. Rev. Lett.* 8:137–40.
Sakurai, J. J., and Wattenberg, A. 1967. *Phys. Rev.* 161:1449–50.
Salam, A. 1980. *Rev. Mod. Phys.* 52:525–38.
Salzman, G. 1955. *Phys. Rev.* 99:973–79.
Sandars, P. G. H. 1967. *Phys. Rev. Lett.* 19:1396–98.
Satchler, G. R. 1958. *Nucl. Phys.* 8:65–68.

Schubert, K. R.; Wolff, B.; Chollet, J. C.; Gaillard, J.-M.; Jane, M. R.; Ratcliffe, T. J.; and Repellin, J.-P. 1970. *Phys. Lett.* 31B:662–65.

Schwarz, J. H. 1985. *Superstrings: The first fifteen years of superstring theory.* Singapore: World Scientific Publications.

Sharman, P.; Hamilton, W. D.; Cavaignac, J.-F.; Charvet, J.-L.; Hungerford, P.; and Vignon, B. 1978. *J. Phys. G.* 4:973–87.

Shifman, M. A.; Vainshtein, A. I.; and Zakharov, V. I. 1977. *Nucl. Phys.* B120:316–24.

Shirokov, M. I. 1957. *JETP* 33:975–81. (Translation 1958, *Soviet Phys. JETP* 6:748–53.)

Shrock, R. E., and Wang, L.-L. 1978. *Phys. Rev. Lett.* 41:1692–95.

Slobodrian, R. J.; Rioux, C.; Roy, R.; Conzett, H. E.; von Rosen, P.; and Hinterberger, F. 1981. *Phys. Rev. Lett.* 47:1803–7.

Smith, K. M.; Booth, P. S. L.; Renshall, H. R.; Jones, P. B.; Salmon, G. L.; and Williams, W. S. C. 1973. *Nucl. Phys.* B60:411–18.

Snow, G. A. 1956. *Phys. Rev.* 103:1111–15.

Sober, D. I.; Cassel, D. G.; Sadoff, A. J.; Chen, K. W.; and Crean, P. A. 1969. *Phys. Rev. Lett.* 22:430–33.

Sprung, D. W. L. 1961. *Phys. Rev.* 121:925–26.

Stech, B. 1983. *Phys. Lett.* 130B:189–93.

Steinberg, R. I.; Liand, P.; Vignon, B.; and Hughes, V. W. 1974. *Phys. Rev. Lett.* 33:41–44.

Stoyonav, D. Tz., and Todorov, I. T. 1968. *J. Math. Phys.* 9:2146–67.

Streater, R. F., and Wightman, A. S. 1964. *PCT, spin and statistics, and all that.* New York: W. A. Benjamin.

Sudarshan, E. C. G., and Marshak, R. E. 1958. *Phys. Rev.* 109:1860–61.

Tanner, N. W., and Dalitz, R. H. 1986. *Ann. Phys.* 171:463–88.

Teichmann, T. 1950. *Phys. Rev.* 77:506–15.

Teichmann, T., and Wigner, E. P. 1952. *Phys. Rev.* 87:123–35.

ter Haar, D. 1954. *Elements of statistical mechanics.* New York: Rinehart.

Thorndike, E. H. 1965. *Phys. Rev.* 138:B586–96.

Thornton, S. T.; Jones, C. M.; Bair, J. K.; Mancusi, M. D.; and Willard, H. B. 1971. *Phys. Rev.* C3:1065–86.

Treiman, S. B.; Jackiw, R.; and Gross, D. J. 1971. *Current algebra and its applications.* Princeton: Princeton University Press.

Treiman, S. B., and Sachs, R. G. 1956. *Phys. Rev.* 103:1545–49.

Trelle, R. P.; Birkhäuser, J.; Hinterberger, F.; Kuhn, S.; Prasuhn, D.; and von Rossen, P. 1984. *Phys. Lett.* 134B:34–36.

Tsinoev, V. G.; Chertov, Yu. P.; Danengirsh, S. G.; Shcherbina, Yu. I.; Stepanov, E. P.; and Veronin, A. A. 1982. *Phys. Lett.* 110B:369–71.

UA-1 Collaboration. 1983. *Phys. Lett.* 122B:103–16; 126B:398–410.

UA-2 Collaboration. 1983. *Phys. Lett.* 122B:476–85; 129B:130–40.

Vainshtein, A. I.; Zakharov, V. I.; and Shifman, M. A. 1975. *Pis'ma Zh. Eksp. Teor. Fiz.* 22:123–26. (Translation, *JETP Lett.* 22:55–56.)

van der Waerden, B. L. 1932. *Die Gruppen theoretische Methode in der Quantenmechanik.* Berlin: Julius Springer.

Vold, T. G.; Raab, F. J.; Heckel, B.; and Forson, E. N. 1984. *Phys. Rev. Lett.* 52:2229–32.

von Witsch, W.; Richter, A.; and von Brentano, P. 1968. *Phys. Rev.* 169:923–32.

Wald, R. M. 1980. *Phys. Rev.* D21:2742–55.

Vold, T. G.; Raab, F. J.; Heckel, B.; and Forson, E. N. 1984. *Phys. Rev. Lett.* 52:2229–32.
von Witsch, W.; Richter, A.; and von Brentano, P. 1968. *Phys. Rev.* 169:923–32.
Wald, R. M. 1980. *Phys. Rev.* D21:2742–55.
———. 1984. *General relativity.* Chicago: University of Chicago Press.
Wang, G. W.; Becker, A. J.; Chirovsky, L. M.; Groves, J. L.; and Wu, C. S. 1978. *Phys. Rev.* C18:476–85.
Watson, K. M. 1952. *Phys. Rev.* 88:1163–71.
Weinberg, S. 1976. *Phys. Rev. Lett.* 37:657–61.
———. 1980. *Rev. Mod. Phys.* 52:515–23.
Weisskopf, V., and Wigner, E. P. 1930. *Z. Physik* 63:54–73; 65:18–29.
Weitkamp, W. G.; Storm, D. W.; Shreve, D. C.; Braithwaite, W. J.; and Bodansky, D. 1968. *Phys. Rev.* 165:1233–44.
Wentzel, G. 1943. *Quantentheorie der Wellenfelder.* Vienna: Franz Deuticke.
Weyl, H. 1931. *The theory of groups and quantum mechanics.* London: Methuen.
White, D. H., and Sullivan, D. 1979. *Physics Today* 32, no. 4:40–47.
Whittaker, E. T. 1944. *Analytical dynamics.* New York: Dover.
Wick, G. C.; Wightman, A. S.; and Wigner, E. P. 1952. *Phys. Rev.* 88:101–5.
Wigner, E. P. 1932. *Nachr. Ges. Wiss. Göttingen Math-Physik* K1, no. 32:546.
———. 1948. *Phys. Rev.* 73:1002–9.
———. 1949. *Am. J. Phys.* 17:99–109.
———. 1955. *Am. J. Phys.* 23:371–80.
———. 1959. *Group theory.* New York: Academic Press.
Wigner, E. P., and Eisenbud, L. 1947. *Phys. Rev.* 72:29–41.
Wigner, E. P., and von Neumann, J. 1954. *Ann. Math.* 59:418–33.
Winter, K. 1972. In *Proceedings of the Amsterdam International Conference on Elementary Particles*, pp. 333–71. Amsterdam: North-Holland.
Wolfenstein, L. 1949. *Phys. Rev.* 75:1664–74; 76:541–42.
———. 1956. *Ann. Rev. Nucl. Sci.* 6:43–76.
———. 1964. *Phys. Rev. Lett.* 13:562–64.
———. 1966. In *Preludes in theoretical physics*, ed. A. De-Shalit, H. Feshbach, and L. van Hove, pp. 170–76. Amsterdam: North-Holland.
———. 1968. *Uspekhi Fiz. Nauk.* 95:428–37. (Translation 1969, *Sov. Phys. Uspekhi* 11:477–81.)
———. 1969a. In *Theory and particle phenomenology in particle physics*, ed. A. Zichichi, pp. 218–61. New York: Academic Press.
———. 1969b. *Nuovo Cim.* 63A:269–78.
———. 1982. *Physics Today* 35, no. 8:66–67.
———. 1983. *Phys. Rev. Lett.* 51:1945–47.
———. 1984a. *Phys. Lett.* 144B:425–26.
———. 1984b. *Nucl. Phys.* B246:45–51.
———. 1985a. *Comments Nucl. Part. Phys.* 14:135–47.
———. 1985b. *Phys. Rev.* D31:2381–82.
———. 1986a. *Phys. Rev.* D34:897–98.
———. 1986b. *Ann. Rev. Nucl. Sci.* 36:137–70.
Wolfenstein, L., and Ashkin, J. 1952. *Phys. Rev.* 85:947–49.
Wu, C. S.; Ambler, E.; Hayward, R. W.; Hoppes, D. D.; and Hudson, R. P. 1957. *Phys. Rev.* 105:1413–15.
Wu, T. T., and Yang, C. N. 1964. *Phys. Rev. Lett.* 13:380–85.

Yang, C. N., and Feldman, D. 1950. *Phys. Rev.* 79:972–78.
Yang, C. N., and Tiomno, J. 1950. *Phys. Rev.* 79:495–98.
Yennie, D. R.; Lévy, M. M.; and Ravenhall, D. G. 1957. *Rev. Mod. Phys.* 29:144–57.
Zocher, H., and Török, C. 1953. *Proc. Nat. Acad. Sci. U.S.A.* 39:681–86.

Index